Computing with Spatial Trajectories

Yu Zheng • Xiaofang Zhou

Editors

Computing with Spatial Trajectories

Foreword by Jiawei Han

 Springer

Editors
Yu Zheng
Microsoft Research Asia
Beijing
China, People's Republic
yuzheng@microsoft.com

Xiaofang Zhou
School of Information Technology
and Electrical Engineering
University of Queensland
Brisbane Queensland
Australia

ISBN 978-1-4899-9105-8   ISBN 978-1-4614-1629-6 (eBook)
DOI 10.1007/978-1-4614-1629-6
Springer New York Dordrecht Heidelberg London

Springer is part of Springer Science+Business Media (www.springer.com)

Foreword

With the rapid development of wireless communication and mobile computing technologies and global positioning and navigational systems, spatial trajectory data has been mounting up, calling for systematic research and development of new computing technologies for storage, preprocessing, retrieving, and mining of trajectory data and exploring its broad applications. Thus, *computing with spatial trajectories* becomes an increasingly important research theme. Although there are many books on spatial databases, mobile computing, and data mining, this is a unique book dedicated to computing with spatial trajectory data, with a broad spectrum of coverage and authoritative overview.

Despite of many years of research on algorithms and methods on general database systems and data mining, spatial trajectory computing deserves dedicated study and in-depth treatment because of its unique nature of data semantics, structures, and applications. Such a unique nature calls for in-depth study of many interesting issues, including spatial trajectory data preprocessing, trajectory indexing and query processing, trajectory pattern mining, uncertainty and privacy in trajectory data, location-based social networks, and application of trajectory computing, such as for driving and other activities. This book, *"Computing with Spatial Trajectories"*, by Yu Zheng and Xiaofang Zhou, provides a comprehensive coverage on the above topics timely, with conciseness and clear organization. The authors of the book are active researchers on different aspects on computing with spatial trajectories, and have made tangible contributions to the progress of this dynamic research frontier. This ensures that the book is authoritative and reflects the current state of the art. Nevertheless, the book gives a balanced treatment on a wide spectrum of topics, well beyond the authors' own methodologies and research scopes.

Computing with spatial trajectories is still a fairly young and dynamic research field. This book may serve researcher and application developers a comprehensive overview of the general concepts, techniques, and applications on trajectory indexing, search and data mining, and help them explore this exciting field and develop new methods and applications. It may also serve graduate students and other interested readers a general introduction to the state-of-the-art of this promising research theme.

I find the book is enjoyable to read. I hope you like it too.

July, 2011

Jiawei Han
University of Illinois at Urbana-Champaign

Preface

A *spatial trajectory* is a trace generated by a moving object in geographical spaces, usually represented by of a series of chronologically ordered points, e.g., $p_1 \rightarrow p_2 \rightarrow \cdots \rightarrow p_n$, where each point consists of a geospatial coordinate set and a timestamp such as $p = (x, y, t)$.

The advances in location positioning and wireless communication technologies have given rise to the prevalence of mobile computing systems and location-based services (LBS), leading to a myriad of spatial trajectories representing the mobility of a variety of moving objects, such as people, vehicles, animals, and natural phenomena, in both indoor and outdoor environments. Below are some examples.

1) Mobility of people: People have been recording their real-world movements in the form of spatial trajectories, passively and actively, for a long time.

- Active recording: Travelers log their travel routes with GPS trajectories for the purpose of memorizing a journey and sharing experiences with friends. Bicyclers and joggers record their trails for sports analysis. In Flickr, a series of geo-tagged photos can formulate a spatial trajectory as each photo has a location tag and a timestamp corresponding to where and when the photo was taken. Likewise, the "check-ins" of a user in Four-square can be regarded as a trajectory, when sorted chronologically.

- Passive recording: A user carrying a mobile phone unintentionally generates many spatial trajectories represented by a sequence of cell tower IDs with corresponding transition times. Meanwhile, transaction records of a credit card also indicate the spatial trajectory of the cardholder, as each transaction contains a timestamp and a merchant ID denoting the location where the transaction occurred.

2) Mobility of vehicles: In recent years, a large number of GPS-equipped vehicles have appeared in our daily life. For instance, many taxis in major cities have been equipped with a GPS sensor, which enables them to report a time-stamped location to a data center with a certain frequency. Such reports formulate a large amount of spatial trajectories that can be used for resource allocation, security management, and traffic analysis.

3) Mobility of animals and natural phenomena: Biologists solicit the moving trajectories of animals like migratory birds for research projects. Similarly, climatologists are busy collecting the trajectories of some natural phenomena, such as hurricanes, tornados, and ocean currents. These trajectories provide scientists with rich information about the objects they are studying.

Overall, spatial trajectories have offered us unprecedented information to understand moving objects and locations, calling for systematic research and development of new computing technologies for the processing, retrieving, and mining of trajectory data and exploring its broad applications. Therefore, *computing with spatial trajectories* has become an increasingly important research theme, attracting extensive attention from numerous areas, including computer science, biology, sociology, geography, and climatology.

Although there are many books on spatial databases, mobile computing, and data mining, this is the first book dedicated to computing with spatial trajectory data, with a broad spectrum of coverage and an authoritative overview. Aimed at advanced undergraduates, graduate students, researchers, and professionals, this book covers the major fundamentals and the key advanced topics that shape the field. Each chapter is a tutorial that provides readers with an introduction to one important aspect of computing with spatial trajectories and also contains many valuable references to relevant research papers. This book provides researchers and application developers a comprehensive overview of the general concepts, techniques, and applications of trajectory indexing, search, and data mining, and helps them explore this exciting field and develop new methods and applications. It also offers graduate students and other interested readers a general introduction to the most recent developments in this promising research area.

We chose 17 active researchers in the field of computing with spatial trajectories to contribute chapters to this book in their areas of expertise. These chapters are organized according to the paradigm of "trajectory preprocessing (prior databases) → trajectory indexing and retrieval (in databases) → advanced topics (above databases)," as illustrated in Figure 1.

-   The first two chapters of the book introduce the foundation of technology dealing with spatial trajectory data: Trajectory Preprocessing (Chapter 1) and Trajectory Indexing and Retrieval (Chapter 2).
-   The second section is comprised of 6 advanced topics: Uncertainty in Spatial Trajectories (Chapter 3), Privacy of Spatial Trajectories (Chapter 4), Trajectory Pattern Mining (Chapter 5), Activity Recognition Based on Spatial Trajectories (Chapter 6), Trajectory Analysis for Driving (Chapter 7), and Location-Based Social Networks (Chapter 8 and 9).

Specifically, the book gradually introduces the concepts and technologies for solving the problems that newcomers will be faced with when exploring this field, starting from the preprocessing and managing of spatial trajectories, then to mining uncertainty, privacy, and patterns of trajectories, and finally ending with some advanced applications based on spatial trajectories including activity recognition,

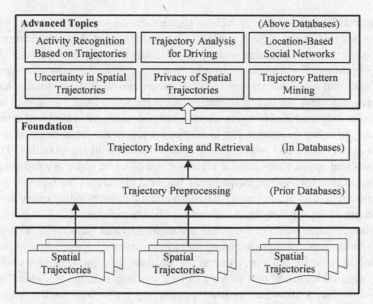

Fig. 1 The framework of the book

driving, and location-based social networks. Below is a brief introduction to the topics that will be covered in each chapter:

Chapter 1: While spatial trajectories carry rich information that can be used in a variety of applications, we have to deal with a number of issues before using them. Generally, the continuous movement of an object is recorded in an approximate form as discrete samples of location points. A high sampling rate of location points generates accurate trajectories, but will result in a massive amount of data leading to enormous overhead in data storage, communications, and processing. Thus, it is vital to design data reduction techniques that compress the size of a trajectory while maintaining the utility of the trajectory. Meanwhile, a trajectory is usually generated with occasional outliers or some noisy points caused by the poor signal of location positioning systems. For example, when traveling in a "city canyon," the satellite signals to a GPS device might be very poor, thereby generating some location points with a significant offset to the real positions. Sometimes, the offsets are more than a mile, creating noise in the trajectories and reducing the effectiveness of techniques and systems that use such trajectories. As a result, techniques for filtering the noisy points are needed for preprocessing spatial trajectories.

To address these two issues, Chapter 1 first presents data reduction techniques that can run in a batch mode after the data is collected or in an online mode as the data is being collected. The second part of the chapter introduces methods for filtering measurement noise from location trajectories, including mean and median filtering, the Kalman filter, and the particle filter. In short, this chapter provides a newcomer with the fundamentals for preprocessing spatial trajectories.

Chapter 2: The prevalence of various location-based services leads to a myriad of trajectories that create a huge burden of computation for these application systems. It is very time-consuming to find information in a trajectory dataset that is not well organized. For instance, retrieving the trajectories that pass a crossroad is a simple task, but it would make online systems unfeasible if these systems have to scan a large trajectory dataset in a direct way. At the same time, we need to search for particular trajectories satisfying certain criteria, such as spatial and temporal constraints, on many occasions. For example, retrieving the trajectories of tourists passing a given region and within a time span can help with trip planning. The requests from real applications call for effective and efficient trajectory indexing and retrieval technologies. As a consequence, Chapter 2 introduces the types of queries that are usually issued in a trajectory database and the corresponding query processing approaches supported by indexing and retrieval techniques.

Chapter 3: After preprocessing and organizing spatial trajectories with corresponding techniques, we can start using them in a variety of applications. However, positioning devices are inherently imprecise, resulting in some uncertainty with regards to acquired locations of a moving object. For instance, the reading of a GPS sensor usually has a 10 meters or more positioning error. With such a reading, it might not be easy to identify the exact point of interest (like a restaurant or a shopping mall) the moving object visited, especially in a dense urban area. At the same time, objects move continuously while their locations can only be updated at discrete times, leaving the location of a moving object between two updates uncertain. Two reasons that lead to long-interval updates are to save energy consumption and communication bandwidth. When the time interval between two updates exceeds several minutes or hours, the uncertainty in a spatial trajectory will reduce its utility and create new challenges when searching for a moving object.

To handle the above-mentioned uncertainty, Chapter 3 presents a systematic overview of the various issues and solutions related to the notion of uncertainty in the settings of spatial trajectories. The problems related to modeling and representing the uncertainty in Moving Objects Databases (MOD) are introduced. The problems of efficient algorithms for processing various spatio-temporal queries are also discussed. Note that the query processing introduced in Chapter 2 does not consider the uncertainty of a spatial trajectory while this is one of the focuses of Chapter 3.

Chapter 4: Although LBSs provide many valuable applications to mobile users, revealing people's private locations to potentially untrusted LBS service providers poses privacy concerns. There is a trade-off between the quality of services offered by a LBS provider and the privacy of a user's location. The more precisely a user reveals her location, the better the services that can be offered, the less privacy is preserved.

In general, there are two types of LBSs, namely, snapshot and continuous LBSs. For a snapshot LBS, a mobile user only needs to report her current location to a service provider once to get the desired information. In fact, it is not necessary for a user to tell a LBS system her exact location when using such services. For instance, finding hotels close to a user only requires a rough geo-region including the user's current location. A bunch of literature has discussed a handful of methods protecting a user's snapshot location, so Chapter 4 will not cover them again.

On the other hand, a mobile user has to report her location to a service provider in a periodic or on-demand manner to obtain continuous LBS (for example, getting real-time traffic conditions around a user or searching for the nearest gas stations while driving). Protecting user location privacy for a continuous LBS is more challenging than a snapshot LBS, because adversaries may use the spatial and temporal correlations between the samples of a user's trajectory to infer the user's location information with a higher degree of certainty. In short, releasing original spatial trajectories to the public or a third party could pose serious privacy concerns. As a result, privacy protection in a continuous LBS and trajectory data publication has increasingly drawn attention from the research community and industry. Under the circumstances, Chapter 4 describes state-of-the-art privacy-preserving techniques for a continuous LBS and trajectory publication.

Chapter 5: The huge volume of spatial trajectories enables opportunities for analyzing the mobility patterns of moving objects, which can be represented by an individual trajectory containing a certain pattern or a group of trajectories sharing similar patterns. It can also be segments of different trajectories holding similar properties (e.g., formulating a cluster according to spatial and temporal constraints), or a set of full trajectories that satisfy the same conditions. These patterns can benefit a broad range of application areas and services, including transportation, biology, sports, and social services. For example, finding clusters of trajectories, in which trajectories have similar spatial shapes, can help detect the popular driving paths of users or migrating routes of birds. Additionally, identifying a group of people moving together may contribute to the exploration of social relationships, enable friend recommendations, or allow for taxi sharing. For these reasons, Chapter 5 provides an overview of trajectory patterns introduced in existing literature, divides these patterns into categories according to different taxonomies, and reviews indexing structures and algorithms for trajectory pattern mining.

Chapter 6: After the preprocessing, managing, and pattern mining of spatial trajectories, people might wonder what kind of advanced applications can be enabled based on these trajectories. Activity recognition is one of the primary applications that can be used with trajectories. Intuitively, spatial trajectories generated by people imply user behavior and activities, offering new insights into the high-level goals and objectives of users based on low-level sensor readings. First, the activities of an individual can be used as a context to trigger services that satisfy the individual's unspoken needs. For example, if it is known that a user is driving, her mobile phone can automatically show the traffic conditions on roads around the user and disable the phone's entertainment functions (for her safety) that would distract a driver's attention. If we know a user is in a meeting or watching a movie, the mobile phone of a user can be switched into quiet mode. Second, the activities of multiple users enable us to mine collective social knowledge that contributes to the analysis of social networks and transportation. With the activity information of multiple users, we can estimate the similarity between two different users more precisely, thereby, offering better services for community discovery and friend and location recommendations.

While providing richer information beyond snapshot location data, spatial trajectories also demand more advanced techniques for trajectory-based activity recognition. To help newcomers address this issue, Chapter 6 overviews the existing research into trajectory-based activity recognition and classifies them into categories according to the number of users involved in the training and inference stage.

Chapter 7: The trajectories of vehicles have a potentially strong connection to transportation, as driving is one of the most central aspects of our lives. Rich knowledge can be learned from these trajectories, such as information about road networks, traffic conditions, and driver behavior, contributing to different aspects of the driving experience. For example, creating a road map from GPS trajectories can be a less expensive way to make up-to-date road maps than traditional methods. Meanwhile, effective route recommendations can be enabled based on the trajectories of one or more experienced drivers.

Chapter 7 describes how a driver benefits from the analysis of spatial trajectories, following the paradigm of "creating road map from GPS trajectories → mapping a single trajectory to road networks → mining effective driving routes from drivers' trajectories → personalizing driving routes for a particular driver based on the preferences learned from her trajectories."

Chapter 8 and 9: The advances in location positioning and wireless communication technologies have led to a myriad of user-generated spatial trajectories, which imply rich information about user behavior, interests, and preferences. Recently, people have started sharing their trajectory data via online social networking services for a variety of purposes, fostering a number of trajectory-centric LBSNs (location-based social networks). For example, users can record travel routes with GPS trajectories to share travel experiences in an online community (e.g., GeoLife), or log jogging and bicycle trails for sports analysis and experience sharing. In addition, the "check-ins" of an individual in Foursquare and the photo trips of a user in Flickr can be regarded as spatial trajectories. These trajectory-centric LBSNs enable us to understand users and locations respectively, and explore the relationship between them.

On the one hand, we can understand an individual and the similarity between two different users with user-generated trajectories, thereby providing a user with personalized services and enabling friend recommendation and community discovery. On the other hand, we are able to understand a location and the correlation between two different locations based upon the information from users, thereby offering users better travel recommendations.

Chapter 8 defines the meaning of location-based social network and discusses the research philosophy behind LBSNs from the perspective of users and locations. Under the circumstances of a trajectory-centric LBSN, this chapter explores two fundamental research points concerned with understanding users in terms of their locations. One is modeling the location history of an individual using the individual's trajectory data. The other is estimating the similarity between two different people according to their location histories. The similarity represents the strength of connection between two users in a location-based social network, and can enable friend recommendations and community discovery. Some possible methods for evaluation of these applications have been discussed, and a number of publically available datasets have been listed in Chapter 8 as well.

While chapter 8 studies the research philosophy behind a location-based social network from the point of view of users, Chapter 9 gradually explores the research into LBSNs from the perspective of locations. A series of research topics is presented with respect to mining the collective social knowledge from many users'GPS trajectories to facilitate travel. First, the generic travel recommendations provide a user with the most interesting locations, travel sequences and travel experts in a region, as well as an effective itinerary conditioned by a user's start location and available time length. Second, the personalized travel recommendations discover the locations matching an individual's interests, which can be gleaned from the individual's historical data.

We hope you will find this book provides a useful overview of and a practical tutorial on the young and evolving field of computing with spatial trajectories.

July, 2011   *Yu Zheng,*
                                   Microsoft Research Asia, Beijing, China

                                                              Xiaofang Zhou,
                          The University of Queensland, Brisbane, Australia

Acknowledgements

Many thanks to Mr. Jing Yuan for helping us compile this book.
Special thanks to the editorial board consisting of Ralf Hartmut Güting, Hans-Peter Kriegel, and Hanan Samet for their constructive suggestion and comments.
Thanks John Krumm for proposing the name of the book.

Yu Zheng and *Xiaofang Zhou*

Contents

List of Contributors

Wang-Chien Lee
The Pennsylvania State University, USA, e-mail: wlee@cse.psu.edu

John Krumm
Microsoft Research, Redmond, WA, USA, e-mail: jckrumm@microsoft.com

Ke Deng
The University of Queensland, Brisbane, Australia,
e-mail: dengke@itee.uq.edu.au

Kexin Xie
The University of Queensland, Brisbane, Australia,
e-mail: kexin@itee.uq.edu.au

Kevin Zheng
The University of Queensland, Brisbane, Australia,
e-mail: kevinz@itee.uq.edu.au

Xiaofang Zhou
The University of Queensland, Brisbane, Australia,
e-mail: zxf@itee.uq.edu.au

Goce Trajcevski
Northwestern University, USA, e-mail: goce@ece.northwestern.edu

Mohamed F. Mokbel
University of Minnesota, USA, e-mail: mokbel@cs.umn.edu

Chi-Yin Chow
City University of Hong Kong, China, e-mail: chiychow@cityu.edu.hk

Hoyoung Jeung
École Polytechnique Fédérale de Lausanne, Switzerland,
e-mail: hoyoung.jeung@epfl.ch

Man Lung Yiu
Department of Computing, Hong Kong Polytechnic University, Hong Kong,
e-mail: csmlyiu@comp.polyu.edu.hk

Christian S. Jensen
Department of Computer Science, Aarhus University, Denmark,
e-mail: csj@cs.au.dk

Yin Zhu
Hong Kong University of Science and Technology, China,
e-mail: yinz@cse.ust.hk

Vincent Wenchen Zheng
Hong Kong University of Science and Technology, China,
e-mail: vincentz@cse.ust.hk

Qiang Yang
Hong Kong University of Science and Technology, China,
e-mail: qyang@cse.ust.hk

Yu Zheng
Microsoft Research Asia, China, e-mail: yuzheng@microsoft.com

Xing Xie
Microsoft Research Asia, China, e-mail: xing.xie@microsoft.com

Acronyms

GPS	Global positioning system
RFID	Radio frequency identification
LBS	Location-based services
EHSID	Euler histogram-based on short IDs
SD	Snapshot disclosure
TD	Trajectory disclosure
DD	Distance deviation
POI	Points of interest
HITS	Hypertext induced topic search
TBHG	Tree-based hierarchical graph
LBSN	Location-based social network
CF	Collaborative filtering
TF-IDF	Term frequency-inverse document frequency
MAP	Mean average precision
nDCG	Normalized discounted cumulative gain

Part I
Foundations

Chapter 1
Trajectory Preprocessing

Wang-Chien Lee and John Krumm

Abstract A spatial trajectory is a sequences of (x,y) points, each with a time stamp. This chapter discusses low-level preprocessing of trajectories. First, it discusses how to reduce the size of data required to store a trajectory, in order to save storage costs and reduce redundant data. The data reduction techniques can run in a batch mode after the data is collected or in an on-line mode as the data is collected. Part of this discussion consists of methods to measure the error introduced by the data reduction techniques. The second part of the chapter discusses methods for filtering spatial trajectories to reduce measurement noise and to estimate higher level properties of a trajectory like its speed and direction. The methods include mean and median filtering, the Kalman filter, and the particle filter.

1.1 Introduction

Owing to the rapid advent of wireless communication and mobile computing technologies, the vision of pervasive computing is becoming a reality. Mobile devices, including smart phones, PDAs, navigational systems on vehicles, and RFIDs on cargos, have played a growing important role in various applications in our daily life. Nowadays, many of these mobile devices have location positioning and wireless communicating capabilities and thus are able to locally log or dynamically report their locations to the server.[1] Indeed, there is a tremendous demand for location tracking of moving objects from various *location-based services (LBS)*,

Wang-Chien Lee
Department of Computer Science and Engineering, The Pennsylvania State University, University Park, PA 16802, USA. e-mail: wlee@cse.psu.edu

John Krumm
Microsoft Research, Redmond, WA 98052, USA. e-mail: jckrumm@microsoft.com

[1] We call these mobile devices (and their carriers) *location-aware moving objects* or *moving objects* in short.

3

ranging from fleet management, traffic information services, transportation logistic-s, location-based games, to location-based social networks. According to forecast of the 2011-2015 LBS market made by Pyramid Research, revenue of the global location-based services market is expected to reach US10.3 billions in 2015, up from 2.8 billions in 2010.[2]

To support LBS applications, the database community has made a tremendous research effort on the development of mobile object databases (MODs) in the past decade [33, 14, 13, 28]. In addition to the conventional search functions of moving objects, *trajectory management* are essential operations of MODs since many LBS applications require analyzing and mining moving phenomenons/patterns of the monitored objects. Thus, trajectories of moving objects, i.e. their geographical-temporal traces, are often treated as first-class citizens in MODs. Due to the need to acquire and preprocess trajectory data before loading them into the MODs, in this book chapter, we review a number of issues and techniques for trajectory prepro-cessing, including trajectory data generation, filtering and reduction.

Based on Wikipedia, a trajectory is the path that a moving object follows through space as a function of time. To capture the accurate and complete trajectory of a moving object, however, is very difficult and expensive due to the inherent lim-itations of data acquisition and storage mechanisms. As a result, the continuous movement of an object is usually obtained in an approximate form as discrete sam-ples of *spatio-temporal location points* (or simply *location points*). Supposedly the more sample points are acquired in a trajectory, the more accurate the trajectory is. However, adopting high sampling rates in acquiring the location points of moving objects to generate the trajectories may result in a massive amount of data lead-ing to enormous overheads in data storage, communications and processing. Take the Taipei eBus system as an example.[3] The system tracks the trajectories of about 4000 buses daily, covering 287 bus routes in the greater Taipei metropolitan area. The locations of buses tracked in the system are transmitted to the system server every 15-25 seconds, generating millions of sampled data points daily. As the posi-tioning technology and processing power of data acquisition mechanisms continue to advance rapidly, the problem of data explosion gets only worse. Hence, it's a mandate to employ the data reduction techniques in trajectory preprocessing.

In addition to data reduction, trajectories can benefit from filtering to reduce noise and estimate higher-level properties like speed and direction. Since trajectories are normally measured by a sensor, they inevitably have some error, including occa-sional outliers. Simple techniques like mean and median filtering can reduce these errors. In addition to error reduction, certain filters like the Kalman filter and particle filter can also give error estimates and inferences on speed and direction.

Figure 1.1 shows a high-level system model for typical location-based services. As shown, the system consists of three components: 1) the location server, 2) mov-ing objects; and 3) LBS applications. As in most pervasive computing applications, we assume wireless communications between the server and moving objects. In

---

[2] See http://www.pyramidresearch.com/store/Report-Location-Based-Services.htm

[3] http://www.e-bus.taipei.gov.tw/

such systems, the locations of tracked moving objects are reported to the location server in accordance with the adopted reporting schemes, e.g. periodically. The location point data (which form trajectories of moving objects) are then uploaded to the moving object databases. On the other hand, the LBS applications submit queries to the location server to retrieve moving objects of interests (as well as their attributes such as locations and other phenomenons/patterns discovered from moving behaviors of objects) to meet various application needs.

As discussed earlier, systems in support of location based services naturally generate enormous volumes of data with measurement noise. Consequently, the data reduction and filtering techniques are particularly important for cleansing, transmission, and storage of trajectory data in location based services. Even though object movement is continuous, the representation of object trajectories is inevitably in a discrete form due to the nature of sampling-based data acquisition approach. Thus, an intuitive strategy to reduce the volumes of trajectory data is to reduce the sampling rate of data acquisition or to reduce the number of sample points in the trajectory representation. However, the question is whether we are able to discard some sample points without sacrificing the quality of trajectory data required for supporting the targeted applications. Additionally, what techniques can be used to effectively filter measurement noise not only in the raw location points of trajectories but also in high-level properties of trajectories such as direction and speed. Fortunately, due to the linear characteristics of the underlying transportation infrastructure, object movements in many LBS applications exhibit predictable patterns. As a result, many redundant and erroneous information can be removed from the trajectory without compromising much of the application requirements.

In the following, we review some trajectory data reduction strategies for LBSs. We first consider the location update scenarios and then review the data reduction strategies under these scenarios. Accordingly, we classify the data reduction strategies into two categories: 1) off-line compression and 2) on-line reporting.

After discussing data reduction, we review filtering techniques, including mean and median filtering, the Kalman filter, and the particle filter.

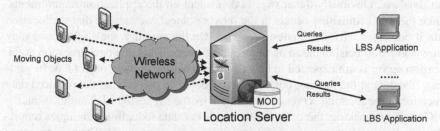

Fig. 1.1 A high-level system model for typical location-based services. The locations of tracked moving objects are reported to the location server via wireless communications. The LBS applications submit queries to the server to retrieve moving object data for analysis or other application needs.

1.2 Trajectory Data Generation

A trajectory is the path that a moving object follows through space as a function of time. Thus, it can be captured as a time-stamped series of location points, denoted as $\{\langle x_1, y_1, t_1 \rangle, \langle x_2, y_2, t_2 \rangle, ..., \langle x_N, y_N, t_N \rangle\}$ where x_i, y_i represent geographic coordinates of the moving object at time t_i and N is the total number of elements in the series. To generate the trajectory, a moving object needs to acquire its coordinates x, y at time t. There are many positioning technologies, e.g. global positioning system (GPS), that can be employed to determine the location of a moving object. For example, most of the smart phones already have a built-in GPS receiver and thus can easily derive their own locations. Augmenting the GPS, some positioning techniques have used wifi access points as the underlying reference system to determine the location of a moving object. Additionally, even in situations where the GPS signal is poor and there is no nearby wifi access point to serve as a positioning reference, the "dead reckoning" techniques can be used to estimate the position of a moving object by advancing from a known position using course, speed, time and distance to be traveled. In other words, where a moving object will be at a certain time can be derived based upon known or estimated speeds over elapsed time, and course. Notice that dead reckoning relies on accurate estimation of speed, elapsed time and direction, which can be measured by using accelerometers and g-sensors built in many mobile devices today. Thus, while the traditional navigational methods of dead reckoning for location acquisition have been replaced by modern positioning technologies, it is still very useful for generating trajectory data, especially when GPS reception is lost, e.g. in a tunnel. Since positions of moving objects calculated by dead reckoning is based on previous positions and estimated distance and directions, the errors in subsequent locations are cumulative. Therefore, the dead reckoning methods only serve as a remedy when the more accurate modern positioning techniques are not applicable.

Given that the time-stamped geographical coordinates can be sampled arbitrarily by a moving object, the next question is whether the moving object needs to report all the sampled trajectory data to the location server for upload to the mobile object database. Obviously the answer is dependent on the application requirements. Since the data acquisition occurs at the moving object, we assume that the location data it possesses have the highest precision. On the contrary, the applications may allow some imprecision based on their requirements. Thus, the data precision at the location server is not expected to be as high as what the moving object has. In summary, the data at the moving object are considered as precise and the required data precision at the location server is determined by the supported LBS applications.

Generally speaking, there are two categories of data reduction techniques reported in the literature of moving object and trajectory management. These approaches aim to reduce the communication and storage overhead of trajectory data representation while not to compromise much precision in the new data representation of trajectory. The basic idea behind data reduction techniques in the first category, called *batched compression techniques*, is to first collect the full set of sampled location data points and then compress the data set by discarding redundant location points for transmission to the location server. Because the full data set is taken into consideration by the compression algorithms, the results tend to approaching the global optimal better than the techniques in the other category. These batched compression techniques are very suitable for off-line uploading and analysis of trajectories at various Web 2.0 sites such as Everytrail[4] and Bikely[5]. Take Everytrail as an example, it provides trajectory logging tools on iphone and Android phones for users to record their trips. By uploading a trip trajectory (usually after the trip is completed), a user can annotate the trajectory with pictures and travelogues for sharing with her friends. Thus, the batched compression techniques can be used to reduce the transmission and storage overheads for the user and the hosting server.

For many LBS applications which require timely updates of the moving objects' locations, e.g. fleet management and traffic monitoring applications, the batched compression techniques may not be applied directly. In these applications, the location server needs to know the whereabouts of moving objects constantly. Since continuous location updates of moving objects is infeasible, data reduction of the sample points in a trajectory is usually achieved on-line by selective updates of the locations based on specified precision requirements. Thus, this category of trajectory data reduction techniques is named as *on-line data reduction techniques*. Two ideas are usually exploited in this category of on-line data reduction techniques: (i) use a line segment to fit as many location points in a trajectory as possible; (ii) predict the object movements and report only those location points deviating significantly from the prediction. Techniques based on (i) are able to capture the geometric properties of trajectories pretty well with linear approximation. On the other hand, techniques based on (ii) may capture additional features, such as speed and headings, of object movements and use them in prediction. Based on previously reported location of a moving object, the server is able to predict its next move even though the predicted location may not be exactly accurate. Later in this chapter we discuss the Kalman filter and particle filter, which can both be used for trajectory prediction. By obtaining a prediction model from the server, the moving object applies the same prediction algorithm on the previously reported object locations to figure out where the server perceives as its current location. As a result, by comparing the location perceived by the server and its true location (acquired from its positioning mechanism locally), the moving object is able to decide whether a location update should be reported to the server in order to calibrate the precision of object location and trajectory.

Figure 1.2 illustrates a simple update policy, called *point policy*, that models the tracked object as a jumping point [5]. This policy assumes that the object jumps to a distant point from its current location, stays around the new location for a while, then jumps to another remote location and stays there for a while. The process repeats in the trajectory of the tracked moving object. As shown in the figure, the object moves to location A and sends a location update to the location server. At this point, a circular neighborhood of radius r is set. As long as the object moves within the neighborhood of location A, no update report is sent to the server. When the object

---

[4] http://www.everytraiil.com

[5] http://www.bikely.com

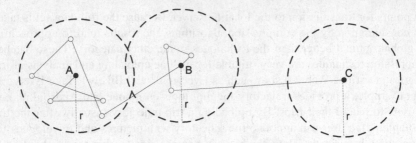

Fig. 1.2 The point update policy. A moving object does not update its new locations as long as they are within the error threshold of the previously reported location.

moves beyond the neighborhood of A to arrive at location B, a new location update is reported to the server. Similarly, a neighborhood of radius r is set at location B. Accordingly, there is no location update issued until the object moves to location C.

In summary, the batched compression techniques rely solely on the moving object to decide which sample points can be discarded. Globally optimized precision in the compressed trajectories may be retained but the required time constraints on reports may be challenging to meet. On the other hand, the on-line reduction techniques require collaboration between the moving object and location server to meet the requirement of updating moving object locations timely while optimizing the precision locally. Notice that the ideas behind the two categories of data reduction techniques can be combined depending on the time and precision requirements.

1.3 Performance Metrics and Error Measures

The primary goal of the trajectory data reduction techniques is to reduce the data size of trajectory representation without compromising much of its precision. Additionally, for the on-line data reduction techniques, the location of an object needs to be reported to the server if the imprecision of the predicted location goes beyond an application-dependent error threshold. Thus, there is a need to find appropriate metrics and error measures for use in algorithms and performance evaluation. The following are the main performance metrics often used to evaluate the efficiency and effectiveness of the trajectory data reduction techniques:

- Processing time: the execution time spent to run a trajectory data reduction algorithm;
- Compression rate: the ratio in the size of an approximate trajectory vs. the size of its original trajectory;
- Error measure: the deviation of an approximate trajectory from its original trajectory.

Among them, the processing time assesses how efficiently a trajectory data reduction technique processing a given trajectory data set. On the other hand, the compression rate and error measure are used to assess the effectiveness of the examined technique. Notice that there may be a tradeoff between these two effectiveness metrics. Thus, trajectory data reduction techniques are usually compared in a plot of these two metrics in order to find the Pareto front.

From the above, we can observe that there is a room to further define different error measures while the definition of compression rate is quite straightforward. For the rest of the section, we discuss two error measures, namely, *perpendicular Euclidean distance* and *time synchronized Euclidean distance*, that are widely used in literature since they have an implication in specifying the imprecision allowed by application and the performance [24, 27, 6].

To specify the allowed imprecision, *distance-based error measure* is a natural choice due to its simplicity and ability to deal with positions of points in multi-dimensional space. Take the error threshold used in on-line data reduction techniques as an example. The distance between a location on the original trajectory acquired from the positioning mechanism and the estimated location on the approximated trajectory intuitively represents how closely the estimated location approximates the original location.[6] With the same reasoning, the aggregated distance between the approximated trajectory and the original trajectory can be used to measure the error introduced by the data reduction process. As mentioned earlier, one of the error measure is to compute the perpendicular Euclidean distances, i.e. the shortest distance, from each of the sampled location points in the original trajectory to the approximated trajectory. As such, we can measure the error by the average or total distances.

Figure 1.3 illustrates the computation of error measure based on the perpendicular Euclidean distance between the original trajectory acquired by a moving object and an approximated trajectory generated by applying one of the trajectory data reduction algorithms. As shown in the figure, the original trajectory is represented by a series of time-stamped location points denoted by $\{p_0, p_1, ..., p_{16}\}$ where p_i is the location of the moving object at time t_i. On the other hand, the approximated trajectory, reduced from the original trajectory, consists of three location points, p_0, p_5 and p_{16}. Notice that the approximated trajectory can also be repre-

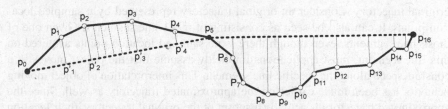

Fig. 1.3 Error measure based on perpendicular Euclidean distance. This error measure takes into account the geometric relationship of the trajectories. However, the temporal factor is not incorporated in this error measure.

---

[6] Note that here we ignore the inherent noises and imprecision from the location acquisition mechanism and assume the measured original location to be precise.

sented as two line segments $\overline{p_0 p_5}$ and $\overline{p_5 p_{16}}$. Thus, the approximated trajectory can be interpreted as the path in which the object moved from p_0 to p_5 and then to p_{16}. Based on this interpretation, the sampled location points on the subtrajectories $\{p_0, p_1, ..., p_5\}$ and $\{p_5, p_6, ..., p_{16}\}$ should be projected to the corresponding line segments $\overline{p_0 p_5}$ and $\overline{p_5 p_{16}}$ for error measurement. Figure 1.3 shows the estimated location points $p_0', p_1', ...$ and p_5' on the line segment $\overline{p_0 p_5}$, corresponding to $p_0, p_1, ...$ and p_5, respectively. Notice that the perpendicular Euclidean distance between a location point on the original trajectory to the approximated trajectory is the shortest distance between the point and the approximated trajectory. Thus, the error between the approximated trajectory and the original trajectory can be calculated by summing up the distances of the projection or computing their average distance. Notice that the error measurements by total or average are actually quite sensitive to the number of sampled location points in the original trajectory [24]. A remedy to this deficiency is to take into consideration all possible location points on the original trajectory instead of limiting the error measure to only the sampled location points. This can be achieved by interpolating some *pseudo sampled points* on the original trajectory. For example, five pseudo sampled points between p_2 and p_3 and their projection on the line segment $\overline{p_0 p_5}$ (indicated by the five dash lines) are illustrated in Figure 1.3. When an infinite number of pseudo sample points are considered, the area between the original trajectory and the approximated trajectory naturally measures the error between them.

The aforementioned approach elegantly captures the error in the approximated trajectory using perpendicular Euclidean distance. The idea of projecting each possible points in the original trajectory onto the line segments of the approximated trajectory, nevertheless, takes only geometric properties of the trajectories into account. The temporal factor of object movement in the trajectories is not considered in the projection. Notice that a sampled data point $\langle x, y, t \rangle$ in the original trajectory denotes the time t when the moving object are located at x, y. Thus, there is a need to also consider the temporal factor in the projection.

The *time synchronized Euclidian distance* has been proposed as a new error measure for approximated trajectories generated by trajectory data reduction algorithms [24, 27]. The intuition is that the movement projected on the approximated trajectory should be synchronized in terms of "time" with the actual movement on the original trajectory. Consider an original trajectory represented by n sampled location points. It can also be seen as consisting of $n - 1$ line segments. Given one of those line segments, even though there is no sampled location points acquired on this line segment, most applications implicitly assume that the object moves in a constant speed along the specific line segment. This interpretation of object moving behavior has been made earlier on the approximated trajectory as well. Since the approximated trajectory is actually a subset of the original trajectory, their location points can be used naturally for time and spatial synchronization of the represented object movement on both trajectories. Consider a line segment on an approximated trajectory and its corresponding subtrajectory on the original trajectory, their end points are the same and thus synchronized. Moreover, the projection of the sampled location points on the original trajectory onto the corresponding line segment on the approximated trajectory can be determined proportionally by using the time interval spent to move from a location point to a subsequent location point as the weight. As such, the time synchronized location points on the approximated trajectory can be easily determined. Finally, the distance between a location point on the original trajectory and the corresponding time synchronized location points can be derived accordingly.

Figure 1.4 illustrates the idea of time synchronized Euclidean distance. As shown, the location points on the approximated trajectory, i.e. p_0, p_5 and p_{16}, are already synchronized by time. The other sampled location points, e.g. p_1, p_2, p_3 and p_4, are projected to time synchronized location points p'_1, p'_2, p'_3 and p'_4, on the line segment $\overline{p_0p_5}$. The projection can be computed easily. For example, the coordinates of p'_1, i.e. x_1, y_1, can be derived as follows.

$$x_1 = x_0 + \frac{t_1 - t_0}{t_5 - t_0} \cdot (x_5 - x_0)$$

and

$$y_1 = y_0 + \frac{t_1 - t_0}{t_5 - t_0} \cdot (y_5 - y_0)$$

To eliminate the sensitivity of the time synchronized Euclidean distance to the number of sampled location points, we can interpolate pseudo sampled points on the original trajectory similar to what we discussed earlier regarding the error measure based on the perpendicular Euclidean distance. As shown in Figure 1.4, five pseudo sampled points and their projection to the line segments $\overline{p_0p_5}$ are shown by dash lines. It is also worth noting that the the lengths of line segments on the approximated trajectory are indicators of the time intervals spent instead of distances moved between sampled location points on the original trajectory. By comparing Figure 1.3 and Figure 1.4, it's easy to observe the difference between these two error measures for approximated trajectories.

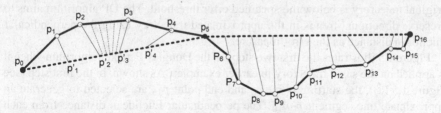

Fig. 1.4 Error measure based on time synchronized Euclidean distance. This error measure takes into account both the geometric relationship and temporal factor of the trajectories.

1.4 Batched Compression Techniques

Given a trajectory that consists of a full series of time-stamped location data points, a batched compression algorithm aims to generate an approximated trajectory by discarding some location points with negligible error from the original trajectory. This is similar to the *line generalization* problem, which has been well studied in the computer graphics and cartography research communities [18, 32, 25, 6]. Works on cartographic line generalization aim to derive small-scale map data from the large-scale high-granularity data. As a result, they can be used to reduce the number of location points in trajectories and thus save storage space.

Some of the line generalization algorithms are very simple in nature. The idea is to retain a fraction of the location points in the original trajectory, without considering the redundancy or other relationships between neighboring data points. For example, the *uniform sampling* algorithm may keep the every i-th location points (e.g. 5th, 10th, 15th, etc) and discard the other points [27]. Since the original trajectory is acquired as a sample of the true trajectory, the new trajectory generated by the uniform sampling process basically is an approximated trajectory with a more coarse granularity. The uniform sampling approach is very efficient computationally, but it may not be useful for certain applications that require better capture of some special trajectory details.

Notice that every location point in the original trajectory may contain different amount of information required to represent the trajectory and that some neighboring location points may contain redundant information, the location points in an approximated trajectory can be selected based on other criteria instead of uniform sampling. A well-known algorithm, called *Douglas-Peucker (DP)*, can be used to approximate the original trajectory [9, 15]. The idea is to replace the original trajectory by an approximate line segment. If the replacement does not meet the specified error requirement, it recursively partitions the original problem into two subproblems by selecting the location point contributing the most errors as the *split point*. This process continues until the error between the approximated trajectory and the original trajectory is below the specified error threshold. The DP algorithm aims to preserve directional trends in the approximated trajectory using the perpendicular Euclidean distance as the error measure.

Figure 1.5 illustrates the first two steps of the Douglas-Peucker algorithm when it is applied on the same trajectory in earlier examples. As shown, in the first step (see Figure 1.5 (a)), the starting point p_0 and end point p_{16} are selected to generate an approximate line segment $\overline{p_0 p_{16}}$. The perpendicular Euclidean distance from each sampled location point on the original trajectory to the approximate line segment $\overline{p_0 p_{16}}$ is derived. Since some of the perpendicular error distances are greater than the pre-defined error distance threshold, the sampled location point deviating the most from $\overline{p_0 p_{16}}$, i.e. p_9 in this example, is chosen as the split point. As a result, in the second step of the algorithm (see Figure 1.5 (b)), a trajectory p_0, p_9, p_{16} is used to approximate the original trajectory. In this step, the original problem is divided into two subproblems where the line segment $\overline{p_0 p_9}$ is to approximate the subtrajectory $\{p_0, p_1, ..., p_9\}$ and the line segment $\overline{p_9 p_{16}}$ is to approximate the other subtrajectory

$\{p_9, p_{10}, ..., p_{16}\}$. As shown, in the first subproblem, several sampled location points have their perpendicular error distances to $\overline{p_0p_9}$ greater than the pre-defined error distance threshold. Therefore, p_5, the sampled location point deviating the most from $\overline{p_0p_9}$, is chosen as the split point and the split subtrajectories are processed recursively until all the sampled location points have perpendicular distances to their approximate line segments within the error threshold. On the other hand, in the second subproblem, the perpendicular distances of all the sample points to the line segment $\overline{p_9p_{16}}$ are smaller than the error threshold. Therefore, further splitting is not necessary.

The Douglas-Peucker algorithm is widely used in cartographic and computer graphic applications. Several studies have analyzed and evaluated various line generalization algorithms mathematically and perceptually and ranked the Douglas-Peucker algorithm highly [18, 32, 25]. Many cartographers considers the Douglas-Peucker algorithm as one of the most accurate line generalization algorithms available but some think it is too costly in terms of processing time. The time complexity of the original Douglas-Peucker algorithm is $O(N^2)$ where N is the number of trajectory location points. Several improvements have been proposed for implementation of the Douglas-Peucker algorithm and reduce its time complexity to $O(NlogN)$ [15].

As we discussed earlier, the error measure of perpendicular Euclidean distance used in the Douglas-Peucker algorithm only takes into account the geometric aspect of the trajectory representation. Unfortunately, it does not capture the important temporal aspect of the trajectories very well. To address this issue, Meratina and de

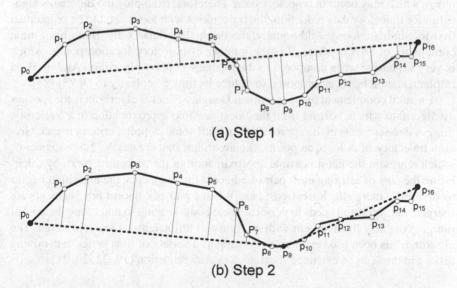

(a) Step 1

(b) Step 2

Fig. 1.5 The Douglas-Peucker algorithm. Line segments are used to approximate the original trajectory. The original trajectory is split into two subtrajectories by selecting the location point contributing the most errors as the split point. In Step (1), p_9 is selected as the split point. In Step (2), p_3 is selected as the split point.

By propose to adopt a new error metrics, called *time-distance ratio metric*, to replace the perpendicular Euclidean distance in the Douglas-Peucker algorithm [24].[7] They claim that the improvement is important because this new error measure is not only more accurate but also taking into consideration both geometric and temporal properties of object movements. A modified Douglas-Peucker algorithm, called the *top-down time-ratio (TD-TR)* algorithm, is proposed in [24] because the Douglas-Peucker algorithm decomposes the trajectory approximation problem in a top-down fashion [19].

The Douglas-Peucker algorithm is based on a heuristic that selects the most deviating location points for inclusion in the approximated trajectory in order to lower the introduced error. However, there is no guarantee that the selected split points are the best choices. To ensure that the approximated trajectory is optimal, dynamic programming technique can be employed even thought its computational cost is expected to be high. The Bellman's algorithm [3] applies the dynamic programming technique to approximate a continuous function $g(x)$ by a finite number of line segments. Even though the algorithm considers a one-dimensional value space, it can be generalized to compute an approximated trajectory in the two-dimensional spatial space. The optimization problem is formulated as to minimize the "area" between the original function and the approximate line segments. In this algorithm, including more line segments in the approximated trajectory fits the original trajectory better but is less effective in terms of compression rate. Thus, the Bellman's algorithm can also adopt a penalty to control the tradeoff between compression rate and quality.

Since Bellman's algorithm approximates a continuous function, it can not handle loops, which may occur in trajectory data. Therefore, to employ the Bellman's algorithm for trajectory data reduction, the trajectories with loops need to be segmented first to eliminate loops. Additionally, the original Bellman's algorithm has a time complexity of $O(N^3)$ where N is the number of trajectory location points, which is very expensive when compared to the Douglas-Peucker algorithm. An improved implementation has been proposed to reduce its time complexity to $O(N^2)$ [23].

A natural complement to the top-down Douglas-Peucker algorithm is the *bottom-up* algorithm which, starting from the finest possible approximation of a trajectory, merges line segments in the approximation until some stopping criteria is met. Given a trajectory of N location points, the algorithm first creates $N/2$ line segments, which represent the finest possible approximation of the trajectory. Next, by calculating the cost of merging each pair of adjacent line segments, the algorithm begins to iteratively merge the lowest-cost pair. When a pair of adjacent line segments are merged, the algorithm needs to perform some book-keeping to make sure the cost of merging the new line segment with its right and left neighbors are considered. The algorithm has been used extensively to support a variety of time series data mining tasks and thus can be extended for trajectory data reduction [19, 22, 20, 21].

---

[7] The time-distance ratio metric is the same as the time synchronized Euclidean distance discussed in Section 1.3.

1.5 On-Line Data Reduction Techniques

The batched compression algorithms, especially the Bellman's algorithm, are expected to produce high-quality approximations due to the access of the whole trajectory. However, they are not as practical as the on-line algorithms in realistic application scenarios. For example, a fleet management application may require trajectory data from tracked moving objects, e.g. trucks, to be reported in a timely fashion back to the fleet control center in order to support multiple continuous queries on truck status in real time. To address the issue of excessive trajectory data continuously generated, there is a demand for on-line trajectory data reduction techniques.

While it's important to reduce the data size of trajectories in order to alleviate storage and communication overheads as well as the computational workload at the location server, there may be certain trajectory properties to be preserved for application needs. Therefore, the on-line trajectory data reduction techniques needs to select some negligible location points intelligently in order to retain a satisfactory approximated trajectory.

One of the essential requirement for on-line processing algorithms is to be able to make efficient on-line decisions when a location point is acquired, i.e. to decide whether to retain the location point in the trajectory or not. The *reservoir sampling* algorithms [30] is well suited for processing trajectory data. The basic idea behind the reservoir sampling algorithms is to maintain a reservoir of size R (or greater than R) which are used to to generate an approximated trajectory of size R. Since the location points in an on-going trajectory are acquired continuously, we do not know in advance the final size of the trajectory. Thus, the key issue is how to select without replacement an approximated trajectory of size R, i.e. once a location point is discarded, there is no way to get it back into the reservoir.

The reservoir algorithm works as follows. It puts the first R location points in the reservoir and decide whether to insert a new location point into the reservoir when it is acquired. Suppose that the k-th location point is acquired (where $k > R$). The algorithm randomly decides, with a probability of R/k, whether this location point should be included as a candidate point in the final approximated trajectory. If the decision is positive, one of the R existing candidates in the reservoir is discarded randomly to make space for the new location point. As such, the algorithm always maintains only R location points in the reservoir, which form a random sample of the original trajectory. Evidently, the reservoir algorithm always maintains a uniform sample of the evolving trajectory without even knowing the eventual trajectory size. Overall, the time complexity is $O(R(1 + logN/R))$, where N is the trajectory size.

While the reservoir sampling algorithm is efficient, it does not consider the sequential, spatial and temporal properties of a trajectory. Since all the location points included in the final trajectory are determined randomly and independently, temporal locality and spatial locality in nearby location points are not considered. The *sliding window* algorithm developed for time series data mining can be adapted for trajectory approximation [19, 24]. The idea is to fit the location points in a growing sliding window with a valid line segment and continue to grow the sliding window (and its corresponding line segment) until the approximation error exceeds some

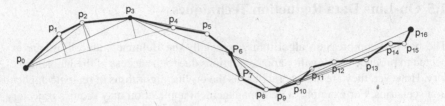

Fig. 1.6 The sliding window algorithm. The idea is to fit the location points in a growing sliding window with a valid line segment and continue to grow the sliding window and its corresponding line segment until the approximation error exceeds some error bound.

error bound. The algorithm first initializes the first location point of a trajectory as the *anchor point* p_a and then starts to grow the sliding window (i.e. by including the next location point in the window). When a new location point p_i is added to the sliding window, the line segment $\overline{p_a p_i}$ is used to fit the subtrajectory consisting of all the location points within the sliding window. As long as the distance errors for all the location points in the sliding window derived against the potential line segment $\overline{p_a p_i}$ are smaller than the user-specified error threshold, the sliding window grows by including the next location point p_{i+1}. Otherwise, the last valid line segment $\overline{p_a p_{i-1}}$ is included as part of the approximated trajectory and p_i is set as the new anchor point. The algorithm continues until all the location points in the original trajectory are visited.

Figure 1.6 illustrates the sliding window algorithm. First, p_0 is set as the anchor point and the initial sliding window is $\{p_0, p_1\}$. Next, p_2 is added into the sliding window. Since $\overline{p_0 p_2}$ fits $\{p_0, p_1, p_2\}$ very well, the sliding window grows into $\{p_0, p_1, p_2, p_3\}$. Again, all the location points within the sliding window do not have error greater than a pre-determined error threshold, i.e. $\overline{p_0 p_3}$ fits $\{p_0, p_1, p_2, p_3\}$ sufficiently well. Thus, the algorithm continues to grow the sliding window into $\{p_0, p_1, p_2, p_3, p_4\}$. This time, the errors for some location points in the sliding window, i.e. p_1, p_2 and p_3, are greater than the error threshold. Thus, the last valid line segment, i.e. $\overline{p_0 p_3}$, is included as a part of the approximated trajectory. Next, the anchor point and the sliding window are reset as p_3 and $\{p_3, p_4\}$, respectively. The algorithm continues to process the rest of the trajectory and then eventually chooses to fit $\{p_3, p_4, p_5, p_6\}$ with $\overline{p_3 p_6}$, $\{p_6, p_7, p_8, p_9\}$ with $\overline{p_6 p_9}$, and $\{p_9, p_{10}, ..., p_{16}\}$ with $\overline{p_9 p_{16}}$. Thus, the final approximated trajectory is $\{p_0, p_3, p_6, p_9, p_{16}\}$.

Meratnia and de By have applied the sliding window algorithm for on-line trajectory data reduction [24]. They consider both of the perpendicular Euclidean distance and time synchronized Euclidean distance as error measures and rename them as *Before Open Window (BOPW)* algorithms, because the location points included in the final approximate trajectory are located before those that result in excessive error. Moreover, Meratnia and de By also apply the heuristic of the Douglas-Peucker algorithm in the open window algorithm. Instead of choosing the location points that result in the longest valid line segments, the new algorithm, called *Normal Opening Window (NOPW)*, chooses location points with the highest error within their sliding window as the *closing point* of the approximating line segment as well as the new anchor point. As it is in the Douglas-Peucker algorithm, this heuristic works very well in reducing the approximation error. With the new anchor point, the NOPW algorithm continues to process the rest of the trajectory.

Figure 1.7 illustrates the NOPW algorithm. First, p_0 is set as the anchor point and the initial open window is $\{p_0, p_1\}$. Next, p_2 is added into the open window. Similar to the illustration in Figure 1.6, $\overline{p_0p_2}$ and $\overline{p_0p_3}$ respectively fits $\{p_0, p_1, p_2\}$ and $\{p_0, p_1, p_2, p_3\}$ sufficiently well, because all the location points within these windows do not have error greater than a pre-determined error threshold. When the opening window grows into $\{p_0, p_1, p_2, p_3, p_4\}$, the errors for p_1, p_2 and p_3 are greater than the error threshold. Instead of choosing p_3 as the closing point, the NOPW algorithm chooses p_2 as the closing point to include the line segment $\overline{p_0p_2}$ as a part of the approximated trajectory. Then, the anchor point and the opening window are reset as p_2 and $\{p_2, p_3, p_4, p_5\}$, respectively. The algorithm continues to process the rest of the trajectory and then eventually chooses $\{p_2, p_2, p_5, p_8, p_{16}\}$ as the approximated trajectory.

1.6 Trajectory Data Reduction Based on Speed and Direction

The data reduction techniques described earlier all use a subset of the location points in the original trajectory as an approximation. In these algorithms, the approximation error, measured by variants of Euclidean distances such as perpendicular Euclidean distance or time synchronized Euclidean distance, are used to select data points that represents the original trajectory as close as possible. In [27], Potamias et. al argue that a data point should be included in the approximated trajectory as long as it reveals changes in the course of a trajectory. As long as the location of an incoming data point can be predicted (e.g. by interpolation or dead reckoning) from the previous movement, this data point can be safely discarded without significant loss in accuracy since it contributes little information. They also argue that, in addition to spatial positions, changes in *speed* and *direction* are key factors for

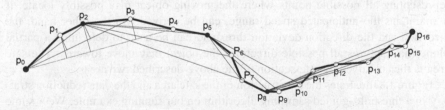

Fig. 1.7 The open window algorithm. The idea is similar to the sliding window algorithm but it applies the heuristic of the Douglas-Peucker algorithm to choose location points with the highest error within their sliding window as the closing point of the approximating line segment as well as the new anchor point.

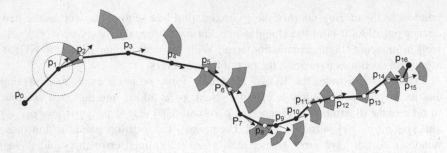

Fig. 1.8 The construction of the safe area and the data reduction strategy in a threshold-guided sampling algorithm. Only the location points fallen out of the projected safe areas are reported to the location server.

predicting locations in a trajectory.[8] Therefore, a decision to include a particular location point or not needs to take into account changes in speed and direction.

In many real life scenario, e.g. driving on a highway, moving objects usually do not make dramatic speed and direction changes. In other words, a moving object may likely to move in the same speed and direction (with some minor changes) for some time. Thus, the current location of a moving object can usually be predicted with little cost by using the speed, direction and time from the last observed location(s).

Based on the specified speed and direction tolerance thresholds, Potamias et. al propose *threshold-guided sampling* algorithms to reduce redundant data points in trajectories [27]. The basic idea is to use a *safe area* derived from the last two locations and a given thresholds to efficiently determine whether a newly acquired location point contains important information. If the new data point, as predicted, is located within the safe area, then this location point is considered as redundant and thus can be discarded. On the other hand, if the location point is fallen outside the safe region, it is included in the approximate trajectory since considerable movement change has happened.

A key issue in the threshold-guided sampling algorithms is the construction of the safe area, which is derived based on the last known location point of the trajectory. Using the last observed speed and the speed tolerance threshold, a circular area representing all possible points where the moving object may possibly locate, if it maintains the anticipated speed range, can be derived. On the other hand, the direction and the direction deviation threshold can be used to determine a partial plane that captures all possible directions the object may move towards. The safe area is then obtained by intersection of the above-described two areas.

Figure 1.8 illustrates the construction of the safe area and the data reduction strategy in a threshold-guided sampling algorithm on our running example. We assume that p_0 and p_1 are included in the approximated trajectory and that the speed and direction of the moving object at p_1 are known. Upon acquisition of the location point

---

[8] To maintain consistency and integrity of discussions in the chapter, here we use the term "direction" to refer to "heading" considered in [27].

p_2, the safe area based on the speed/direction at p_1, the specified speed/direction thresholds, and the time interval between p_2 and p_1, a fan-shape safe area for p_2 is derived. Since p_2 has fallen in its safe area as predicted, the information carried by p_2 is not considered as important. Thus, p_2 is discarded. Next, upon acquisition of the location point p_3, the safe area for p_3 is derived. This time, p_3 has fallen outside the safe area and thus been considered as an important location point for the trajectory. The same decision process has been performed on all the rest of the trajectory location points. The final approximated trajectory in this example is $\{p_0, p_1, p_3, p_4, p_7, p_9, p_{10}, p_{16}\}$.

Observed that the decisions made based on the above-described safe areas are vulnerable to the problem of error propagation, which also exists in dead reckoning, Potamias et. al consider an alternative scheme of constructing a safe area. Instead of constructing the safe area using the last two points included in the approximated trajectory to derive the speed and direction, the new scheme derive the speed and direction from the last two actual location points acquired. Since this scheme is also susceptible to error propagation, when the object movement exhibits a smooth but significant change in the object's direction, Potamias et. al further adopt the joint safe area intersect by the two safe area schemes described earlier.

While the threshold-guided sampling algorithms may achieve significant trajectory data reduction, they may not be effective under the constraint of limited memory. Therefore, Potamias et. al propose another on-line sampling algorithm, called *STTrace*, to obtain an approximated trajectory under a given memory of known and constant size [27]. The idea is to insert data points into the sample memory based on the movement features (e.g., speed and direction) as in the aforementioned threshold-guided sampling algorithms. However, once the memory used to maintain the approximated trajectory is full, we need to decide whether to evict an existing data point (and which one) in order to accommodate a new data point. To address this issue, SSTrace adopts a deletion scheme based on time-synchronous Euclidean distance to discard a data point that results in the least distortion to the maintained approximated trajectory.

In addition to [27], Meratnia and de By also exploit the speed information hidden in the trajectories [24]. By analyzing the derived speeds at subsequent segments of a trajectory, they propose to use the speed difference of two subsequent segments as a criteria to decide whether the location point between the two segments should be retained in the approximated trajectory. Accordingly, a new class of spatio-temporal algorithms are obtained by integrating the speed difference threshold and the time synchronized Euclidean distance with the top-down algorithms and open window algorithms.

Finally, Hung and Peng propose a model-driven data acquisition technique that reports the speeds of a moving object [17]. They develop a kernel regression algorithm and derive a set of kernel functions to model a time series of speeds readings.

1.7 Trajectory Filtering

Spatial trajectories are never perfectly accurate, due to sensor noise and other factors. Sometimes the error is acceptable, such as when using GPS to identify which city a person is in. In other situations, we can apply various filtering techniques to the trajectory to smooth the noise and potentially decrease the error in the measurements. This section explains and demonstrates some conventional filtering techniques using sample data.

It is important to note that filtering is not always necessary. In fact, we rarely use it for GPS data. Filtering is important in those situations where the trajectory data is particularly noisy, or when one wants to derive other quantities from it, like speed or direction.

1.7.1 Sample Data

To demonstrate some of the filtering techniques in this chapter, we recorded a trajectory with a GPS logger, shown in Figure 1.9. The GPS logger recorded 1075 points at a rate of one per second during a short walk around the Microsoft campus in Redmond, Washington USA. For plotting, we converted the latitude/longitude points to (x, y) in meters. While the walk itself followed a casual, smooth path, the recorded trajectory shows many small spikes due to measurement noise. In addition, we manually added some outliers to simulate large deviations that sometimes appear in recorded trajectories. These outliers are marked in Figure 1.9. We will use this data to demonstrate the effects of the filtering techniques we describe below.

1.7.2 Trajectory Model

The actual, unknown trajectory is denoted as a sequence of coordinates $\mathbf{x}_i = (x_i, y_i)^T$. The index i represents time increments, with $i = 1...N$. The boldface $\mathbf{x}_i$ is a two-element vector representing the x and y coordinates of the trajectory coordinate at time i.

Due to sensor noise, measurements are not exact. This error is usually modeled by adding unknown, random Gaussian noise to the actual trajectory points to give the known, measured trajectory, whose coordinates are given as vectors $\mathbf{z}_i$ as

$$\mathbf{z}_i = \mathbf{x}_i + \mathbf{v}_i \tag{1.1}$$

The noise vector $\mathbf{v}_i$ is assumed to be drawn from a two-dimensional Gaussian probability density with zero mean and diagonal covariance matrix R, i.e.

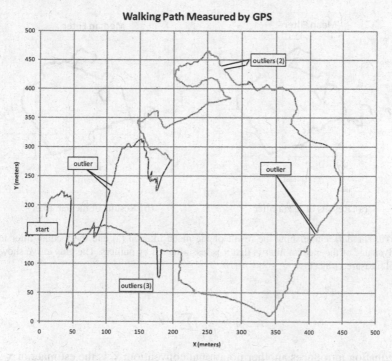

Fig. 1.9 This is a trajectory recorded by a GPS logger. The outliers were inserted later for demonstration.

$$\mathbf{v}_i \sim N(\mathbf{0}, R) \qquad R = \begin{bmatrix} \sigma^2 & 0 \\ 0 & \sigma^2 \end{bmatrix} \qquad (1.2)$$

With the diagonal covariance matrix, this is the same as adding random noise from two different, one-dimensional Gaussian densities to x_i and y_i separately, each with zero mean and standard deviation σ. It is important to note that Equation (1.1) is just a model for noise from a location sensor. It is not an algorithm, but an approximation of how the measured sensor values differ from the true ones. For GPS, the Gaussian noise model above is a reasonable one [7]. In our experiments, we have observed a standard deviation σ of about four meters.

1.8 Mean and Median Filters

One simple way to smooth noise is to apply a mean filter. For a measured point $\mathbf{z}_i$, the estimate of the (unknown) true value is the mean of $\mathbf{z}_i$ and its $n-1$ predecessors in time. The mean filter can be thought of as a sliding window covering n temporally adjacent values of $\mathbf{z}_i$. In equation form, the mean filter is

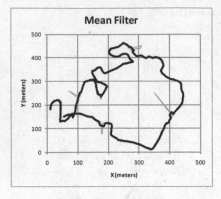

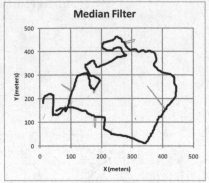

(a) Results of mean filter   (b) Results of median filter

Fig. 1.10 The dark curve show the result of the mean filter in (a) and the median filter in (b). One advantage of the median filter is that it is less affected by outliers. The gray curve shows the original measured trajectory.

$$\hat{\mathbf{x}}_i = \frac{1}{n} \sum_{j=i-n+1}^{i} \mathbf{z}_j \qquad (1.3)$$

This equation introduces another notational convention: $\hat{\mathbf{x}}_i$ is the estimate of $\mathbf{x}_i$.

Figure 1.10(a) shows the result of the mean filter with $n = 10$. The resulting curve is smoother.

The mean filter as given in Equation (1.3) is a so-called "causal" filter, because it only depends on values in the past to compute the estimate $\hat{\mathbf{x}}_i$. In fact, all the filters discussed in this chapter are causal, meaning they can be sensibly applied to real time data as it arrives. For post-processing, one could use a non-causal mean filter whose sliding window takes into account both past and future values to compute $\hat{\mathbf{x}}_i$.

One disadvantage of the mean filter is that it introduces lag. If the true underlying value $\mathbf{x}_i$ changes suddenly, the estimate from the mean filter will respond only gradually. So while a larger sliding window (larger value of n) makes the estimates smoother, the estimates will also tend to lag changes in $\mathbf{x}_i$. One way to mitigate this problem is to use a weighted mean, where more recent values of $\mathbf{z}_i$ are given more weight.

Another disadvantage of the mean filter is its sensitivity to outliers. From Figure 1.10(a), it is clear that the artificially introduced outliers noticeably pull away the estimated curve from the data. In fact, it is possible to find an outlier value to pull the mean to any value we like.

One way to mitigate the outlier problem is to use a median filter rather than a mean filter. The median filter simply replaces the mean filter's mean with a median. The equation for the median filter that corresponds to the mean filter in Equation (1.3) is

$$\hat{\mathbf{x}}_i = median\{\mathbf{z}_{i-n+1}, \mathbf{z}_{i-n+2}, ..., \mathbf{z}_{i-1}, \mathbf{z}_i\} \qquad (1.4)$$

Figure 1.10(b) shows the result of the median filter, where it is clear that it is less sensitive to outliers and still gives a smooth result.

The mean and median filters are both simple and effective at smoothing a trajectory. They both suffer from lag. More importantly, they are not designed to help estimate higher order variables like speed. In the next two sections, we discuss the Kalman filter and the particle filter, two more advanced techniques that reduce lag and can be designed to estimate more than just location.

1.9 Kalman Filter

The mean and median filters use no model of the trajectory. More sophisticated filters, like the Kalman and particle filters, model both the measurement noise (as given by Equation (1.1)) and the dynamics of the trajectory.

For the Kalman filter, a simple example is smoothing trajectory measurements from something arcing through the air affected only by gravity, such as a soccer ball. While measurements of the ball's location, perhaps from a camera, are noisy, we can also impose constraints on the ball's trajectory from simple laws of physics. The trajectory estimate from the Kalman filter is a tradeoff between the measurements and the motion model. Besides giving estimates that obey the laws of physics, the Kalman filter gives principled estimates of higher order motion states like speed.

The subsections below develop the model for the Kalman filter for the example trajectory from above. We use notation from the book by Gelb, which is one of the standard references for Kalman filtering [1].

1.9.1 Measurement Model

While the mean and median filters can only estimate what is directly measured, the Kalman filter can estimate other variables like speed and acceleration. In order to do this, the Kalman formulation makes a distinction between what is measured and what is estimated, as well as formulating a linear relationship between the two.

As above, we assume that the measurements of the trajectory are taken as noisy values of x and y:

$$\mathbf{z}_i = \begin{pmatrix} z_i^{(x)} \\ z_i^{(y)} \end{pmatrix} \qquad (1.5)$$

Here $z_i^{(x)}$ and $z_i^{(y)}$ are noisy measurements of the x and y coordinates.

The Kalman filter gives estimates for the state vector, which describes the full state of the object being tracked. In our case, the state vector will include both the object's location and velocity:

$$\mathbf{x}_i = \begin{pmatrix} x_i \\ y_i \\ s_i^{(x)} \\ s_i^{(y)} \end{pmatrix} \tag{1.6}$$

The elements x_i and y_i are the true, unknown coordinates at time i, and $s_i^{(x)}$ and $s_i^{(y)}$ are the x and y components of the true, unknown velocity at time i. The Kalman filter will produce an estimate of $\mathbf{x}_i$, which includes velocity, even though this is not directly measured. The relationship between the measurement vector $\mathbf{z}_i$ and the state vector $\mathbf{x}_i$ is

$$\mathbf{z}_i = H_i \mathbf{x}_i + \mathbf{v}_i \tag{1.7}$$

where H_i, the measurement matrix, translates between $\mathbf{x}_i$ and $\mathbf{z}_i$. For our example, H_i expresses the fact that we are measuring x_i and y_i to get $z_i^{(x)}$ and $z_i^{(y)}$, but we are not measuring velocity. Thus,

$$H_i = \begin{bmatrix} 1 & 0 & 0 & 0 \\ 0 & 1 & 0 & 0 \end{bmatrix} \tag{1.8}$$

H_i also neatly accounts for the dimensionality difference between $\mathbf{x}_i$ and $\mathbf{z}_i$. While the subscript on H_i means it could change with time, it does not in our example.

The noise vector $\mathbf{v}_i$ in Equation (1.7) is the same as the zero-mean, Gaussian noise vector in Equation (1.2). Thus Equation (1.7) is how the Kalman filter models measurement noise. In fact, Gaussian noise has been proposed as a simple model of GPS noise [7], and for our example it would be reasonable to set the measurement noise σ to a few meters.

1.9.2 Dynamic Model

If the first half of the Kalman filter model is measurement, the second half is dynamics. The dynamic model approximates how the state vector $\mathbf{x}_i$ changes with time. Like the measurement model, it uses a matrix and added noise:

$$\mathbf{x}_i = \Phi_{i-1} \mathbf{x}_{i-1} + \mathbf{w}_{i-1} \tag{1.9}$$

This gives $\mathbf{x}_i$ as a function of its previous value $\mathbf{x}_{i-1}$. The system matrix Φ_{i-1} gives the linear relationship between the two. For the example problem, we have

$$\Phi_{i-1} = \begin{bmatrix} 1 & 0 & \Delta t_i & 0 \\ 0 & 1 & 0 & \Delta t_i \\ 0 & 0 & 1 & 0 \\ 0 & 0 & 0 & 1 \end{bmatrix} \tag{1.10}$$

Here Δt_i is the elapsed time between the state at time i and time $i-1$. Recalling the state vector from Equation (1.6), the top two rows of the system matrix say that $x_i = x_{i-1} + \Delta t_i s_i^{(x)}$ and similarly for y_i. This is standard physics for a particle with constant velocity.

The bottom two rows of system matrix say $s_i^{(x)} = s_{i-1}^{(x)}$ and $s_i^{(y)} = s_{i-1}^{(y)}$, which means the velocity does not change. Of course, we know this is not true, or else the trajectory would be straight with no turns. The dynamic model accounts for its own inaccuracy with the noise term $\mathbf{w}_{i-1}$. This is another zero-mean Gaussian noise term. For our example, we have

$$\mathbf{w}_i \sim N(\mathbf{0}, Q_i) \qquad Q_i = \begin{bmatrix} 0 & 0 & 0 & 0 \\ 0 & 0 & 0 & 0 \\ 0 & 0 & \sigma_s^2 & 0 \\ 0 & 0 & 0 & \sigma_s^2 \end{bmatrix} \qquad (1.11)$$

With the first two rows of zeros, this says that the relationship between location and velocity (e.g. $x_i = x_{i-1} + \Delta t_i s_i^{(x)}$) is exact. However, the last two rows say that the assumption in the system matrix about constant velocity is not quite true, but that the velocity is noisy, i.e. $s_i^{(x)} = s_i^{(x)} + N(0, \sigma_s^2)$. This is how the Kalman filter maintains its assumption about the linear relationship between the state vectors over time, yet manages to account for the fact that the dynamic model does not account for everything.

1.9.3 Entire Kalman Filter Model

The Kalman filter requires a measurement model and dynamic model, both discussed above. It also requires assumptions about the initial state and uncertainty of the initial state. Here are the all the required elements:

H_i – measurement matrix giving measurement $\mathbf{z}_i$ from state $\mathbf{x}_i$, Equation (1.8).
R_i – measurement noise covariance matrix, Equation (1.2).
Φ_{i-1} – system matrix giving state $\mathbf{x}_i$ from $\mathbf{x}_{i-1}$, Equation (1.10).
Q_i – system noise covariance matrix, Equation (1.11).
$\hat{\mathbf{x}}_0$ – initial state estimate.
P_0 – initial estimate of state error covariance.

The initial state estimate can usually be estimated from the first measurement. For our example, the initial position came from $\mathbf{z}_0$, and the initial velocity was taken as zero. For P_0, a reasonable estimate for this example is

$$p_0 = \begin{bmatrix} \sigma^2 & 0 & 0 & 0 \\ 0 & \sigma^2 & 0 & 0 \\ 0 & 0 & \sigma_s^2 & 0 \\ 0 & 0 & 0 & \sigma_s^2 \end{bmatrix} \qquad (1.12)$$

The value of σ is an estimate of the sensor noise for GPS. For our example, we set $\sigma = 4$ meters based on earlier experiments with our particular GPS logger. We set $\sigma_s = 6.62$ meters/second , which we computed by looking at the changes in velocity estimated naively from the measurement data.

1.9.4 Kalman Filter

For the derivation of the Kalman filter, see [1]. The result is a two-step algorithm that first extrapolates the current state to the next state using the dynamic model. In equations, this is

$$\hat{\mathbf{x}}_i^{(-)} = \Phi_{i-1}\hat{\mathbf{x}}_{i-1}^{(+)} \tag{1.13}$$

$$P_i^{(-)} = \Phi_{i-1}P_{i-1}^{(+)}\Phi_{i-1}^T + Q_{i-1} \tag{1.14}$$

The terms in Equation (1.13) should be familiar. The $*^{(-)}$ superscript refers to the extrapolated estimate of the state vector, and the $*^{(+)}$ superscript refers to the estimated value of the state vector. Equation (1.14) is interesting in that it concerns an extrapolation P_i of the covariance of the state vector, giving some idea of the error associated with the state vector.

The first step of the Kalman filter is pure extrapolation, with no use of measurements. The second step incorporates the current measurement to make new estimates. The equations are

$$K_i = P_i^{(-)}H_i^T(H_iP_i^{(-)}H_i^T + R_i)^{-1} \tag{1.15}$$

$$\hat{\mathbf{x}}_i^{(+)} = \hat{\mathbf{x}}_i^{(-)} + K_i(\mathbf{z}_i - H_i\hat{\mathbf{x}}_i^{(-)}) \tag{1.16}$$

$$P_i^{(+)} = (I - K_iH_i)P_i^{(-)} \tag{1.17}$$

Equation (1.15) gives the Kalman gain matrix K_i. It is used in Equation (1.16) to give the state estimate $\hat{\mathbf{x}}_i^{(+)}$ and in Equation (1.17) to give the state covariance estimate $P_i^{(+)}$.

Applying these equations to the example trajectory gives the plot in Figure 1.11(a).

1.9.5 Kalman Filter Discussion

One of the advantages of the Kalman filter over the mean and median filters is its lack of lag. There is still some intrinsic lag, because the Kalman filter depends on previous measurements for its current estimate, but it also includes a dynamic model to keep it more current. Another advantage is the richer state vector. In our example, it includes velocity as well as location, making the Kalman filter a principled way to estimate velocity based on a sequence of location measurements. It is easy to add acceleration as another part of the state vector. Yet another advantage is the Kalman filter's estimate of uncertainty in the form of the covariance matrix $P_i^{(+)}$ given in Equation 1.17. Knowledge of the uncertainty of the estimate can be used by a higher-level algorithm to react intelligently to ambiguity, possibly by invoking another sensor or asking a user for help.

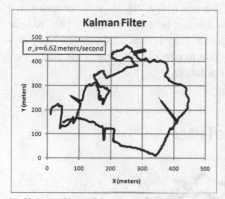 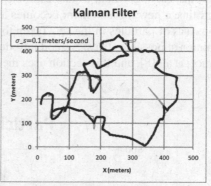

(a) Kalman filter with $\sigma_s = 6.62$ meters/second (b) Kalman filter with $\sigma_s = 0.1$ meters/second

Fig. 1.11 The dark curve show the result of the Kalman filter. In (a), the process noise σ_s comes from an estimate on the original noisy data. The process noise in (b) is much smaller, leading to a smoother filtered trajectory.

One of the mysteries of the Kalman filter is the process noise, which is embodied as σ_s in our example. In our example, this represents how much the tracked object's velocity changes between time increments. In reality, this is difficult to estimate in many cases, including our example trajectory of a pedestrian. A larger value of σ_s represents less faith in the dynamic model relative to the measurements. A smaller value puts more weight on the dynamic model, often leading to a smoother trajectory. This is illustrated in Figure 1.11(b) where we have reduced the value of σ_s from our original value of 6.62 meters/second to 0.1 meters/second. The resulting trajectory is indeed smoother and less distracted by outliers.

One of the main limitations of the Kalman filter is the requirement that the dynamic model be linear, i.e. that the relationship between x_{i-1} and x_i be expressed as a matrix multiplication (plus noise). Sometimes this can be solved with an extended Kalman filter, which linearizes the problem around the current value of the state. But this can be difficult for certain processes, like a bouncing ball or an object constrained by predefined paths. In addition, all the variables in the Kalman filter model are continuous, without a convenient way to represent discrete variables like the mode of transportation or goal. Fortunately, the particle filter fixes these problems, and we discuss it next.

1.10 Particle Filter

The particle filter is similar to the Kalman filter in that they both use a measurement model and a dynamic model. The Kalman filter gains efficiency by assuming linear models (matrix multiplication) plus Gaussian noise. The particle filter relaxes these assumptions for a more general, albeit generally less efficient, algorithm. But, as shown by Hightower and Borriello, particle filters are practical for tracking even on mobile devices [16].

The particle filter gets its name from the fact that it maintains a set of "particles" that each represent a state estimate. There is a new set of particles generated each time a new measurement becomes available. There are normally hundreds or thousands of particles in the set. Taken together, they represent the probability distribution of possible states. A good introduction to particle filtering is the chapter by Doucet et al. [8], and this section uses their notation.

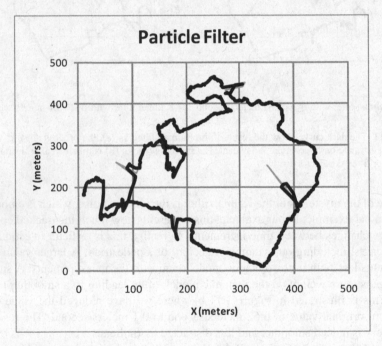

Fig. 1.12 This is the result of the particle filter. It is similar to the result of the Kalman filter in Figure 3(a) since it uses the same measurement model, dynamic model, and noise assumptions.

As in the previous section on Kalman filtering, this section shows how to apply particle filtering to our example tracking problem.

1.10.1 Particle Filter Formulation

As with the Kalman filter discussed above, the particle filter makes estimates $\hat{x}_i$ of a sequence of unknown state vectors $\mathbf{x}_i$, based on measurements $\mathbf{z}_i$. For our example, these vectors are formulated just as in the Kalman filter. As a reminder, the state vector $\mathbf{x}_i$ has four scalar elements representing location and velocity. The measurement vector $\mathbf{z}_i$ has two elements representing a location measurement with some degree of inaccuracy.

The particle filter's measurement model is a probability distribution $p(\mathbf{z}_i|\mathbf{x}_i)$ giving the probability of seeing a measurement given a state vector. This distribution must be provided to the particle filter. This is essentially a model of a noisy sensor which might produce many different possible measurements for a given state. It is much more general than the Kalman filter's measurement model, which is limited to $\mathbf{z}_i = H_i\mathbf{x}_i + \mathbf{v}_i$, i.e. a linear function of the state plus added Gaussian noise.

To stay consistent with the example, however, we will use the same measurement model as in the Kalman filter, writing it as

$$p(\mathbf{z}_i|\mathbf{x}_i) = N((x_i, y_i)^T, R_i) \tag{1.18}$$

This says that the measurement is a Gaussian distributed around the actual location with a covariance matrix R_i from Equation (1.2). The measurement ignores the velocity components in $\mathbf{x}_i$.

While this is the same measurement model that we used for the Kalman filter, it could be much more expressive. For instance, it might be the case that the location sensor's accuracy varies with location, such as GPS in an urban canyon. The particle filter's model could accommodate this.

In addition to the measurement model, the other part of the particle filter formulation is the dynamic model, again paralleling the Kalman filter. The dynamic model is also a probability distribution, which simply gives the distribution over the current state $\mathbf{x}_i$ given the previous state $\mathbf{x}_{i-1}$: $p(\mathbf{x}_i|\mathbf{x}_{i-1})$. The analogous part of the Kalman filter model is $\mathbf{x}_i = \Phi_{i-1}\mathbf{x}_{i-1} + \mathbf{w}_{i-1}$ (Equation (1.9)). The particle filter version is much more general. For instance, it could model the fact that vehicles often slow down when climbing hills and speed up going down hills. It can also take into account road networks or paths through a building to constrain where a trajectory can go. This feature of the particle filter has proved useful in many tracking applications.

It is not necessary to write out the dynamic model $p(\mathbf{x}_i|\mathbf{x}_{i-1})$. Instead, it is sufficient to sample from it. That is, given a value of $\mathbf{x}_{i-1}$, we must be able to create samples of $\mathbf{x}_i$ that adhere to $p(\mathbf{x}_i|\mathbf{x}_{i-1})$. For our example, we will use the same dynamic model as the Kalman filter, which says that location changes deterministically as a function of the velocity and that velocity is randomly perturbed with Gaussian noise:

$$x_{i+1} = x_i + v_i^{(x)} \Delta t_i$$
$$y_{i+1} = y_i + v_i^{(y)} \Delta t_i$$
$$v_{i+1}^{(x)} = v_i^{(x)} + w_i^{(x)} \qquad w_i^{(x)} \sim N(0, \sigma_s^2) \qquad (1.19)$$
$$v_{i+1}^{(y)} = v_i^{(y)} + w_i^{(y)} \qquad w_i^{(y)} \sim N(0, \sigma_s^2)$$

The above is a recipe for generating random samples of $\mathbf{x}_{i+1}$ from $\mathbf{x}_i$. Contrary to the Kalman filter, the particle filter requires actually generating random numbers, which in this case serve to change the velocity.

Finally, also paralleling the Kalman filter, we need an initial distribution of the state vector. For our example, we can say the initial velocity is zero and the initial location is a Gaussian around the first measurement with a covariance matrix R_i from Equation (1.2).

1.10.2 Particle Filter

The particle filter maintains a set of P state vectors, called particles: $\mathbf{x}_i^{(j)}, j = 1...P$. There are several versions of the particle filter, but we will present the Bootstrap Filter from Gordon [12]. The initialization step is to generate P particles from the initial distribution. For our example, these particles would have zero velocity and be clustered around the initial location measurement with a Gaussian distribution as explained above. This is the first instance of how the particle filter requires actually generating random hypotheses about the state vector. This is different from the Kalman filter which generates state estimates and uncertainties directly. We will call these particles $\mathbf{x}_0^{(j)}$.

With a set of particles and $i > 0$, the first step is "importance sampling," which uses the dynamic model $p(\mathbf{x}_i|\mathbf{x}_{i-1})$ to probabilistically simulate how the particles change over one time step. This is analogous to the extrapolation step in the Kalman filter in that it proceeds without regard to the measurement. For our example, this means invoking Equation (1.19) to create $\tilde{\mathbf{x}}_i$. The tilde ($\sim$) indicates extrapolated values. Note that this involves actually generating random numbers for the velocity update.

The next step computes "importance weights" for all the particles using the measurement model. The importance weights are

$$\tilde{w}_i^{(j)} = p(\mathbf{z}_i|\tilde{\mathbf{x}}_i^{(j)}) \qquad (1.20)$$

Larger importance weights correspond to particles that are better supported by the measurement. The important weights are then normalized so they sum to one.

The last step in the loop is the "selection step" when a new set of P particles $\mathbf{x}_i^{(j)}$ is selected at random from the $\tilde{\mathbf{x}}_i^{(j)}$ based on the normalized importance weights. The probability of selecting an $\tilde{\mathbf{x}}_i^{(j)}$ for the new set of particles is proportional to its importance weight $\tilde{w}_i^{(j)}$. It is not unusual to select the same $\tilde{\mathbf{x}}_i^{(j)}$ more than once if it has a larger importance weight. This is the last step in the loop, and processing then returns to the importance sampling step with the new set of particles.

While the particles give a distribution of state estimates, one can compute a single estimate with a weighted sum:

$$\tilde{\mathbf{x}}_i = \sum_{j=1}^{P} \tilde{w}_i^{(j)} \tilde{\mathbf{x}}_i^{(j)} \tag{1.21}$$

Applying the particle filter to our example problem gives the result in Figure 1.12. We used $P = 1000$ particles. This result looks similar to the Kalman filter result, since we used the same measurement model, dynamic model, and noise in both.

1.10.3 Particle Filter Discussion

One potential disadvantage of the particle filter is computation time, which is affected by the number of particles. More particles generally give a better result, but at the expense of computation. Fox gives a method to choose the number of particles based on bounding the approximation error [10].

Even though the particle filter result looks similar to the Kalman filter result in our example, it is important to understand that the particle filter has the potential to be much richer. As mentioned previously, it could be made sensitive to a network of roads or walking paths. It could include a discrete state variable representing the mode of transportation, e.g. walking, bicycling, in a car, or on a bus.

While it is tempting to add many variables to the state vector, the cost is often more particles required to make a good state estimate. One solution to this problem is the Rao-Blackwellized particle filter [26]. It uses a more conventional filter, like Kalman, to track some of the state variables and a particle filter for the others.

1.11 Summary

In this chapter, we discussed two low-level preprocessing tasks for spatial trajectory computing and data management: 1) how to reduce the data size for representing a trajectory; and 2) how to filter spatial trajectories to reduce measurement noise and to estimate higher level properties of a trajectory. For task 1, the data reduction techniques can run in a batch mode after the data is collected or in an on-line mode as the data is collected. Due to the inherent spatio-temporal characteristics in spatial trajectories, conventional error measure, e.g. the perpendicular Euclidean distance that has been widely used in many line generalization algorithms, does not work well in determining the location points to be included in the approximated trajectory. On the other hand, the time synchronized Euclidean distance, providing a more precise error measurement for approximated trajectories, has been incorporated into several trajectory data reduction techniques recently. Moreover, research on trajectory data reduction techniques have been extended from focusing on location information to high-level properties of trajectories such as speed and directions. For task 2, trajectory filtering techniques are employed to reduce measurement noise. Additionally, they can be used to estimate high-level properties of a trajectory like its speed and direction and thus can possibly be integrated with trajectory data reduction techniques. Trajectory filtering methods, including mean and median filtering, the Kalman filter, and the particle filter, are important techniques for trajectory data preprocessing.

Many transportational and recreational activities have left very useful information in form of trajectories. In recent years, researchers have started to explore the semantics, e.g. activity types and transportation modes, behind various trajectories and thus proposed the notion of *semantic trajectories* [31, 2, 34, 11]. Accordingly, semantic compression techniques, while in its infancy, have been proposed [29, 4]. We anticipate more advanced data reduction and filtering techniques for semantic trajectories to be developed in the coming years as we obtain more in-depth understanding of these concepts.

References

1. A. Gelb, et al.: Applied Optimal Estimation. The MIT Press (1974)
2. Alvares, L., Bogorny, V., Kuijpers, B., de Macelo, J., Moelans, B., Palma, A.: Towards semantic trajectory knowledge discovery. Tech. rep., Hasselt University (2007)
3. Bellman, R.: On the Approximation of Curves by Line Segments Using Dynamic Programming. Communications of the ACM **4**(6), 284 (1961)
4. Bogorny, V., Valiati, J., Alvares, L.: Semantic-based Pruning of Redundant and Uninteresting Frequent Geographic Patterns. Geoinformatica **14**(2), 201–220 (2010)
5. Civilis, A., Jensen, C., Nenortaite, J., Pakalnis, S.: Efficient Tracking of Moving Objects with Precision Guarantee. In: IEEE International Conference on Mobile and Ubiquitus Systems: Networking and Services (MobiQuitous) (2004)
6. C.T. Lawson and S.S. Rvi and J.-H. Hwang: Compression and Mining of GPS Trace Data: New Techniques and Applications. Technical Report. Region II University Transportation Research Center
7. van Diggelen, F.: GNSS Accuracy: Lies, Damn Lies, and Statistics. GPS World (2007)
8. Doucet, A., Freitas, N., Gordon, N.: An Introduction to Sequential Monte Carlo Methods. In: Sequential Monte Carlo Methods in Practice., pp. 3–13. Springer: New York (2001)
9. Douglas, D., Peucker, T.: Algorithms for the Reduction of the Number of Points Required to Represent a Line or its Caricature. The Canadian Cartographer **10**(2), 112–122 (1973)
10. Fox, D.: Adapting the Sample Size in Particle Filters Through KLD-Sampling. The International Journal of Robotics Research **22**(12), 985–1003 (2003)
11. Giannotti, F., Nanni, M., Pedreschi, D., Renso, C., Trasarti, R.: Mining Mobility Behavior from Trajectory Data. In: International Conference on Computational Science and Engineering (CSE)., pp. 948–951 (2009)
12. Gordon, N.: Bayesian Methods for Tracking. Imperial College, University of London, London (1994)
13. Guting, R., M.H.Bohlen, Erwig, M., Jensen, C., Lorentzos, N., M, S., Vazirgiannis, M.: A Foundation for Representing and Querying Moving Objects. ACM Transaction on Database Systems (TODS) **25**, 1–42 (2000)

14. G"uting, R., Schneider, M.: Moving Object Databases. Morgan Kaufmann, San Francisco, CA (2005)
15. Hershberger, J., Snoeyink, J.: Speeding up the Douglas-Peucker Line simplification Algorithm. In: International Symposium on Spatial Data Handling, pp. 134–143 (1992)
16. Hightower, J., Borriello, G.: Particle Filters for Location Estimation in Ubiquitous Computing: A Case Study. In: in 6th International Conference on Ubiquitous Computing., pp. 88–106 (2004)
17. Hung, C.C., Peng, W.C.: Model Driven Traffic Data Acquisition in Vehicle Sensor Networks. In: International Conference of Parallen Processing (ICPP)., pp. 424–432 (2011)
18. Jenks, G.: Lines, Computers, and Human Frailties. Annuals of the Association of American Geographers **71**, 1–10 (1981)
19. Keogh, E., Chu, S., Hart, D., Pazzani, M.: An On-Line Algorithm for Segmenting Time Series. In: International Conference on Data Mining (ICDM), pp. 289–296 (2001)
20. Keogh, E., Pazzani, M.: An Enhanced Representation of Time Series which Allows Fast and Accurate Classification, Clustering and Relevance Feedback. In: International Conference of Knowledge Discovery and Data Mining (KDD)., pp. 239–241 (1998)
21. Keogh, E., Pazzani, M.: An On-Line Algorithm for Segmenting Time Series. In: Annual International ACM-SIGIR Conference on Research and Development in Information Retrieval (SIGIR). (1999)
22. Keogh, E., Smyth, P.: A Probabilistic Approach to Fast Pattern Matching in Time Series Databases. In: International Conference of Knowledge Discovery and Data Mining (KDD)., pp. 24–30 (1997)
23. Kleinberg, J., Tardos, E.: Algorithm Design. Addison Wesley, Reading, MA (2005)
24. Maratnia, N., de By, R.: Spatio-Temporal Compression Techniques for Moving Point Objects. In: International Conference on Extending Database Technology (EDBT), pp. 765–782 (2004)
25. McMaster, R.: Statistical Analysis of Mathematical Measures of Linear Simplification. The American Cartographer **13**, 103–116 (1986)
26. Murphy, K., Russell, S.: Rao-Blackwellised Particle Filtering for Dynamic Bayesian Networks. In: Sequential Monte Carlo Methods in Practice., pp. 499–515. Springer: New York (2001)
27. Potamias, M., Patroumpas, K., Sellis, T.: Sampling Trajectory Streams with Spatio-Temporal Criteria. In: International Conference on Scientific and Statistical Database Management (SSDBM), pp. 275–284 (2006)
28. Saltenis, S., Jensen, C., Leutenegger, S., Lopez, M.: Indexing the Positions of Continuously Moving Objects. In: ACM International Conference on Management of Data (SIGMOD), pp. 331–342 (2000)
29. Schmid, F., Richter, K.F., Laube, P.: Semantic Trajectory Compression. In: International Symposium on Advances in Spatial and Temporal Databases (SSTD). (2009)
30. Vitter, J.: Random sampling with a reservoir. ACM Transactions on Mathematical Software (TOMS) **11**(1) (1985)
31. Wang, X., Tieu, K., Grimson, E.: Learning Semantic Scene Models by Trajectory Analysis. In: European Conference on Computer Vision (ECCV), pp. 110–123 (2006)
32. White, E.: Assessment of Line Generalization Algorithms. The American Cartographer **12**, 17–27 (1985)
33. Wolfson, O., Sistla, P., Xu, B., Zhou, J., Chamberlain, S., Yesha, Y., Rishe, N.: Tracking Moving Objects Using Database Technology in DOMINO. In: The Fourth Workshop on Next Generation Information Technologies and Systems (NGITS), pp. 112–119 (1999)
34. Yan, Z.: Towards Semantic Trajectory Data Analysis: A Conceptual and Computational Approach. In: Internation Conference on Very Large Data Base (VLDB) PhD Workshop. (2009)

Chapter 2
Trajectory Indexing and Retrieval

Ke Deng, Kexin Xie, Kevin Zheng and Xiaofang Zhou

Abstract The traveling history of moving objects such as a person, a vehicle, or an animal have been exploited in various applications. The utility of trajectory data depends on the effective and efficient trajectory query processing in trajectory databases. Trajectory queries aim to evaluate spatiotemporal relationships among spatial data objects. In this chapter, we classify trajectory queries into three types, and introduce the various distance measures encountered in trajectory queries. The access methods of trajectories and the basic query processing techniques are presented as another component of this chapter.

2.1 Introduction

Trajectories are the traveling history of moving objects such as a person, a vehicle, or an animal. Trajectories can be used for complex analysis across different domains. For example, public transport systems may go back in time to any particular instant or period to analyze the pattern of traffic flow and the causes of traffic congestions; in biological studies movements of animals may be analyzed when considering road networks to reveal the impact of human activity on wild life; or the urban planning of a city council may analyze the trajectories to predict the development of suburbs and provide support in decision making. Other applications include path optimization of

Ke Deng
The University of Queensland, Brisbane, Australia, e-mail: dengke@itee.uq.edu.au

Kexin Xie
The University of Queensland, Brisbane, Australia, e-mail: kexin@itee.uq.edu.au

Kevin Zheng
The University of Queensland, Brisbane, Australia, e-mail: kevinz@itee.uq.edu.au

Xiaofang Zhou
The University of Queensland, Brisbane, Australia, e-mail: zxf@itee.uq.edu.au

logistics companies, improvement of public security management, and personalized location-based services, etc.

Performing this complex analysis requires trajectory databases to support trajectory queries effectively. Trajectory queries can be classified into three types according to their spatiotemporal relationships: 1) trajectories and points (e.g. find all trajectories within 500m of a gas station between 9:00pm-9:30pm), 2) trajectories and regions (e.g. find the region which is passed by τ trajectories between 9:00pm-9:30pm), and 3) trajectories and trajectories (e.g. travelers who may take a similar path in the coming 30mins). Clearly, the spatiotemporal relationship in trajectory queries concerns not only the topological relationship such as passing a region, but also the distance measures between spatial objects from the simple such as Euclidean distances to the complex such as the similarities between trajectories. The background diversity of applications determines that the definition of distance measure is variable and should have a sound interpretation in certain application contexts. Several popular distance functions are introduced in this chapter.

Due to their historical nature, the size of trajectory databases is supposed to be very large. Thus, a critical aspect of trajectory databases is to support an effective trajectory index to accelerate query processing. In general, the spatiotemporal data index technique is an extension of the spatial data index with augmentation of a time dimension. However, the trajectory data have specific requirements to index techniques due to unique data characteristics, i.e. continuous long period of time, and due to unique query characteristics, i.e. often asking for information in an instantaneous/continuous time window. We present several representative trajectory indexes in this chapter with processing techniques for various trajectory query types.

This chapter is organized as follows. In section 2.2, we classify trajectory queries into three types. Section 2.3 focuses on the trajectory similarity measures/distance matrices used in trajectory queries. A number of trajectory indexes are discussed in section 2.4 and in section 5 we present the processing of trajectory queries with the support of index. This chapter is summarized in section 2.6.

2.2 Trajectory Query Types

Trajectories are the historical spatiotemporal data which are the foundation of many important and practical applications such as traffic analysis, behavior analysis and so on. The utility of trajectory data is built on various queries in a trajectory database. A typical trajectory query asks for the information against spatiotemporal relationships between trajectories and with other spatial data objects, i.e. points (P), regions (R) and trajectories (T). We generally classify the trajectory queries into three types:

- P-Query   1) asks for points of interest (POI) which satisfy the spatiotemporal relationship to specified trajectory segment(s) (e.g. top k nearest neighbors); or 2) asks for trajectories which satisfy the spatiotemporal relationship to a specified point/points.

- R-Query   1) asks for trajectories passing a given spatiotemporal region *R*; or 2) asks for spatiotemporal regions which are overlapped or frequently passed by trajectories.
- T-Query   1) asks for similar trajectories in a set of trajectories; or 2) asks for trajectories within a distance threshold (i.e. nearest approaching points of two trajectories).

In addition to spatiotemporal relationships, the query in a trajectory database may ask for trajectories which satisfy extra navigational conditions (e.g. top speed and direction within a certain area during a given time interval). To process such queries, the relevant trajectory segments extracted using R-query are examined against the navigational conditions. In trajectory queries, the absence of specification in time (time stamp or interval) implies that any character of object within a time dimension is acceptable. But, we treat this situation as a special case in trajectory queries.

2.2.1 P-Query

2.2.1.1 Ask for POI

P-query aims to find spatiotemporal point/points of interest which satisfy expected spatiotemporal relationships given segments of a trajectory/trajectories. A basic type of P-query is to retrieve the nearest neighbor of every point in a trajectory segment [5]. For example, you want to find all nearest gas stations along the specified trajectory segment from point *s* to point *e*. The result contains a set of < *POI*, *trajectory_interval* > tuples, such that *POI* is the nearest neighbor of all points in the corresponding trajectory interval. Some variants of P-query consider more practical factors in the context of a road network, that is, the traveling direction and destination along the trajectory [30, 29, 9]. The aim of such P-queries is to find one point from many, such as a set of gas stations, via which the detour from the original route (i.e. leave the route and come back after visiting the gas station) on the way to the destination is the smallest.

P-Query can also be an envelop query (or range query) which asks for all POIs (i.e. gas stations) whose distances (according to a distance metric) to a given trajectory is less than a threshold in a specified time interval.

2.2.1.2 Ask for Trajectory

P-query may ask for segment(s) of trajectory/trajectories when a given spatiotemporal point/points are specified. The single point based query [11, 26] looks for the nearest trajectories to only one point (e.g. a supermarket). Similarly, P-Query for trajectory may also ask for all trajectories which are within a proximity of a point such as within 500m in a time interval.

In the multiple-points trajectory query [10], given a small set of points with or without an order specified, the target is to find the trajectories from a trajectory database such that these trajectories best connect the designated points geographically. This query is useful when planning a traveling path passing several must-go locations such as sightsee points.

2.2.2 R-Query

2.2.2.1 Ask for Trajectory

Given a three dimensional spatiotemporal region (or block), R-query searches for trajectory segments that belong to the specified spatiotemporal region. This type of query can be used to support traffic analysis, and located-based services. As a basic operation in various applications, this query type has been studied extensively and various indexes have been proposed (TB, 3D-R tree, STR tree).

2.2.2.2 Ask for Regions

Even though trajectories are independent of each other, they often show common behavior such as passing a small region within a certain period of time. Identifying these regions with R-query is important in trajectory clustering [19, 17]. An application is to identify the regions which are more likely to be passed by a given user in a time window based on the many trajectories relevant to that user.

2.2.3 T-Query

2.2.3.1 Ask for Similar Trajectories

T-Query usually asks for similar trajectories in a trajectory database with attempt to classify/cluster trajectories. The trajectory classification/clustering can be used in many applications such as to predict the class labels of moving objects based on their trajectories and other features, to identify the traffic flow in road networks, or for instance, to identify interesting behavior patterns, etc. One version of T-query is to discover common sub-trajectories [20, 19] and another version targets the problem of long trajectories spreading over large geographic areas [18].

2.2.3.2 Ask for Close Trajectories

Another task of T-trajectory is to find the closest trajectories. While the similarity of trajectories is based on the distance measure at a sequence of points, the closeness of trajectories refers to the shortest distance. Using this query, scientists may, for instance, investigate the behaviors of animals near the road where the traffic rate has been varied.

2.2.4 Applications

The booming industry of location-based services has accumulated a huge collection of users' location trajectories of driving, cycling, hiking, etc. In trajectory databases, a pattern can represent the aggregated abstraction of many individual trajectories of moving object(s). The trajectory based pattern has been widely used in many important and practical applications including traffic flow monitoring, path planning, behavior mining and predictive queries.

Traffic Flow Patterns Discovery of traffic flow patterns is important for applications requiring classification/profiling based on monitored movement patterns, such as targeted advertising, or resource allocation. The problem of maintaining hot motion paths, i.e. routes frequently followed by multiple objects over the recent past, has been investigated in recent years [28, 21].

To achieve this goal, [28] delegate part of the path extraction process to objects, by assigning to them adaptive lightweight filters that dynamically suppress unnecessary location updates and, , and thus, help reduce the communication overhead. [21] proposes a new density-based algorithm named FlowScan which finds hot routes in a road network. Instead of clustering the moving objects, road segments are clustered based on the density of common traffic they share. Another interesting problem concerns the road speed pattern based on the the traffic database that records observed road speeds under a variety of conditions (e.g., weather, time of day, and accidents) [12]. For example, if a road segment is a certain condition, then a speed factor is assigned, where speed factor is the speedup or slowdown for the road segment with respect to the base-speed. Speed factors are clustered to increase rule support. The concise set of rules is mined through a decision tree induction algorithm.

Behavior Mining In many applications, objects follow the same routes (approximately) over regular time intervals. For example, people wake up at the same time and follow more or less the same route to work every day and go shopping weekly. Thus, trajectory patterns can be viewed as concise descriptions of frequent behaviors, in terms of both space (i.e., the regions of space visited during movements) and time (i.e., the duration of movements). The discovery of hidden periodic movement patterns in spatiotemporal data, apart from unveiling important information to the data analyst, can facilitate data management substantially. The basic task is to map the trajectories to human activities based on the spatiotemporal relationship, e.g., the trajectory of a person frequently remaining at a location from 9.00am to 7.00pm usually means that they work at that location.

Path Planning An effective path planning (or route recommendation) solution is beneficial to travelers who are asking directions or planning a trip. Historical traveling experiences can reveal valuable information on how other people usually choose routes between locations. The rationale behind this is that people must have good reasons to choose these routes, e.g., they may want to avoid those routes that are likely to encounter accidents, road construction, or traffic jams.

A problem is how to mine frequently traveled road-segment-sequences in order to obtain important driving hints that are hidden in the data [12]. An area-level mining computes frequent path-segments with support relative to the traffic in the area, e.g., lower support thresholds for edges in rural areas compared to those in big cities. While the road-segment-sequence may help in suggesting a drive turn at some intersection in a general case, they are not sufficient and accurate to discover a popular route to some specified destination. This is because simply counting the number of trajectories is not enough to discover the popular route between any two locations, for instance, there may be no records of direct trajectories. To overcome this difficulty, a so-called *transfer network* is generated from raw trajectories [8] developed a transfer network from raw trajectories as an intermediate result to capture the moving behaviors between locations, and to facilitate the search for the popular route.

Another interesting problem presented in [39] mines the history of travelers to find the sound traveling path when passing several locations, like culturally important places such as the Sydney Opera House, or frequented public areas like shopping malls and restaurants, and so on.

Predictive Query Given the recent movements of an object and the current time, predictive queries ask for the probable location of the object at some future time. A simple approach is based on the assumption that objects move according to linear functions. In practice movement is more complex, and individual objects may follow drastically different motion patterns. In order to overcome these problems, a non-linear function can be modeled to accurately capture its movement in the recent past [32].

While [32] is more suitable for prediction of short time, [16, 23] accurately forecast locations when the forecast time is far away from the current time. The long term prediction uses previously extracted movement patterns named Trajectory Patterns, which are a concise representation of behaviors of moving objects as sequences of regions frequently visited within a typical travel time. It has been shown that prediction based on the trajectory patterns of an object is a powerful method.

2.3 Trajectory Similarity Measures

An important consideration in similarity-based retrieval of moving object trajectories is the definition of a distance function.

2.3.1 Point to Trajectory

The similarity between a query point q and a trajectory A is usually measured by the distance from q to the nearest point of A, i.e.,

$$D(q,A) = \min_{p' \in A} d(q,p')$$

where $d(,)$ is some distance function between two points, which can be either L_p-norms (if they are in Euclidean space) or network distance (if they are on spatial networks).

We can also extend a single query point to multiple query points. In this case we need a similarity function to score how well a trajectory connects the query locations, and which considers the distance from the trajectory to each query point. The work [10] has studied this kind of query and propose a novel similarity function

$$Sim(Q,A) = \sum_{q \in Q} e^{D(q,A)}$$

The intuition of using the exponential function is to assign a larger contribution to a closer matched pair of points while giving much lower value to those far away pairs. As a result, only the trajectory that is reasonably close to all the query locations can be considered as 'similar'.

2.3.2 Trajectory to Trajectory

The similarity between two trajectories is usually measured by some kind of aggregation of distances between trajectory points. Along this line, several typical similarity functions for different applications include Closest-Pair Distance, Sum-of-Pairs Distance [1], Dynamic Time Warping (DTW) [38], Longest Common Subsequence (LCSS) [39], and Edit Distance with Real Penalty (ERP) [10], Edit Distance on Real Sequences (EDR) [7]. It is worth noting that some of those similarity functions were originally proposed for time series data. But as trajectories can be regarded as a special kind of time series in multi-dimensional space, these similarity functions can also be applied to trajectory data.

2.3.2.1 Closest-Pair Distance

A simple way to measure the similarity between two trajectories is to use their minimum distance. To do so, we find the closest pair of points from two trajectories and calculate the distance between them. More formally, given two trajectories A and B, their Closest-Pair distance can be computed as follows:

$$CPD(A,B) = \min_{a_i \in A, b_j \in B} d(a_i, b_j)$$

2.3.2.2 Sum-of-Pairs Distance

Another similarity function between two trajectories was proposed by Agrawal et al. [1], who simply use the sum of distances between the corresponding pairs of points to measure the similarity. Let A, B be two trajectories with the same number of points, their distance is defined as follows:

$$SPD(A,B) = \sum_{i=1}^{n} d(a_i, b_i)$$

2.3.2.3 Dynamic Time Warping Distance

As we have seen, an obvious limitation of Euclidean distance is that, it requires the two trajectories to be of the same length, which is very unlikely for real life applications. More ideal similarity measures should have a certain flexibility on the lengths of two trajectories. Dynamic time warping (DTW) distance is the first one based on this motivation. The basic idea of DTW is to allow 'repeating' some points as many times as needed in order to get the best alignment.

Given a trajectory $A = \langle a_1, ..., a_n \rangle$, let $Head(A)$ denote a_1 and $Rest(A)$ denote $\langle a_2, ..., a_n \rangle$. The time warping distance between two trajectories A and B with lengths of n and m is defined as [38]:

$$DTW(A,B) = \begin{cases} 0, & \text{if } n = 0 \text{ and } m = 0 \\ \infty, & \text{if } n = 0 \text{ or } m = 0 \\ d(Head(A), Head(B)) + \min \begin{cases} DTW(A, Rest(B)) \\ DTW(Rest(A), B) \\ DTW(Rest(A), Rest(B))) \end{cases} \end{cases}$$

where $d(,)$ can be any of the distance functions defined on points.

2.3.2.4 Longest Common Subsequence

A common drawback of the previous two similarity functions is that they are relatively sensitive to noise, since all points including noises are required to match. Therefore it is possible to accumulate an overly large distance merely due to a single noisy point. To address this issue, Longest Common Subsequence (LCSS) has been proposed to measure the distance of two trajectories in a more robust manner. Its basic idea is to allow skipping over some points rather than just rearranging them. As a consequence, far-away points will be ignored, making it robust to noises. The advantages of the LCSS method are twofold: 1) some points can be unmatched, while in Euclidean distance and DTW distance all points have to be matched, even when they are outliers. 2) The LCSS distance allows a more efficient approximate computation.

Let A and B be two trajectories with lengths of n and m respectively. Given an integer δ and a distance threshold ε, the LCSS between A and B is defined as follows:

$$LCSS(A,B) = \begin{cases} 0, \ if \ n = 0 \ or \ m = 0 \\ 1 + LCSS(Rest(A),Rest(B)), \ if \ d(Head(A),Head(B)) \leq \varepsilon, \\ \qquad\qquad\qquad\qquad\qquad\qquad and \ |n - m| < \delta \\ \max(LCSS(Rest(A),B),LCSS(A,Rest(B))), otherwise \end{cases}$$

The parameter δ is used to control how far in time we can go in order to match a given point from one trajectory to a point in another trajectory. ε is a matching threshold to determine whether to take the point into account.

Based on the concept of LCSS, two similarity functions $S1$ and $S2$ have been proposed in [39].

Definition 2.1. The similarity function $S1$ between two trajectories A and B, given δ and ε, is defined as follows:

$$S1(\delta,\varepsilon,A,B) = \frac{LCSS(A,B)}{min(n,m)}$$

$S2$ is defined based on $S1$, but it allows the translations of trajectories. A translation of trajectory is a movement on all the points with the same offset. Let $\mathscr{F}$ be the set of all possible translations.

Definition 2.2. Given δ, ε and the family $\mathscr{F}$ of translations, $S2$ between two trajectories A and B is defined:

$$S2(\delta,\varepsilon,A,B) = \max_{f\in\mathscr{F}} S1(\delta,\varepsilon,A,f(B))$$

So both similarity functions take the range from 0 to 1. It is worth noting that $S2$ is an improvement over the $S1$, because by allowing translations, similarities between movements that are parallel in space but not identical can be detected.

2.3.2.5 EDR Distance

Although LCSS can handle trajectories with noises, but it is a very coarse measure since it does not differentiate trajectories with similar common subsequences, but focuses on different sizes of gaps in between. This has motivated the proposal of a new distance function, called Edit Distance on Real Sequence [7], to be proposed.

Definition 2.3. Given two trajectories A, B with lengths of n and m respectively, a matching threshold ε, the Edit Distance on Real sequence (EDR) between A and B is the number of insert, delete, or replace operations that are needed to change A into B.

$$EDR(A,B) = \begin{cases} n & if\ m = 0 \\ m & if\ n = 0 \\ \min\{EDR(Rest(R),Rest(S) + subcost, & otherwise \\ EDR(Rest(R),S) + 1, EDR(R,Rest(S)) + 1\} \end{cases}$$

where

$$subcost = \begin{cases} 0, & if\ d(Head(A),Head(B)) \leq \varepsilon \\ 1, & otherwise \end{cases}$$

Compared to Euclidean distance, DTW, and LCSS, EDR has the following virtues:

- In EDR, the matching threshold reduces effects of noise by quantizing the distance between a pair of elements to two values, 0 and 1 (LCSS also performs the same quantization). Therefore, the effect of outliers on the measured distance is much less in EDR than that in Euclidean distance, DTW.
- Contrary to LCSS, EDR assigns penalties to the gaps between two matched sub-trajectories according to the lengths of gaps, which makes it more accurate than LCSS.

2.3.2.6 ERP Distance

All the similarity functions discussed before can be classified into two categories. The first one is the Euclidean distance, which is a metric but cannot support local time shifting. The second category includes DTW, LCSS, EDR which are capable of handling local time shifting but are non-metric. To tackle this problem, Edit distance with Real Penalty (ERP for short) [6] has been proposed, representing a marriage of L1-norm and the edit distance.

By analyzing DTW distance more carefully, it can be observed that the reason DTW is not metric is because, when a gap needs to be added, it repeats the previous points. Thus the difference between a point and a gap depends on the previous point. On the contrary, ERP uses real penalty between two matched points, but a constant value for computing the distance for unmatched points. As a consequence, ERP can support local time shifting, and is a metric.

Given two trajectories A, B with lengths of n and m respectively, a random point g, ERP distance is defined as follows:

$$ERP(A,B) = \begin{cases} \sum_1^n |s_i - g|, \; if \; m = 0 \\ \sum_1^m |r_i - g|, \; if \; n = 0 \\ \min \begin{cases} ERP(Rest(A), Rest(B)) + d(Head(A), Head(B)), \\ ERP(Rest(A), B) + d(Head(A), g), \; otherwise \\ ERP(A, Rest(B)) + d(Head(B), g) \end{cases} \end{cases} \quad (2.1)$$

2.4 Trajectory Indexes

It is important that trajectory databases can support the retrieval of trajectories of an arbitrarily large size. To extract qualitative information by querying databases containing very large numbers of trajectories, the efficiency depends crucially upon an appropriate index of trajectories. The trajectory database has specific requirements to index techniques due to both unique data characteristics, i.e. a continuous long period of time, and unique query characteristics, i.e. often asking for information in an instantaneous/continuous time window.

There are mainly three index approaches. The first is to use any multidimensional access method like R-tree indexes with augmentation in temporary dimensions such as 3D R-tree [25], or STR-tree [25]. The second approach uses multiversion structures, such as MR-tree [37], HR-tree [24], HR+-tree [33], and MV3R-tree [34]. This approach builds a separate R-tree for each time stamp and shares common parts between two consecutive R-trees. The third approach divides the spatial dimension into grids, and then builds a separate temporal index for each grid. This category includes SETI [4], and MTSB-tree [40].

Here, we are only interested in the data structure indexing the trajectory in history. We ignore the indexes, like [31] [31] used, to update a large number of moving objects in a system, since they concern the latest location information about the moving objects.

2.4.1 Augmented R-tree

R-Tree We first briefly introduce R-tree and then two extended versions of R-tree with augmentation of temporal dimension. R-tree [13] is efficient and simple. It has been widely recognized in the spatial database research community and found its way into commercial spatial database management systems. In addition, many trajectory indexing structures are variants or based on R-trees.

R-tree is a height-balanced data structure. Each node in R-tree represents a region which is the minimum bounding box (MBB) of all its children nodes. Each data entry in a node contains the information of the MBB associated with the referenced children node. The search key of an R-tree is the MBB of each node. Figure 2.1 shows two views of an R-tree. In Figure 2.1(b), we see a tree structure, and in Figure 2.1(a), we see how the data objects and bounding boxes are distributed in space.

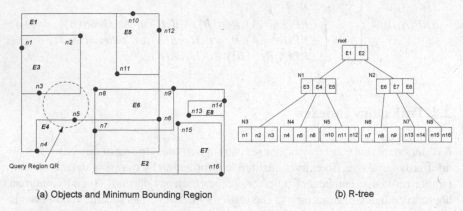

(a) Objects and Minimum Bounding Region   (b) R-tree

Fig. 2.1 Two views of an R-tree Example.

The root has two data entries $E1$, $E2$ referring to the children nodes $N1$ and $N2$ separately. $N1$ represents the minimum bounding box of its children nodes $N3$, $N4$ and $N5$ and the information of the bounding box is contained in $E1$, as with $N2$. The spatial objects are referred by the entries of the leaf nodes in R-tree.

R-tree can be used in range query to retrieve all spatial objects within a specified query region, e.g., a sphere of 5km radius around a location. R-tree is traversed from the root. The query region is examined against the MBB in each entry visited. If they are contained/overlapped, this entry is accessed and its referring child node is visited. In this way, the nodes are visited only if its MBB is relevant (e.g., contained/overlapped) to the query region. When all relevant nodes have been visited, query processing terminates. For example in Figure 2.1, QR is a query region and $N1$ is visited since its MBB overlaps with QR; then for the same reason $N3$ and $N4$ are visited and $n5$ is found from $N4$. To search for a certain point using R-tree, is the same as with processing range query. For example, to search for $n7$, $N1$ and $N2$ are visited and then $n7$ is found in $N6$. Note that $N2$ is visited although $n7$ is not in the subtree of $N2$. The reason is that MBBs of R-tree nodes overlap each other and therefore multiple searching paths may be followed. A variant of R-tree, called R*-tree [2], has been developed to minimize node access by reducing overlapping of MBBs.

R-tree can also be used to process a nearest neighbor query for a given query point. There are two traverse strategies, depth-first and best-first [14]. In both strategies, the minimum distance from each MBB to the query point is used and denoted as *mindist*. Using the best-first traverse, the root node of R-tree is visited first. Each entry in the root node computes its *mindist* to the query point and is kept in a heap sorted in ascending order of *mindist*. Each time the first entry in the heap is expanded and replaced by its child nodes. Once the first entry is a leaf node, the object is the first nearest neighbor. Using the depth-first traverse, a leaf node is visited first in some way and its *mindist* to the query point is used as a threshold T. Then, other nodes of R-tree which probably contain objects closer to the query point (i.e., *mindist* $< T$) are accessed and checked. If an object has *mindist* $< T$, T is updated with *mindist*. The query is returned when all such nodes have been visited. R-tree is also used in processing spatial join queries. More details can be found in [15, 3].

Inserting a new data point p to a R-tree takes the following steps. First, a leaf node L is selected in which to place p. The selection starts from the root of the R-tree and each time a subtree is chosen until the lead node is reached. Among several subtrees, we always chose the one which needs the least enlargement to include p. Second, we insert p into L if L has room for anther entry, otherwise a structure change happens by splitting L to two nodes. Third, the structure change is propagated upward. If the node split propagation caused the root to split, a new root is created and the R-tree grows taller as a consequence.

Deleting a data point p from an R-tree takes following steps. First, the leaf node L containing p is found and remove p from L. Second, L is eliminated if it has too few entries and relocate its entries. The change is propagated upward as necessary. Third, if the root node has one child after the tree has been adjusted, the child becomes the new root and R-tree is shortened as a consequence.

3D R-Tree In trajectory databases, traditional spatial queries need to consider the temporal aspect of the spatial objects. Two basic queries in temporal databases are *timestamp queries*, which retrieve all objects at a specific timestamp, and *interval queries*, which retrieve all objects lasting for several continuous timestamps. As a result, common queries for trajectory databases include finding the spatial objects according to some spatial requirement (e.g. trajectories within a region) for a given timestamp or interval.

In order to capture the temporal aspect of the spatial objects, R-tree can be simply modified to model the temporal aspect of the trajectory as an additional dimension. As a result, a 2 dimensional trajectory line segment can be bounded by a 3 dimensional MBB (i.e.minimum bounding box) as shown in Figure 2.4.1(a). The problem with using MBB to bound 3D line segments is that the MBB used to bound the actual line segment covers a much larger portion of space, whereas the line segment itself is small. As a result, the 3D R-tree have smaller discrimination capability because of the high overlap between MBBs.

Another aspect not captured in R-tree is knowledge about the specific trajectory to which a line segment belongs. To smoothen these inefficiencies, the 3 dimensional R-tree can be modified as follows. As can be seen in Figure 2.4.1(b), a line segment can only be contained in four different ways. This extra information is stored in four different ways in an MBB. This extra information is stored at the leaf level by simply modifying the entry format to $(id, MBB, orientation)$, where the orientation's domain is $\{1, 2, 3, 4\}$. Assuming the trajectories are numbered from 0 to n, a leaf node entry is then of the form $(id, trajectory\#, MBB, orientation)$.

In 3D R-tree, the insertion and deletion of data points follows R-tree as discussed above.

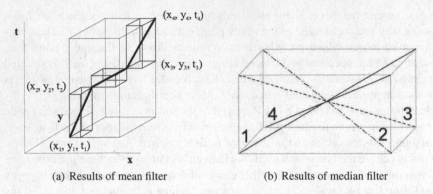

(a) Results of mean filter   (b) Results of median filter

Fig. 2.2 (a) Approximating trajectories using MBBs. (b) Mapping of line segments in a MBB [25]

STR-Tree The STR-tree (Spatio-Temporal R-Tree) is an extension of the 3D R-tree to support efficient query processing of trajectories of moving points. STR-tree and 3D R-tree differ in insertion/split strategy.

The insertion process is considerably different from the 3D R-tree. As already mentioned, the insertion strategy of the R-tree is based purely on the spatial least enlargement criterion. STR-tree improves this by not only considering *spatial closeness*, but also partial *trajectory preservation*, by trying to keep line segments belonging to the same trajectory together. As a result, when inserting a new line segment, the goal is to insert it as close as possible to its predecessor in the same trajectory. In order to achieve this, the STR-tree uses an algorithm *FindNode* to find which node contains the predecessor. As for the insertion, if there is room in this node, the new segment is inserted there, otherwise, a node split strategy has to be applied.

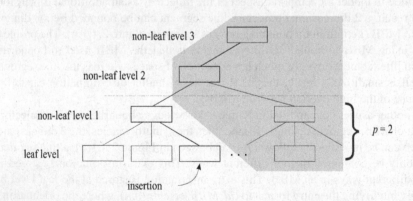

Fig. 2.3 STR-tree Insertion [25]

The ideal characteristics for an index suitable for object trajectories would be to decompose the overall space according to time, the dominant dimension in which "growth" occurs, while simultaneously preserving trajectories. An additional parameter is introduced by [25], called the *preservation parameter*, p, that indicates the number of levels to be "reserved" for the preservation of trajectories. When a leaf node returned by *FindNode* is full, the algorithm checks whether the $p-1$ parent nodes are full (in Figure 2.3, for $p=2$, only the node draw in bold at non-leaf level 1 has to be checked). In case one of them is not full, the leaf node is split. In case all of the $p-1$ parent nodes are full, another leaf node is selected on the subtree including all the nodes further on the right of the current insertion path (the gray shaded tree in Figure 2.3). A smaller p decreases the trajectory preservation and increases the spatial discrimination capabilities of index. The converse is true for a larger p. The experimental results show that the best choice of a preservation parameter is $p=2$ [25].

TB-Tree The TB-tree (Trajectory-Bundle tree) [21] is a trajectory bundle tree that is based on the R-tree. The main idea of the TB-tree indexing method is to bundle segments from the same trajectory into the leaf nodes of the R-tree. The structure of the TB-tree is a set of leaf nodes, each containing a partial trajectory, organized in a tree hierarchy. In other words, a trajectory is distributed over a set of disconnected leaf nodes. Figure 2.4 shows a part of a TB-tree structure and a trajectory illustrating this approach. The trajectory symbolized by the gray band is fragmented across six nodes $c1$, $c3$, etc. In the TB-tree, these leaf nodes are connected through a linked list.

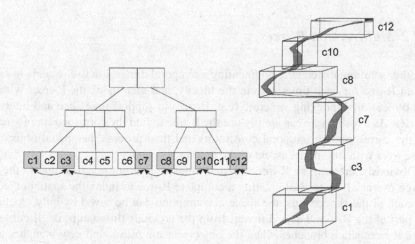

Fig. 2.4 The TB-tree structure [25]

To insert a trajectory into TB-tree, the goal is to "cut" the whole trajectory of a moving object into pieces, where each piece contains M line segments, with M being the fanout. The process starts by traversing the tree from the root and steps into every child node that overlaps with the MBB of the new line segment. The leaf node containing a segment connected to the new entry (stage 1 in Figure 2.5) is chosen. In case the leaf node is full, in order to preserve the trajectory, instead of a split, a new leaf node is created. As shown in the example, the tree is traversed until a non-full parent node is found (stages 2 through 4). The right-most path (stage 5) is chosen to insert the new node. If there is room in the parent node (stage 6), the new leaf node is inserted as shown in Figure 2.5.

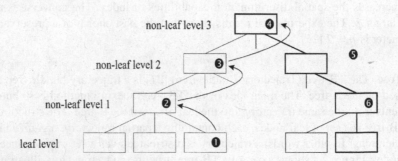

Fig. 2.5 Insertion into TB-tree [25]

2.4.2 Multiversion R-trees

Another solution, different to augmenting a temporal dimension to R-tree, is to create an R-tree for each timestamp in the history, and then index the R-trees with a one dimensional indexing structure (e.g. B-tree) to support time slice and interval queries. As a result, one can simply use the B-tree to find the corresponding R-trees for the corresponding temporal constraints and then process the spatial object in those trees with the spatial aspect of the queries.

Obviously, creating an R-tree for each timestamp is not practical due to the storage overhead. Instead of creating a complete R-tree to index the spatial objects for each of the timestamps, the space consumption can be saved by only creating the part of the R-tree that is different from the previous timestamp, i.e. if consecutive R-trees share branches when the objects do not move, and new branches are only created otherwise. The indexing structures that employ this strategy are MR-tree [37] (Multiversion R-tree), HR-tree [24] (Historical R-tree) and HR+-tree [33].

Figure 2.6 shows an example of HR-tree which stores spatial objects for timestamp 0 and 1. Since all spatial objects in A_0 do not move, the entire branch is shared by both trees R_0 and R_1. In this case, it is not necessary to recreate the entire branch in R_1, instead, a pointer is created to point branch A_0 in R_0.

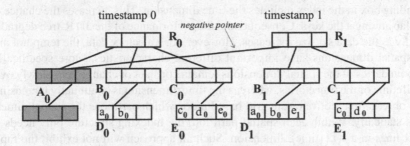

Fig. 2.6 Example of HR-tree [33]

HR+-tree improves HR-tree by allowing entries of different timestamps to be placed in the same node. This means that objects within the same node with a smaller change of position may still be shared with different R-trees. According to [33], HR+-tree consumes less than 20% of the space required by HR-tree yet answers interval queries several times faster, with similar query processing time for timestamp queries.

Both HR-tree and HR+-tree are efficient in timestamp queries. Although, they are better than using an R-tree for each timestamp, however, the extensive duplication of objects still leads to considerable space cost, and relatively poor performance on interval queries. MV3R-tree [34] utilizes both multiversion B-Trees and 3D R-trees to overcome disadvantages. A MV3R-tree consists of two structures: a multiversion R-tree (MVR-tree) and a small auxiliary 3D R-tree built on the leaves of the MVR-tree in order to process interval queries. Figure 2.7 shows an overview of MV3R-tree.

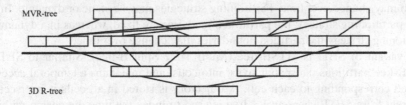

Fig. 2.7 MV3R-tree [34]

2.4.3 Grid Based Index

In a 3D R-tree, the temporal and the spatial dimensions are treated equally, i.e. a bounding box in the index includes the time dimension. This increases the chance of overlap amongst the keys. Consequently, the performance of the 3D R-tree degrades rapidly as the data set size increases. However, for trajectory data, the temporal and the spatial dimensions have important different characteristics. More specifically, the boundaries of the spatial dimensions remain constant or change very slowly over the lifetime of the trajectories, whereas the time dimension is continually increasing. This observation motivates the grid based index which partitions the spatial dimensions statically. Within each spatial partition, the indexing structure only needs to index lines in a 1-D (time) dimension. Such an approach will not exhibit the rapid degradation in index performance observed for 3-D indexing techniques. The SETI indexing mechanism (Scalable and Efficient Trajectory Index) [4] is the first grid based index.

In SETI, spatial discrimination is maintained by logically partitioning the spatial extent into a number of non-overlapping spatial cells. Each cell contains only those trajectory segments that are completely within the cell. If a trajectory segment crosses a spatial partitioning boundary then that segment is split at the boundary, and inserted into both cells. Each trajectory segment is stored as a tuple in a data file, with the restriction that any single data page only contains trajectory segments that belong to the same spatial cell. The lifetime of a data page is defined as the minimum time interval that completely covers the time-spans of all the segments stored in that page. The lifetime values of all pages that are logically mapped to a spatial cell are indexed using an R*-tree. These temporal indices are sparse indices as only one entry for each data page is maintained instead of one entry for each segment. Using sparse indices has two distinct advantages: smaller index overheads and improved insert performance. The temporal indices also provide the temporal discrimination in searches. A good spatial partitioning is one in which the number of moving objects per cell is fairly uniform. Producing a good partitioning strategy is challenging as the distribution of the objects may be non-uniform, and the distribution may change over time. Partitioning strategies may be static or dynamic. In a static partitioning strategy, the partition boundaries are fixed, whereas in a dynamic partitioning strategy the partition boundaries may change over time.

A variant of SETI is MTSB-tree (Multi Time Split B-tree). Similar to SETI, MTSB-tree partitions the spatial space into cells, and maintains a temporal access method corresponding to each cell. A trajectory is stored in all cells it intersects. Different from SETI where an R*-tree is used to index the time dimension within each cell, MTSB-tree uses the TSB-tree (Time Split B-tree) [22] to index the time dimension within each cell. Compared to R*-tree, the advantage of using TSB-tree is that it provides results in order of time. So, MTSB is more suitable for processing queries where trajectories are close in spatial space as well as in time. Another variant of SETI is CSE-tree (Compressed Start-End tree) [36] where different indexes are used to index time dimension within each cell; $B+$-tree for frequently updated data and sorted dynamic array for rarely updated data.

2.5 Query Processing

2.5.1 Query Processing in Spatial Databases

Since trajectories are spatial objects themselves, in this section, we first review the traditional query processing techniques for spatial objects, in particular, the spatial range search as well as nearest neighbor search.

As introduced earlier, spatial objects are usually organized by R-trees in spatial databases. Since computing the spatial relationship (e.g. distances, containments, etc) between spatial objects is expensive, a query processing algorithm typically employs a filter-and-refine approach. The *filter* step uses relatively cheap computation cost to find a set of candidate objects that are likely to be the results. The *refine* step is to further identify the actual query result from the small set of candidates. The overall efficiency of a spatial query processing algorithm mainly depends on the effectiveness of the filtering step.

$N \leftarrow$ the root node of the R-tree;
$Q \leftarrow$ a priority queue initialized with one element $\langle N \rangle$;
$C \leftarrow \emptyset$;
while Q *is not empty* **do**
    $E \leftarrow Q.pop\_first()$;
    if E *is a leaf node* **then**
        if $filter(E, O_q)$ **then**
            | $C \leftarrow C \cup \{E\}$;
        end
    else
        foreach $E' \in E.children$ **do**
            if $filter(E', O_q)$ **then**
                | $Q.append(E')$;
            end
        end
    end
end
return $refine(C)$;

Algorithm 1: Breadth-First Spatial Query Processing

Algorithm 1 and 2 show the general framework of the filter-and-refine approach for spatial query processing. The only difference is how the nodes of the R-tree are traversed. Generally speaking, a depth-first search will consume a relatively small amount of memory, while a breadth-first search can find all the candidates more efficiently. The function $refine(C)$ is usually implemented to evaluate the actual spatial relationship between each element of C and the query object. The function $filter(E, O_q)$ is a function based on $refine(C)$ to eliminate the spatial objects that are impossible to be returned. For example, given a spatial region QR, a *spatial range* query is to search for all the objects that overlaps with QR. The function The

$N \leftarrow$ the root node of the R-tree;
$Q \leftarrow$ a priority queue initialized with one element $\langle N \rangle$;
$C \leftarrow \emptyset$;
while Q *is not empty* **do**
    $E \leftarrow Q.pop\_first()$;
    if E *is a leaf node* **then**
        if $filter(E, O_q)$ **then**
            $C \leftarrow C \cup \{E\}$;
        end
    else
        foreach $E' \in E.children$ **do**
            if $filter(E', O_q)$ **then**
                $Q.insert\_first(E')$;
            end
        end
    end
end
return $refine(C)$;

Algorithm 2: Depth-First Spatial Query Processing

function $filter()$ is to check whether the MBB of the node and the query region overlaps, while $refine()$ is to examine whether the candidate actually overlaps with the query region. As shown in figure 2.1, only the spatial objects n_1, n_2, n_3, n_4, n_5 and n_6 need to be passed to the refine process, since the MBBs of the other objects do not overlap with QR and would thus be eliminated in the filter step.

2.5.2 P-query

2.5.2.1 Querying trajectories by a given point

This kind of query aims to find a set of trajectories whose nearest distance to a given query point is below a certain threshold (i.e., range query) or belongs to the top-k (i.e., k nearest neighbor query), based on some distance function. To answer such a query efficiently, one can index either the line segments or the sampled positions of trajectories using any multidimensional indexing structure (e.g. R-trees), and then search for the line segments or sampled points in the filter-and-refine paradigm. But it is worth noting a subtle difference compared to the traditional query processing for point objects. Whenever we encounter a line segment or a point, the corresponding trajectory should be marked as "visited" so that it will not be processed repeatedly when the other parts of the same trajectory are also reached.

2.5.2.2 Querying points by a given trajectory

A well known query that falls into this category is the Continuous Nearest Neighbor Search [35]. Given a set $\mathscr{P}$ of points, and a line segment $q = \overline{s,e}$, a continuous nearest neighbor (CNN) query retrieves the nearest neighbor from $\mathscr{P}$ for every position on q. The resulting output by such a query contains a set of $\langle R,T \rangle$ tuples, where R is a point in $\mathscr{P}$, and T is an interval of $\overline{s,e}$ within which R is the nearest neighbor of q. Figure 2.8 shows an example of the CNN query. Let $\mathscr{P} = \{a,b,c,d,f,g,h\}$ be the set of points to be considered. Given a query line segment $\overline{s,e}$, the CNN returns $\{\langle a,\overline{s,s_1}\rangle, \langle c,\overline{s_1,s_2}\rangle, \langle f,\overline{s_2,s_3}\rangle, \langle h,\overline{s_3,e}\rangle\}$ as the result. The objective of a CNN query is to split the line segments such that each of them consists of only one nearest neighbor.

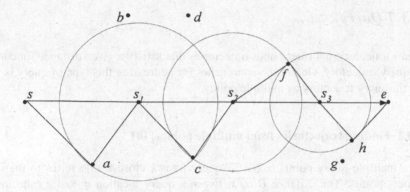

Fig. 2.8 Example of Continuous Nearest Neighbor[35]

Let SL be the list of split points. It can be seen that the start and end points of the line segment constitute the first and last elements in SL. To avoid multiple database scans, the strategy is to start with an initial SL that contains only two split points s and e with their covering points set, and then incrementally updates SL during query processing.

As illustrated in figure 2.9, the initial SL consists of s and e only. The nearest neighbor of the segment starting with s must be a nearest neighbor of s. The process starts by finding the nearest neighbor (i.e. a) of s and then draws a circle centered at s and e with the radius of $\overline{s,a}$ and $\overline{e,a}$, respectively. At this stage, only the data points within the circle need to be processed, because the data points outside the circles have greater distances to the line segment than a. The next step is to find the nearest neighbor of e from the data points within the circles and then draw a circle centered at e with radius $\overline{e,c}$. After that, the perpendicular bisector of $\overline{a,c}$ is computed and intersects with $\overline{s,e}$ at s_1. The above process will repeat, i.e., finding the nearest neighbor of s_1 and performing the perpendicular bisection, until no new point within the circles becomes the nearest neighbor again.

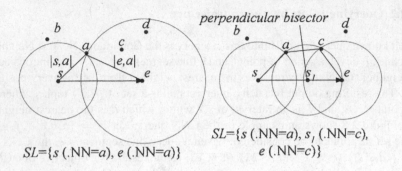

$SL=\{s \ (.NN=a), e \ (.NN=a)\}$

$SL=\{s \ (.NN=a), s_1 \ (.NN=c), \\ e \ (.NN=c)\}$

Fig. 2.9 Updating SL [35]

2.5.3 T-Query

Given a trajectory, a T-query finds trajectories that satisfy a given distance function to a query trajectory. One of the approaches for addressing this type of query is to treat the query trajectory as multiple points.

2.5.3.1 Finding trajectories from multiple points [10]

Given multiple query points, a k-BCT query finds k closest trajectories to the set of query points. The distance $Dist_q$ between a query location q_i and a trajectory $R = \{p_1, p_2, \ldots, p_l\}$ is defined as:

$$Dist_q(q_i, R) = \min_{p_j \in R} \{Dist_e(q_i, p_j)\} \tag{2.2}$$

Given a set of query locations Q, the distance between Q and a trajectory R is defined as follows:

$$Dist(Q, R) = \sum_{i=1}^{m} e^{-Dist_q(q_i, R)} \tag{2.3}$$

The approach adopted to evaluate this query is to search the nearby trajectories for each query location separately by using the k-Nearest Neighbor search [27, 7] algorithm, and then merge the results for the exact k-BCT [10]. Given a set of query locations $\{q_1, q_2, \ldots, q_m\}$, first of all the λ-NN of each query location is retrieved:

$$\lambda - NN(q_1) = \{p_1^1, p_1^2, \ldots, p_1^\lambda\}$$

$$\lambda - NN(q_2) = \{p_2^1, p_2^2, \ldots, p_2^\lambda\}$$

$$\cdots$$

$$\lambda - NN(q_3) = \{p_3^1, p_3^2, \ldots, p_3^\lambda\}$$

The set of scanned trajectories that contain at least one point in λ-NN(q_i) form a candidate set C_i for the k-BCT results. Note the cardinality $|C_i| \le \lambda$, as there may be several λ-NN points belonging to the same trajectory. By merging the candidate sets generated by all the λ-NN(q_i), we get totally f different trajectories as candidates:

$$C = C_1 \cup C_2 \cup \cdots \cup C_m = \{R_1, R_2, \ldots, R_f\}$$

For each trajectory $R_x(x \in [1, f])$ within C, it must contain at least one point whose distance to the corresponding query location is determined. For example, if $R_x \in C_i (C_i \subseteq C)$, then the λ-NN of q_i must include at least one point of R_x, and the shortest distance from R_x to q_i is known. Therefore, at least one matched pair of points between R_x and some q_i can be discovered, and then a lower bound LB of the distance for each candidate $R_x(x \in [1, f])$ can thereafter be computed by using the found matched pairs:

$$LB(R_x) = \sum_{i \in [1,m] \wedge R_x \in C_i} (\max_{j \in [1,\lambda] \wedge p_i^j \in R_x} \{e^{-Dist_e(q_i, p_i^j)}\}) \qquad (2.4)$$

Here $\{q_i | i \in [1, m] \wedge R_x \in C_i\}$ denotes the set of query locations that have already been matched with some point on R_x, and the p_i^j which achieves the maximum $e^{-Dist_e(q_i, p_i^j)}$ with respect to q_i is the point on R_x that is closest to q_i, i.e., $\max_{j \in [1,\lambda] \wedge R_x \in C_i} (e^{-Dist_q(q_i, R_x)})$. Obviously it is not greater than $\sum_{i=1}^{m} e^{-Dist_q(q_i, R_x)}$, because it only takes those matched pairs found so far into account. Thus $LB(R_x)$ must lower bound the exact distance $Dist(Q, R_x)$ defined in equation 2.3. On the other hand if $R_x \notin C_i$, then none of the trajectory points has been scanned by $\lambda - NN(q_i)$ yet.

The upper bound UB of the distances for candidate trajectories in C can be derived by following the same rational:

$$UB(R_x) = \sum_{i \in [1,m] \wedge R_x \in C_i} (\max_{j \in [1,\lambda] \wedge p_i^j \in R_x} \{e^{-Dist_e(q_i, p_i^j)}\}) + \sum_{i \in [i,m] \wedge R_x \notin C_i} (e^{-Dist_e(q_i, p_L^\lambda)}) \qquad (2.5)$$

Having both the lower and upper bound, a filter and refine strategy can be adapted by using λ-NN to find the results. In case that λ-NN returns enough candidates for finding the k nearest trajectories to the query point, the refine process will be invoked to evaluate the exact result. Otherwise, λ is modified to λ^2 in order to include more candidates.

2.5.4 R-Query

R-query searches for trajectories or trajectory segments that belong to a specified space-time window. This type of query can be further classified into timestamp and time window (interval) queries. In general, timestamp queries find the spatial objects within a region at a particular time instant in the past, whereas time window queries find spatial objects within a region for a period of continuous time instants.

2.5.4.1 R-Query with multi-version R-trees

There are many indexing structures that organize the movements of spatial objects by building an R-tree for each timestamp. In this section, we choose HR+-tree [33] as a representative for processing this type of query.

For timestamp queries, the process search is firstly directed to the root whose jurisdiction interval covers the timestamp. After that, it proceeds to the appropriate branches considering both the spatial and temporal extents (lifespan). The processing of interval queries is more complicated; since a node can be shared by multiple branches, it may be visited many times during the search. Sometimes, it is not necessary to traverse the same node if it has already been visited.

One approach to avoid duplicate visits is to store the old spatial extents into the new entry. However, storing such information, will significantly lower the fanout of the R-tree. Another solution is to perform (interval) queries in a breadth-first manner. Specifically, we start with the set of roots whose associated logical trees will be accessed. By examining the entries in these nodes, we can decide the nodes that need to be visited at the next level. Instead of accessing these nodes immediately, we save their block addresses and check for duplicates. Only when we have finished all the nodes at this level, will the ones at the next level be searched via the address information saved. Obviously the duplicate visits will be naturally avoided in this way.

2.5.4.2 R-Query with 3D R-trees

With 3D R-tree, the timestamp and interval queries can be treated as 3D range queries. A timestamp query is to find the spatial objects that overlap with a circle parallel to the x-y plane, while an interval query is to find spatial objects that are contained by a cylinder. In such cases, traditional spatial query processing approaches can then be adopted to answer these queries.

2.6 Summary

In trajectory databases, trajectory queries play an important role in complex analysis in various applications. According to the spatial data types involved, the trajectory queries can be classified into three types: point related trajectory query (P-query), region related trajectory query (R-query) and trajectory related trajectory query (T-query). The evaluation of trajectory queries has to address problems from two aspects. The first aspect is on effectiveness where a proper definition of distance measure between spatial objects is essential in order to accurately capture the spatiotemporal relationships. The second aspect is about efficiency which requires a proper index to speed up the query processing. This chapter introduces fundamental knowledge about these two aspects and discusses the processing techniques for different types of queries. These techniques and knowledge form the background for further study of this book.

References

1. Agrawal, R., Faloutsos, C., Swami, A.N.: Efficient similarity search in sequence databases. FODO pp. 69–84 (1993)
2. Beckmann, N., Kriefel, H., Schneider, R., Seeger, B.: The r^* tree: An efficient and robust access method for points and rectangles. In 9th ACM-SIGMOD Symposium n Principles of Database Systems **6**(1), 322–331 (1990)
3. Brinkhoff, T., Kriegel, H.P., Seeger, B.: Efficient processing of spatial joins using r-trees. ACM SIGMOD Conference pp. 237–246 (1993)
4. Chakka, V.P., Everspaugh, A., Patel, J.M.: Indexing large trajectory data sets with seti. In Proc. of the Conf. on Innovative Data Systems Research(CIDR) (2003)
5. Chen, L.: Robust and fast similarity search for moving object trajectories. VLDB
6. Chen, L., Ng, R.: On the marriage of lp-norms and edit distance. In: VLDB, pp. 792–803 (2004)
7. Chen, L., Ozsu, M.T., Oria, V.: Robust and fast similarity search for moving object trajectories. SIGMOD (2005)
8. Chen, Z., Shen, H.T., Zhou, X.: Discovering popular routes from trajectories. ICDE (2011)
9. Chen, Z., Shen, H.T., Zhou, X., Yu, J.X.: Monitoring path nearest neighbor in road networks. SIGMOD (2009)
10. Chen, Z., Shen, H.T., Zhou, X., Zheng, Y., Xie, X.: Searching trajectories by locations - an efficiency study. SIGMOD (2010)
11. Frentzos, E., Gratsias, K., Pelekis, N., Theodoridis, Y.: Algorithms for nearest neighbor search on moving object trajectories. Geoinformatica **11**(2), 159–193 (2007)
12. Gonzalez, H., Han, J., Li, X., Myslinska, M., Sondag, J.P.: Adaptive fastest path computation on a road network: a traffic mining approach. VLDB (2007)
13. Guttman, A.: R-trees: a dynamic index structure for spatial searching. In: Proceedings of the 1984 ACM SIGMOD international conference on Management of data, SIGMOD '84, pp. 47–57. ACM, New York, NY, USA (1984). DOI http://doi.acm.org/10.1145/602259.602266. URL http://doi.acm.org/10.1145/602259.602266
14. Hjaltason, G.R., Samet, H.: Distance browsing in spatial databases. TODS **24**(2), 265–318 (1999)
15. Huang, Y.W., Jing, N., Rundensteiner, E.A.: Spatial joins using r-trees: Breadth-first traversal with global optimizations. VLDB **24**(2), 396–405 (1997)
16. Jeung, H., Liu, Q., Shen, H.T., Zhou, X.: A hybrid prediction model for moving objects. ICDE (2008)
17. Jeung, H., Yiu, M.L., Zhou, X., Jensen, C.S., Shen, H.T.: Discovery of convoys in trajectory databases. VLDB (2008)
18. Lee, J.G., Han, J., Li, X., Gonzalez, H.: Traclass: trajectory classification using hierarchical region-based and trajectory-based clustering. PVLDB **1**(1), 1081–1094 (2008)
19. Lee, J.G., Han, J., Whang, K.Y.: Trajectory clustering: A partition-and-group framework. SIGMOD (2007)
20. Lee, J.G., Han, J., Whang, K.Y.: Trajectory clustering: a partitionand-group framework. SIGMOD (2007)

21. Li, X., Han, J., Lee, J.G., Gonzalez, H.: Traffic density-based discovery of hot routes in road networks. SSTD (2007)
22. Lomet, D., Salzberg, B.: The performance of a multiversion access method. SIGMOD (1990)
23. Monreale, A., Pinelli, F., Trasarti, R., Giannotti, F.: Wherenext: a location predictor on trajectory pattern mining. SIGKDD (2009)
24. Nascimento, M., Silva, J.: Towards historical r-trees. In: Proceedings of the 1998 ACM symposium on Applied Computing, pp. 235–240. ACM (1998)
25. Pfoser, D., Jensen, C., Theodoridis, Y.: Novel approaches to the indexing of moving object trajectories. VLDB (2000)
26. Pfoser, D., Jensen, C.S., Theodoridis, Y.: Novel approaches in query processing for moving object trajectories. VLDB pp. 395–406 (2000)
27. Roussopoulos, N., Kelley, S., Vincent, F.: Nearest neighbor queries. In: Acm Sigmod Record, vol. 24, pp. 71–79. ACM (1995)
28. Sacharidis, D., Patroumpas, K., Terrovitis, M., Kantere, V., Potamias, M., Mouratidis, K., Sellis, T.: On-line discovery of hot motion paths. EDBT (2008)
29. Shang, S., Deng, K., Xie, K.: Best point detour query in road networks. ACM GIS (2010)
30. Shekhar, S., Yoo, J.S.: Processing in-route nearest neighbor queries: a comparison of alternative approaches. ACM GIS (2003)
31. Song, Z., Roussopoulos, N.: Seb-tree: An approach to index continuously moving objects. Proceedings of International Conference of Mobile Data Management (2003)
32. Tao, Y., Faloutsos, C., Papadias, D., Liu, B.: Prediction and indexing of moving objects with unknown motion patterns. SIGMOD (2004)
33. Tao, Y., Papadias, D.: Efficient historical r-trees. In: ssdbm, p. 0223. Published by the IEEE Computer Society (2001)
34. Tao, Y., Papadias, D.: Mv3r-tree: A spatio-temporal access method for timestamp and interval queries. In: VLDB, pp. 431–440 (2001)
35. Tao, Y., Papadias, D., Shen, Q.: Continuous nearest neighbor search. In: Proceedings of the 28th international conference on Very Large Data Bases, pp. 287–298. VLDB Endowment (2002)
36. Wang, L., Zheng, Y., Xie, X., Ma, W.Y.: A flexible spatio-temporal indexing scheme for large-scale gps track retrieval. MDM (2008)
37. Xu, X., Han, J., Lu, W.: Rt-tree: An improved r-tree indexing structure for temporal spatial databases. In: Int. Symp. on Spatial Data Handling
38. Yi, B.K., Jagadish, H., Faloutsos, C.: Efficient retrieval of similar time sequences under time warping. ICDE (1998)
39. Zheng, Y., Zhang, L., Xie, X., Ma, W.Y.: Mining interesting locations and travel sequences from gps trajectories. WWW (2009)
40. Zhou, P., Zhang, D., Salzberg, B., Cooperman, G., Kollios, G.: Close pair queries in moving object databases. Proceedings of ACM GIS (2005)

Part II
Advanced Topics

Chapter 3
Uncertainty in Spatial Trajectories

Goce Trajcevski

Abstract This chapter presents a systematic overview of the various issues and solutions related to the notion of *uncertainty* in the settings of moving objects trajectories. The sources of uncertainty in this context are plentiful: from the mere fact that the positioning devices are inherently imprecise, to the pragmatic aspect that, although the objects are moving continuously, location-based servers can only be updated in discrete times. Hence come the problems related to modelling and representing the uncertainty in Moving Objects Databases (MOD) and, as a consequence, problems of efficient algorithms for processing various spatio-temporal queries of interest. Given the ever-presence of uncertainty since the dawn of philosophy through modern day nano-level science, after a brief introduction, we present a historic overview of the role of uncertainty in parts of the evolution of the human thought in general, and Computer Science (CS) and databases in particular, which are relevant to this chapter. The focus of this chapter, however, will be on the impact that capturing the uncertainty in the syntax of the popular spatio-temporal queries has on their semantics and processing algorithms. We also consider the impact of different models in different settings – e.g., free motion; road-network constrained motion – and discuss the main issues related to exploiting such semantic dimension(s) for efficient query processing.

3.1 Introduction

Historically, the impact of the imperfect knowledge on the reasoning and belief has been a topic that has attracted a lot of research interest among both philosophers and logicians [41, 63, 44, 123]. With the advent of the computing technologies, as various domains of Computer Science (CS) have emerged, the importance of

Goce Trajcevski
Dept. of EECS, Northwestern University, 2145 Sheridan Rd. Evanston, Il 60208 USA (research supported by NSF-CNS-0910952), e-mail: goce@eecs.northwestern.edu

capturing the uncertain/probabilistic nature of the data has been recognized in many of them:

- Artificial Intelligence which, in a sense popularized the Possible-Worlds semantics [6, 40, 158].
- Knowledge Representation and Reasoning along with Logic Programming and Deductive Databases [1, 7, 98, 173].
- Incorporating it on top of the traditional database technology [16].

to list but a few.

Due to the novel application domains along with advancements in database technology, a lot of recent research has been undertaken, addressing problems in modelling and efficient querying of imprecise/uncertain data [2, 20, 111, 125, 133, 134].

In the past two decades, the advances in sensing and communication/networking technologies, along with the miniaturizations of computing devices and development of variety of embedded systems have spurred the recognition of the importance of Location Based Services (LBS) [126] in a plethora of applications. From military, through structural and environmental monitoring, disaster/rescue management and remediation, to tourist information-providing systems – the efficient management of large amount of *(location, time)* data pertaining to mobile entities over (large) periods of time is a paramount. After several works and development of some ad-hoc solutions [89], the field of Moving Objects Databases (MOD) [56, 163] emerged in the late 1990's as an enabling technology for the LBS-related applications, providing formal foundations and bringing about development of prototype systems [51, 69].

Contrary to the typical assumptions in:

1. Spatial databases [19, 76, 130, 142], where the data items may have dimensionality and extent, but are (relatively) static over time;
2. Temporal databases [35, 72, 132], where the main objective is capturing the time-varying nature of the data in various application domains; and
3. Time-series [79, 78, 116, 177], where the values of the data samples over time often pertain to a single dimension,

in MOD-settings, the objects are assumed to move, either freely in the 2D (or even 3D) space [90, 52, 131], or constrained by a road network [38, 50, 28, 154]. The main features of spatio-temporal data sets:

1. The discrete data samples are expected to represent a continuous motion over the given space, thereby necessitating some type of an interpolation; and
2. The typical queries of interest (e.g., whereabouts-in-time, range, (k)nearest-neighbor, reverse nearest-neighbor, skyline) are *continuous* – which is, their answers need to be re-evaluated in time, or even *persistent* (cf. [131]) – which is, in addition to re-evaluating the answers over time, one may need to take into consideration the entire history of the motion;

have influenced a large body of works addressing issues related to modelling/ representation, indexing and querying such data [9, 14, 24, 80, 37, 99, 109, 59, 112, 71, 93, 60, 97, 105, 164, 102, 119, 106, 140, 139, 141, 155, 169, 172]

In practice, the location data at different time-instants is obtained by some positioning devices like, for example, a GPS-enabled device on-board a moving objects, which eventually transmits the location to a MOD server(s). However, GPS receivers only approximate the actual position of the respective sensor or object due to physical limitations and measurement errors of the sensing hardware [27].

The position data may be obtained by some other (collaborating) tracking-devices e.g., roadside sensors [26]. Even more so in application settings like intrusion-detection and environmental monitoring, where the tracking of the moving object of interest is based on collaborative trilateration and continuous hand-off among the participating sensors [39, 137, 62, 110, 136, 176, 179, 180]. In addition to the interpolation in-between consecutive location samplings, the imprecision of the devices involved, both GPS-based as well as sensor-based, is yet another source of uncertainty. But another example of location imprecision is the investigation of trajectories of various particles in physical and chemical processes [121].

Motivated by these observations, many researchers have focused on addressing the problem of uncertainty management in MOD settings [5, 21, 31, 30, 67, 66, 83, 85, 86, 91, 94, 108, 113, 115, 114, 143, 149, 145, 159, 160], also considering the impact of the restriction of the motions to road networks [5, 31, 30, 45, 84, 83, 178]. While it is often the case to assume that the possible locations of a given object at a particular time-instant is obeying a uniform distribution within certain bounds, recent works have addressed the impact of the different *pdf*s [21, 67, 159].

An important observation when it comes to incorporating uncertainty into the processing of spatio-temporal queries is that, in order to relieve the user from factoring it out from the answers, it needs to be incorporated in the very syntax of the given query [18, 88, 103, 115, 149, 145, 148, 171]. Although many of the existing works have focused on uncertain point-objects with (mostly) linear motion with constant speed in-between updates, some recent results have addressed the uncertainty aspects of points/lines with extent [157], and even uncertain fields [43, 170]. Such models are necessary to capture, for example, the trajectory of the "eye" of a given hurricane [152] – however, in addition to the "eye" being uncertain, the (moving) spatial zone affected by that hurricane is also uncertain.

This chapter gives an overview of the research results in the field of managing and querying uncertain trajectories data, in a manner that will strike a balance between the breadth and the depth of the different topics presented in the existing literature. The intended goal is to present a body of materials in a manner that will be suitable for both non-specialists to get introduced to the field, and specialists to get a coherent presentation that could help influence the selection of research directions.

In the rest of this chapter, in section 3.2.3, we will overview the historic evolution of the incorporation and treatment of uncertainty in the philosophy and logic[1], as well as certain CS-areas related to AI and databases, along with time-geography and geometry. Subsequently, Section 3.3.2 will address the role and impact of uncertainty in spatial databases and temporal databases, paving a way for the crux of the material of this chapter which will follow in Sections 3.4 and 3.5. After a

---

[1] It is well beyond the scope of this chapter to discuss the importance and the treatment of uncertainty in all the different scientific fields like, for instance, physics, chemistry, etc.

brief formal overview of the notion of trajectories and spatio-temporal queries[2], we will focus on a thorough analysis of the issues related to incorporating the uncertainty into the trajectories' data model, queries-syntax, and the corresponding processing algorithms. Specifically, Section 3.4 will present uncertainty models for free-space motion as well as models of uncertainty for motion restricted to road network. Subsequently, in Section 3.5, we consider the issues that uncertainty brings in the query processing, and we present examples of different (type_of _query, uncertainty_model) couplings. Section 3.6 concludes this chapter and gives a brief overview of the role of trajectories uncertainty in a broader context and application settings.

3.2 Uncertainty Throughout the History

We now discuss the evolution of the treatment of the uncertainty along different aspects of the evolution of the human thought. Firstly, we will review the uncertainty of the knowledge/belief and how philosophers and logicians throughout the history have addressed its formalization(s). Subsequently, we follow up with discussing the role and treatment of the uncertainty in the fields of Artificial Intelligence and Databases. The last part of the section touches upon the fields of time-geography and inexact geometries.

3.2.1 Philosophy and Logic

As part of the philosophy focusing on principles of valid reasoning, inference and demonstration, *logic* has had its presence in many of the ancient civilizations – Babylon, Egypt, China and India – clearly demonstrated in some inference rules related to geometric and astronomical calculations. However, the philosophical form of logic that is likely the most influential one for the Western and Islamic cultures, bringing about a symbolic and purely formal axiomatic treatment – is the one developed in ancient Greece, as first formalized by Aristotle. Even the earliest works, however, observed that some aspects of formalizing the thought required the concepts of *knowledge* and/or *belief*, leading to the so called *epistemic logic* [13] which focuses on their systematic properties. The syntax of epistemic logic extends the propositional logic with the unary operator K_a (or B_a) applied to the traditional propositions. Thus, B_aP denotes *"the agent a Believes P"*, where P is any proposition, and its meaning/semantics is: *in all possible worlds compatible with what the agent a believes, it is the case that P* [64].

Although the epistemic logic has found numerous applications in fields like CS (AI, Databases) and economics, it was the *modal logic* [17, 42] that provided a

---

[2] Addressed in greater detail in Chapter 2.

perspective for incorporating the uncertainty into a systematic framework, enabling a qualification of the truth/falsity. Intuitively, a modality is any word or phrase that can be applied to a given statement S which, in a sense, creates a new statement that makes an assertion *about the mode of truth of S* – when, where or how S is true (or about the circumstances under which S may be true) [42]. For a given proposition P, the two main operators of the modal logic are:

1. $\square$: denoting *necessity* – e.g., $\square P \equiv$ *it is necessary that P.*
2. $\Diamond$: denoting *possibility* – e.g., $\Diamond P \equiv$ *it is possible that P.*

Several approaches have been undertaken towards axiomatization of the modal logic, however, it was not until the work of Kripke [81] that the semantics and model-theoretic aspects were fully considered. Given that a particular semantics is as good as its entailment relation (within a given model) can "mimick" the consequence relation in terms of syntactic derivability, the main novelty of the, so called, Kripke structures (or frames) is that they provided a foundation for connecting a particular modal logic to a corresponding class of frames, thereby enabling reasoning about its (in)completeness. As a specific example, instead of evaluating a composite formula based solely on the *true/false* value of its primitive constituents, one may consider the behavior of the composite formula when the truth values of the constituents are *changing gradually* from false to true according to some "scenario" (cf. [29]). A thorough treatment of the topic is well beyond the scope of this chapter – however, the main influence of this line of works is that they brought *in agnitio* one concept that has been widely used since – the *possible worlds*[3]

3.2.2 Uncertainty in AI and Databases

Due to its close relationship with logic and, for that matter, extensive use of the Logic Programming paradigm, AI is one of the very first CS fields that have adopted the concept of possible worlds. The Possible Worlds Approach (PWA) is a powerful mechanism for incorporating new information into logical theories, studied by philosophers interested in belief revision and scientific theory formation [3], as well as database theorists [1, 36]. The basic premise of PWA is to keep a single model of the world that is updated when actions are performed. The update procedure involves constructing the nearest world to the current one, in which the consequences of the actions under consideration hold. As explained in [158], the PWA-based revision of a theory can be summed up as: To incorporate a set S of formulae into an existing theory T, take the maximal subset T of T that is *consistent with* S, and add S to T. This is one of the approaches undertaken for the problem of minimality of view updates in databases [46].

Although it aimed at bringing about computationally efficient procedures for reasoning about actions, PWA was shown to have problems when it comes to, so called,

---

[3] We respectfully note that philosophers and logicians are likely to disagree that semantics based on Kripke frames are model-equivalent to the one based possible worlds.

frame, ramification, or qualification issues, due to the fact that it did not distinguish between the *state of the world* and the *description of the state of the world* [158]. As a remedy, the Possible Models Approach (PMA) was introduced, which observed the *models* of a given theory $\mathbf{T}$, rather than its formulae. The goal of PMA is to change as little as possible the models of $\mathbf{T}$ in order to make the new set of formulae $\mathbf{S}$ true. Once the focus has shifted on the models, reasoning about actions became more amenable to incomplete information.

The archetypical example of uncertainty in traditional relational databases was the one of an absence of value for a particular attribute, denoted as NULL. This value is not associated with a particular type and, more importantly, it implies involvement of the three-valued logic, adding the *unknown* value in the picture and disturbing the "cushy" Closed World Assumption (CWA) model of relational databases. With NULL value, one is no longer justified to assume that the values stored within a database correspond to a complete version of the world and everything not stored in the database is false – thereby imposing the Open World Assumption and demanding extra caution when using SQL in practice.

A plethora of novel application domains such as Location-Based services, health and environmental monitoring based on sensor data analysis, biological image analysis, market analysis and economics – generate a vast amount of data which is inherently uncertain due to the imprecision of measuring devices, randomness and delays in data updates. This has spurred a tremendous research interest in probabilistic databases [2, 11, 10, 20, 92, 125, 133, 167]. In these settings, one typical feature is that some attributes are probabilistic, in the sense that their values are given by a probabilistic density function (*pdf*) – however, in practice, one cannot hope to have the *pdf* available and must rely on samples instead. In general, a probabilistic database can be thought of as finite set of probabilistic tables – one for each plausible value of the uncertain tuples, associated with membership probability. If the probability of a particular instantiation for the objects in the database is greater zero, then that particular instantiation constitutes one of the possible world. The main problem is that the cardinality of the set of all the possible worlds is exponential in the number of uncertain objects [6]. In addition to complicating the issue of the semantics to the answers of the queries, the large number of possible worlds clearly imposes computational costs in their processing – enumerating the answers in all the possible worlds is infeasible in practice. Hence, the researchers have resorted to balancing tradeoff between accuracy and computational cost, e.g., retrieving only objects with highest likelyhood to be in the result; reporting only answers the probability of which exceeds a given threshold; returning approximate answers, etc. A recent approach addressing a generic query optimization for uncertain databases, introducing a threshold operator (τ-operator) to the query plan and demonstrating that it is generally desirable to push it "down" as much as possible, is presented in [118].

Getting into a detailed discussion on the topic of probabilistic databases is beyond the scope of this Chapter, and for a comprehensive overview the reader is referred to the recent tutorials [134, 111, 120], along with a cohesive recent collection of works with an extensive list of references available in [58].

3.2.3 Time-Geography and Inexact Geometries

Time geography [61] addresses questions like:
Given a location of a mobile agent at time t_0, where is the agent at a later time $t_1 >$
t_0, or where was the agent at a previous time $t_2 < t_0$? [160]. Assuming the agent can
move in any direction and is limited only by a maximum speed v_{max}, time geography
represents the reachable locations of this agent by a right cone in $(X, Y, Time)$-space.
The cone apex represents the agents location at t_0, and the aperture represents the
maximum speed of the agent – specifically, a cone base B_i represents the set of
locations the agent may settle at a time $t_i > t_0$.

Fig. 3.1 Possible Whereabouts of an Agent: different
shades correspond to the *pdf*'s
emanating among consecutive
time-instants (cf. [160], with
permission)

Focusing on the discrete probabilistic space-time cone approximation of the continuous *pdf* of the location of a mobile agent, in [160] three approaches are undertaken for deriving that approximation: (1) from a random walk simulation, (2)
from combinatorics, and (3) from convolution The results are targeting some basic
questions of interest for time-geography like, for example, *what is the most prob-*
able arrival time of an agent A at a particular location B? We will discuss the
space-time prisms and their implications to trajectories uncertainty in more detail in
Section 3.4.

While the interest of time geographers is on the uncertainty of location of mobile
agents as the time evolves, a specific type of handling imprecision was considered
by the GIS (Geographic Information Systems) researchers, focusing on the spatial
properties of the basic primitives. Namely, in a vector GIS, the representation is
based on the type of an infinitely small point, in accordance with Euclidean. However, more often than not it is the case that GIS maps are representing geographic
entities that have spatial extent. In [123], an axiomatic *tolerance geometry* was developed, aiming at formalizing the limited capability of distinguishing stimuli in
visual perception. Intuitively, the work "blurred" the concepts of *proximity* and *i-*
dentity, and developed corresponding primitives for a formalism that can substitute
the traditional concept of "between"-ness, with an ε-between-ness.

Recently, an attempt of formalizing the geometric reasoning (and computing) for uncertain object was presented in [174], offering an approach based on multiple modalities of uncertainty in position.

The Euclidian geometry is an axiomatic well-founded logical theory, and it has interpretation of its primitives satisfying its axioms. For example, the Cartesian model of Euclidean geometry provides an interpretation of the geometric primitives point, line, equality, incidence, congruency, etc..., in the real 2D plane. However, once the uncertainty is allowed for the basic primitives, the logical foundations need to be revisited and novel derivation rules (for developing theorems based on valid proofs) are needed. A very recent formalization of the uncertainty into the foundation of the Euclidian geometry – the first postulate – and giving interpretations that capture the GIS-intuition of *points with extension* and *lines with extension* is presented in [157]. The work extends the Boolean-based reasoning by translating it into fuzzy logic, providing means of approximating and propagating positional tolerance within sound inference system.

3.3 Uncertainty in Spatial and Temporal Databases

Two fields that have emerged in the mid/late 1980s – *Spatial Databases* and *Temporal Databases* – became, in some sense, precursors to the spatio-temporal databases. In the rest of this section, we present some issues addressed in each field, which are of relevance to the context of this Chapter.

3.3.1 Spatial Databases

Spatial databases [49, 122, 130] deal with efficient storage and retrieval of objects in space that have identity and well-defined extents, locations, as well as certain geometric and/or topological relationships among them, owing to developments in application fields (GIS, VLSI design, CAD) that needed to deal with large quantities of geometric, geographic, or spatial entities. In addition to some stable and mature prototypes prototypes based on solid algebraic type-foundation [48, 55] commercial Database Management System (DBMS) vendors have provided extensions to their products, supporting spatial types and operations (Oracle Spatial, DB2 Spatial Extender, PostgresGIS, Microsoft SQL server, MySQL). Without a doubt, the results in spatial databases have spurred several important research avenues in MOD settings, e.g.:

- Many popular types of MOD queries (e.g., range, nearest-neighbor) have variants that were studied in spatial databases context [65, 124].
- Indexing structures developed for facilitating the efficient data access for processing spatial queries [8, 57] served as foundations for spatio-temporal indexes.

- Topological properties of and relationships among spatial types [32], as well as generalization issues [156], have also found their "counter-parts" in spatio-temporal data.

What is of a most specific interest for this Chapter, is that the various concepts of *uncertainty* that were investigated in the context of spatial databases have been, in one way or another, applied and/or adopted in the context of uncertain spatio-temporal data.

Fig. 3.2 Processing of Spatial Range Query for Uncertain Objects: a.) crisp range; b.) uncertain range (cf. [142], with permission)

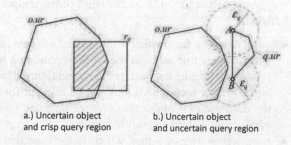

a.) Uncertain object and crisp query region

b.) Uncertain object and uncertain query region

The first such concept is the one of *location uncertainty*. Namely, if one cannot specifically determine the values of the coordinates of a given point in a reference coordinate system, then the specification of that point must incorporate the accompanying uncertainty. We already touched upon the issue of tolerance geometry in Section 3.2.3, which generalized the concept of a point into a point with extension and investigated the impacts on the formal reasoning in such geometries. In practice, however, in addition to capturing the uncertainty – e.g., via *pdf*, or histogram, alongside with some boundary – an important aspect is how to incorporate it in the query processing techniques. The first observation is that the answer to the query must somehow reflect it. An illustrating example is shown in Figure 3.2, pertaining to processing of spatio-temporal range queries for objects with uncertain locations. Part a.) of Figure 3.2 shows the uncertain object *o.ur* whose possible locations are bounded by a heptagonal region. For as long as the query region *r.q* is crisp, one can determine the probability that *o.ur* is inside the range *r.q* – e.g., if the *pdf* of *o.ur* is uniform, the probability is: $|o.ur \cap r.q|/|o.ur|$. However, once the query region itself is uncertain – e.g., its boundaries have some ε bound of possible whereabouts (cf. Figure 3.2.a.)), then the calculations of the probability become more computationally expensive. To cater for this, it was observed in [142] that if one is interested only in objects whose probability of being inside the range exceeds certain threshold, then a pruning could be applied, for which the U-Tree indexing structure was introduced.

Many entities such as regions of toxic spread, temperature maps, water-to-soil boundaries and boundaries among different types of soil, cannot be exactly determined. One of the approaches to address the storing and querying of such data was to introduce the concepts of *fuzzy points*, *fuzzy lines*, and *fuzzy regions* in the Euclidian space. Along with that, fuzzy spatial set operations like union, intersection, and difference, as well as fuzzy topological predicates were introduced to manage spatial joins and selections over fuzzy objects [77, 114, 127, 128].

3.3.2 Temporal Databases

Many applications of databases like, for example, accounting, portfolio management, medical-record and inventory management record information that is time-varying in nature [35, 72]. At the heart of the temporal databases is the distinction between:

- *valid time* of a fact, which is the time at which a particular data item is collected and becomes true as far as the world represented by the database is concerned – possibly spanning the past, present, and future. However, the valid time may not be known, or recording it may not be relevant for the applications supported by the database or – in the case that the database models possible worlds – it may vary across different possible worlds.
- *transaction time* of a fact, which is the time that a given fact is current in the database. Transaction time may be associated with different database entities like, e.g., objects and values that are *not* facts because they cannot be true or false in isolation. Thus, all database entities have a transaction-time aspect, which has a duration: from insertion to (logical) deletion of a given entity.

Capturing the time-varying nature using traditional data models and query languages can be a cumbersome activity and, as a consequence, constructs are needed that will enable capturing the valid and transaction times of the facts, leading to temporal relations. In addition, query languages [23, 132] are needed with syntactic extensions that enable database operations on temporal models.

As an example of uncertainty in temporal databases, consider the following scenario (cf. [12]):

Transportation companies (e.g., UPS, DHL) have massive databases containing information about the various packages they are shipping or have previously shipped and, most importantly, how long it takes for packages to get from a given origin to a given destination. In such cases, based on the existing data about the valid times, it may be the case that the database has the following information regarding the arrival time for packages departing from o_i at 10AM and arriving at d_j:

$$\{(12:30[0.4, 0.6]), (12:35[0.3, 0.5])\}$$

indicating that the probability of a package arriving at 12:30PM is between 0.4 and 0.6, and the probability of a package arriving at 12:35PM is between 0.3 and 0.5.

A similar scenario in the context of trajectories uncertainty occurs, for example, when the *(location,time)* data is obtained via tracking. In addition to the location imprecision, due to the clocks-skew among the sensors participating in tracking [135]. In such cases, even if a crisp location is detected after trillateration, the value of the *time* attribute will be bound to an interval instead of an instant.

3.4 Modelling Uncertain Trajectories

We now focus on the first part of the main topic of this chapter – modelling of uncertainty in spatial trajectories. After a brief overview of some basic spatio-temporal concepts and definitions, we proceed with detailed discussion of the main aspects of some of the existing models for capturing the uncertainty of spatio-temporal data. The last part of this section is dedicated to the uncertainty aspects when the motion of the objects is constrained to a road network.

As mentioned in Section 3.1, the *(location, time)* data capturing the motion of moving objects is subject to uncertainty for a variety of reasons, at every stage of its generation. The GPS receivers only approximate the actual position [27] hardware. The precision of motion sensors deteriorates with the distance from their own location and, moreover, typically the localization of a tracked object is done via trillateration, without guaranteeing that every participating node is reliable [62, 73, 175, 179]. Aside from the location determination per se' additional issues arise due to timing synchronization [135], as well as the protocols used for transmitting the *(location,time)* information from sensors to MOD or LBS servers [37, 162]. Last, but not the least – since it is impossible to record the location for every single time-instant, the interpolation in-between consecutive records yields an uncertainty of the trajectory [74, 88].

In a similar spirit to the works that have developed formal models for representing and querying "crisp" trajectories – i.e., ones without any uncertainty of the moving objects whereabouts(e.g., [53, 154]), researchers have addressed the problem of generic representation of uncertainty, along with a framework for syntactic categorization of spatio-temporal queries [88, 103, 171].

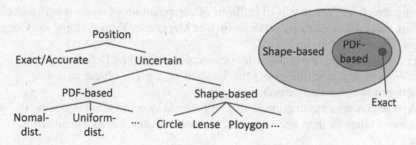

Fig. 3.3 Categorization of Uncertainty Models (cf. [88], with permission)

An example is shown in Figure 3.3 where, in addition to the exact/accurate model in which the location as a given time is assumed to be the actual one without error, two broad categories of location uncertainty models are identified [88].

1. *pdf-based models*: Motivated by the GPS-based location uncertainty, within these models the position/location at time t is described by a two-dimensional probability density function (pdf) $\ell_t : R2 \to [0, +\infty)$.

2. *shape-based models*: Bounding the possible locations by geometric shapes (e.g., circle, lens, polygon), these models may have associated probability values. However, in contrast to the pdf-based models, no claims are made about the spatial pdf within a shape.

Once a model of uncertainty is established, its impact on the syntax of the queries needs to be considered, in conjunction with the type of a particular query – e.g., position, range, nearest neighbor (cf. [88, 103]).

In practice, there is a coupling between selecting the model for the motion-plan of the moving objects – affecting the choice of the representation [54, 52] of trajectories in a MOD and, consequently, the overall strategy of query processing – and the uncertainty model. One of the common definitions of a moving object trajectory is as follows:

Definition 3.1. A *trajectory Tr* is a function *Time* $\rightarrow \mathscr{R}^2$, represented as a sequence of 3D (2D spatial plus time) points, accompanied by a unique ID of the moving object:

$$Tr_i = (oid_i, (x_{i_1}, y_{i_1}, t_{i_1}), \ldots, (x_{i_k}, y_{i_k}, t_{i_k})),$$

where $t_{i_1} < t_{i_2} < \ldots < t_{i_k}$.

We note that Definition 3.1 can be used to represent both *past* trajectories (i.e., ones whose motion is completed relative to the *now*-time) as well as *future* trajectories. In future-trajectories settings, users transmit to the MOD server: (1) the *beginning location*; (2) the *ending location*; (3) the *beginning time*; and (4) possibly a set of points to be visited. Based on the information available from electronic maps and traffic patterns, the MOD server will construct and transmit the *shortest travel time* or *shortest path trajectory* to the user. This model is applicable to the routing of commercial fleet vehicles (e.g., FedEx and UPS) as well as to web services for driving directions, where tens of millions of computations of shortest path trajectories are executed monthly by services such as MapQuest, Yahoo! Maps, and Google Maps [70].

Following are the two noteworthy observations regarding Definition 3.1:
O1: What is (the description of) the location of a given object at a time instant in-between two consecutive points $(t_{i_j} \leq t \leq t_{i_{j+1}})$?

A very common assumption is that in-between two consecutive points, the objects move along straight line segments and with constant speed, calculated as:

$$v_{i_k} = \frac{\sqrt{(x_{i_k} - x_{i_{(k-1)}})^2 + (y_{i_k} - y_{i_{(k-1)}})^2}}{t_{i_k} - t_{i_{(k-1)}}} \tag{3.1}$$

Thus, the coordinates of an object oid_i at time $t \in (t_{i_{(k-1)}}, t_{i_k})$ can be obtained by linear interpolation:

$$\begin{aligned} x_i(t) &= x_{i_{(k-1)}} + v_{i_k} \cdot (t - t_{i_{(k-1)}}) \\ y_i(t) &= y_{i_{(k-1)}} + v_{i_k} \cdot (t - t_{i_{(k-1)}}) \end{aligned} \tag{3.2}$$

However, researchers have observed that the linear interpolation assumption need not be suitable for certain applications, especially if prediction of future locations is needed. Hence, techniques have been proposed for using different (hybrid) models based on representing the objects whereabouts with other algebraic functions [74, 138].

O2: Are the points arriving at a MOD server in a batch manner, i.e., portions of, or the entire trip – as opposed to streams of individual (location,time) data values [100, 102, 39, 162]?

Catering to observations **O1** and **O2**, researchers have proposed several models of uncertainty of motion, which we address in detail next.

3.4.1 Cones, Beads and Necklaces

An idea discussed in the works of Hägerstrand in the early 1970s in time-geography [61], was the first one to have found its way into MOD research. The first consideration of the implications of the fact that the object's motion was constrained by some maximal speed v_{max} *in-between* two updates was presented in [113]. Based on its definition as a geometric set of 2D points, it was demonstrated that the objects whereabouts are bound by an ellipse, with foci at the respective point-locations of the consecutive updates, as illustrated by the spatial (X-Y) projection in Figure 3.4. Subsequently, [66], presented a spatio-temporal version of the model, naming the volume in-between two update points a *bead*[4], and the entire uncertain trajectory, a *necklace*. Note that, in a sense, a bead is a "backward-extension" of the concept of spatio-temporal cone as discussed in Section 3.2.3. Namely, the assumption is that for as long as there is no new *(location, time)* update, the object can be located anywhere inside the cone emanating from the last-known update. However, once a new update arrives, in addition to the possible-future locations, it also constraints the possible locations from the past (since the previous update). A thorough analysis of the properties of the beads was recently done in [85, 86, 108].

Definition 3.2. Let v_{max}^i denote the maximum *speed* that an object can take within the time-interval (t_i, t_{i+1}). A *bead* $B_i = ((x_i, y_i, t_i), (x_{i+1}, y_{i+1}, t_{i+1}), v_{max}^i)$ is defined as the set of all the points (x, y, t) satisfying the following constraints:

$$t_i \leq t \leq t_{i+1}$$
$$(x - x_i)^2 + (y - y_i)^2 \leq [(t - t_i)v_{max}^i]^2$$
$$(x - x_{i+1})^2 + (y - y_{i+1})^2 \leq [(t_{i+1} - t)v_{max}^i]^2 \tag{3.3}$$

The first and the second constraint of Equation 3.3, when taken together describe a cone emanating upwards from (x_i, y_i, t_i), with a vertical axis and with circles whose radius value at time t is $(t - t_i)v_{max}^i$, whereas the first and the third constraint together, specify a cone emanating downwards from $(x_{i+1}, y_{i+1}, t_{i+1})$, with a vertical axis and with circles whose radius at time t is $(t_{i+1} - t)v_{max}^i$. Hence, the bead B_i can be

---

[4] We note that, more recently, this model is also called *space-time prism*.

Fig. 3.4 Spatio-temporal
Beads and their (X,Y) Projection

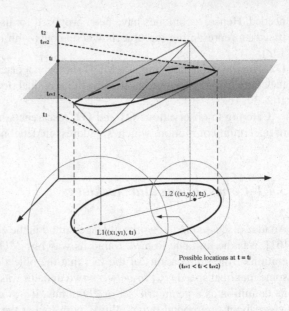

viewed as volume defined by the intersection of those two cones. We note that at
$t = t_i$ (resp. $t = t_{i+1}$) the locations of the object are crisp (i.e., no uncertainty).

For a given bead B_i, let $d_i = \sqrt{(x_{i+1} - x_i)^2 + (y_{i+1} - y_i)^2}$ denote the distance
between locations of the starting location (at t_i) and ending location (at t_{i+1}). Also,
let $t_{sv_i} = (t_i + t_{i+1})/2 - d_i/2v^i_{max}$ and $t_{sv_{i+1}} = (t_i + t_{i+1})/2 + d_i/2v^i_{max}$. We observe
that each bead had two distinct types of volumes:

1. *Single disk volumes:* For every $t \in [t_i, t_{sv_i}]$, the spatial boundary of the bead at
   t is a circle with radius $r(t) = v^i_{max}(t - t_i)$, centered at (x_i, y_i). Similarly, for
   every $t \in [t_{sv_{i+1}}, t_{i+1}]$, the spatial boundary of the bead at t is a circle with radius
   $r(t) = (t_{i+1} - t)v^i_{max}$, centered at (x_{i+1}, y_{i+1}). Hence, throughout $[t_i, t_{sv_i}]$, the 3D
   volume of the bead consists of a single cone, with a vertex at (x_i, y_i, t_i) (similarly
   for $[t_{sv_{i+1}}, t_{i+1}]$).
2. *Two-disks volume:* In-between t_{sv_i} and $t_{sv_{i+1}}$, (i.e., $t \in [t_{sv_i}, t_{sv_{i+1}}]$), the boundary
   of the bead at t is an intersection of two circles: $C^i_{down}(t)$, centered at (x_i, y_i),
   with radius $r_{down}(t) = (t - t_i)v^i_{max}$, and $C^i_{up}(t)$, centered at (x_{i+1}, y_{i+1}), with radius $r(t) = (t_{i+1} - t)v^i_{max}$.

We conclude this section with noting one more property of the beads: the projection of the bead B_i onto the the (X, Y) plane is an ellipse (cf. [86, 113]), with foci at
(x_i, y_i), and (x_{i+1}, y_{i+1}), and with equation:

$$\frac{(2x - x_i - x_{i+1})^2}{(v^i_{max})^2(t_{i+1} - t_i)^2} +$$

$$\frac{(2y - y_i - y_{i+1})^2}{(v^i_{max})^2(t_{i+1} - t_i)^2 - (x_{i+1} - x_i)^2 - (y_{i+1} - y_i)^2} = 1 \qquad (3.4)$$

We will use El_i to denote the ellipse resulting from projecting the bead B_i in the (X,Y) plane. Figure 3.4 provides an illustration of the different components of the (volume of the) bead and its corresponding shapes at different time-points, as projected on the horizontal (X,Y) plane.

Definition 3.3. Given a trajectory *Tr*, its corresponding *uncertain trajectory UTr* is a *sequence of beads*, $B_1, B_2, \ldots, B_{n-1}$.

A *Possible Motion Curve PMC(Tr)* of UTr is any function $f: Time \rightarrow \mathbf{R}^2$ for which every point (x,y,t), is either a vertex of the polyline of Tr, or it satisfies $(x,y) = f(t)$ and is inside the corresponding bead – i.e., $(\forall t)(t_i < t < t_{i+1}) \Rightarrow ((x,y,t) \in B_i)$.

The concept of a possible motion curve is illustrated in Figure 3.4 and we note that, in a sense, each possible motion curve corresponds to a "possible world" of the object's motion in-between two updates.

We note that in a recent work [94], an analogy is used between the expected-trajectory (i.e., the line segment between consecutive points) where necklace is reserved for the "known" part of the motion, and the uncertain part termed "pendant".

3.4.2 Sheared Cylinders

Another uncertainty model is the one in which an uncertain trajectory is represented as a *sheared cylinder* in the 2D-space + Time coordinate system. This is obtained by associating a fixed uncertainty threshold r at each time-instant with each line segment of the trajectory. Formally (cf. [149]):

Definition 3.4. Let r denote a positive real number and *Tr* denote a trajectory between the times t_1 and t_n. An *uncertain trajectory UTr* is the pair (Tr, r). r is called the *uncertainty threshold*.

For each point (x,y,t) along *Tr*, its *r-uncertainty area* (or the *uncertainty area* for short) is a horizontal disk (i.e. the circle and its interior) with radius r centered at (x,y,t), where (x,y) is the expected location at time $t \in [t_1,t_n]$.

Let $UTr = (T,r)$ be an uncertain trajectory between t_1 and t_n.

A *Possible Motion Curve PMC(Tr)* of UTr is any *continuous function* $f_{PMC^{Tr}}$: $Time \rightarrow R^2$ defined on the interval $[t_1,t_n]$ such that for any $t \in [t_1,t_n]$, the 3D point $(f_{PMC(Tr)}(t),t)$ is inside the uncertainty disk around the expected location at time t.

For a given uncertain trajectory $UT_r = (Tr,r)$ and two end-points (x_i,y_i,t_i), $(x_{i+1},y_{i+1},t_{i+1}) \in$ *Tr*, the trajectory volume of UTr between t_i and t_{i+1} is the u-nion of all the disks with radius r centered at the points along the line segment $\overline{(x_i,y_i,t_i),(x_{i+1},y_{i+1},t_{i+1})}$. This volume is actually what defines the "sheared cylinder" in the (X,Y,T) coordinate system. The XY projection of the trajectory volume is called an *uncertainty zone*. Figure 3.5 illustrates the basic concepts associated with the motions uncertainty under this model.

Fig. 3.5 Uncertain Trajectory bounded by Sheared Cylinders, and Possible Motion Curves

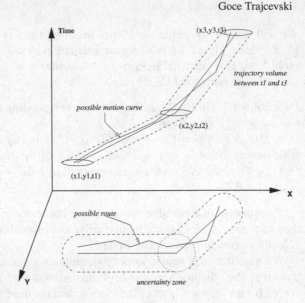

3.4.3 Uncertainty on Road Networks

Fig. 3.6 Possible Positions of an Uncertain Object Moving Along a Road Segment: in-between two location samples, the geometry consists of a union of two triangles, two trapezoids and a parallelogram (cf. [5], with permission)

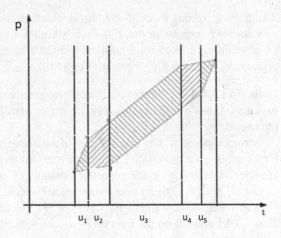

When the motion of an object is restricted by a road network [5, 31, 30, 45, 84, 83, 178], the models described so far (cones/beads and sheared cylinders) become inadequate for representing the moving objects uncertainty. To begin with, road networks are most often represented as (undirected) graph $G(V,E)$, where the vertices correspond to intersections and edges correspond to road segments in-between intersections. Often, a given edge e_{sk} is accompanied with some additional attributes, e.g.,:

– the *length* of e_{sk}, denoted $l(e_{sk})$; and

– the *maximum speed* of e_{sk}, denoted v_{sk}^{max}, which is the upper bound on how fast an object can move along edge e_{sk}.

Rich algebraic specification (types, operators) for representing and querying moving objects on road networks has been presented in [50]. In [5], the authors have extended the corresponding framework by adding data types to represent static (`unqpoint`) and mobile (`munqpoint`) point objects with location uncertainty, along with a detailed specification of the the respective set of predicates/operators. As shown in Figure 3.6, assuming that the location samples at given time-instants are "crisp", the geometric shape bounding the possible whereabouts of the object is a union of three "zones".

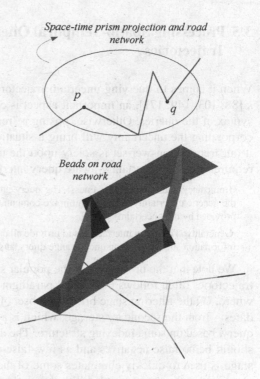

Fig. 3.7 Uncertainty on Road Networks: The possible whereabouts of the moving objects, contrary to the space-time prisms, is now only a subset of the 2D ellipse – the one intersecting the edges. The cones/beads are reduced to unions of vertical line-segments, "sweeping" along, and perpendicular to road network edges (cf. [84], with permission).

The connection (and restrictions) with the beads model is illustrated in Figure 3.7. Note that in road network settings, one cannot consider the entire ellipse (the 2D projection of the bead [113]) as a spatial range of the objects possible locations. Instead, only a subset of it intersecting the edges of the graph can be taken into account [84, 83].

An important consequence of the model of road network trajectories is that the distance between two moving objects can *no longer* be measured using the 2D Euclidian distance (L_2-norm) since the objects are constrained to move along the edges of the road network. Instead we need to rely on the *shortest network distance* which, in turn, may have a two-fold interpretation ([67, 105, 129, 168]):

– shortest path distance, or
– shortest travel-time distance.

Once the distance function is selected, one can proceed with determining the earliest-possible (respectively, the latest-possible) times that an object can be at a given location along an edge, taking the minimum and maximum speed limits along that edge. The bottom part of Figure 3.7 illustrates the shape of the uncertainty "volume" of a given object. Essentially, for each location (x_ℓ, y_ℓ) along a given edge, we have a vertical line-segment bounded by the earliest-possible and the latest-possible time that the moving object can be at that location.

3.5 Processing Spatio-Temporal Queries for Uncertain Trajectories

When it comes to querying uncertain trajectories, as pointed out in several works [88, 103, 149, 171], an important aspect is capturing the uncertainty in the very syntax of the queries. Otherwise, posing a "regular" query to a server without incorporating the uncertainty will bring a situation in which the burden of factoring it out from the answer-set is solely upon the user. As specified in [88], the basic requirements for uncertainty-aware query interface are:

Immediacy and comprehensiveness: The query interface should immediately build upon the generic uncertainty model to minimize computational effort and exploit all information provided by the uncertainty model.

Generality: The query interface has to provide all prevalent spatial query types for position information such as position query, range query, and next neighbor query

We note that the processing of the popular spatio-temporal queries for uncertain trajectories often follows the typical paradigm of *filtering* + *pruning* + *refinement*, where: (1) the filtering stage brings a subset of the total set of trajectories – candidates – from the secondary storage, which is a superset of the relevant data for the query, based on some indexing structure. The desiderata for this stage are that there should be no false-negatives and as few false-positive as possible; (2) the pruning stage is used to quickly eliminates some of the candidates – e.g., based on the assurance that the desired probability threshold cannot be met by a trajectory which satisfies some properties that are computationally easier to evaluate than the refinement algorithm; (3) the refinement stage, which eliminates all the false positives from among the candidates.

In the rest of this section, we will present examples of solutions to the problem of processing spatio-temporal range queries and nearest neighbor queries for uncertain trajectories for different models of uncertainty discussed in Section 3.4. Since explaining the details of the approaches exceeds the scope of this Chapter, we will try to highlight some of the main intuitive features of the existing results and point out to the body of references where more detailed exposition of various topics is available. We finalize this section with an overview of some miscellaneous queries for uncertain moving objects.

Fig. 3.8 Probability of an
Uncertain Moving Object
being Inside a given Region
(adapted from [21], with
permission)

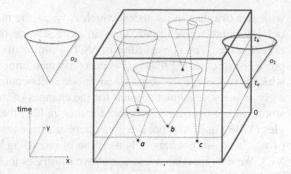

3.5.1 Range Queries for Uncertain Trajectories

The basic form of spatio-temporal range query is:

Q_R: *Retrieve all the trajectories inside the region R between t_1 and t_2.*

where R is typically a bounded region, and t_1 and t_2 denote the begin-time and end-time of interest for the query.

3.5.1.1 Instantaneous Range Query for Cones

If the model of uncertainty is the one of a *cone* and the moving objects are assumed to send *(location, time)* updates, along with a given restriction of the velocity then, for a given spatial *pdf*, one can evaluate the probability of a particular object being inside the region R at each $t \in [t_1, t_2]$. For a given t, the generic formula for calculating whether a given object o_i is inside R would be (cf. [21]):

$$\frac{Area(U_{o_i}(t)) \cap Area(R)}{Area(U_{o_i}(t))}$$

where $U_{o_i}(t)$ denotes the shape and/or *pdf* of the uncertainty zone of o_i at time t. As illustrated in Figure 3.8, all the dark objects have 100% probability of being inside the given rectangular region R for each t of interest, however: (1) the blue object (o_1) has non-zero probability of being inside R starting slightly later than the begin-time of interest for the query; (2) the red object (o_2) always has a zero probability of being inside R – hence, from scalability perspective it should be pruned out of the computation in the refinement phase. The evaluation step(s) taken throughout the refinement stage may typically involve expensive numerical integration, consequently, eliminating objects that should not be evaluated yields benefits in terms of the overall execution time.

We re-iterate that the topic of efficient processing of spatio-temporal queries for trajectories is addressed in greater detail in Chapter 2, however, as but one example, we note that [21] specifically uses the VCI index [117] to aid the elimination of candidates for processing range queries. VCI is an index structure based on R-tree [57], with an extra data in its nodes, which is v_{max} - the maximum possible speed of the objects under a given node, with an extra storage of the overall maximum speed at leaf nodes. The construction of VCI is similar to the R-tree, with an additional provision of ensuring that v_{max} is correctly maintained at the root of each sub-tree – which is properly considered when a node split occurs. When VCI is used to process a given query, one must account for the changes of the position (with respect to the stored value) over time. To cater for this, in [21] the Minimum Bounding Rectangles (MBR) used to process a give query at a time instant t are expanded by a factor of $v_{max} \times (t - t_0)$, where t_0 is the time of recording the entry for a given object at VCI. We note that the discussion above illustrates techniques that can be applied for processing range queries over uncertain trajectories at a particular time-instant.

3.5.1.2 Continuous Range Queries for Sheared Cylinders

In Section 3.4, we introduced the concept of a possible motion curve for a given trajectory (PMC(Tr)) and hinted that, in some sense, it describes a "possible world" of a particular trip taking place. However, the generic form of a range query Q_R discussed above does not reflect this anywhere in its syntax. Towards that, the works in [150, 149] have identified the different qualitative relationships that an uncertain trajectory (i.e., the family of its PMC's) could have with the spatial aspect (region R) and temporal aspect (time-interval of interest $[t_1, t_2]$) of the range query.

Firstly, since the location of the object changes continuously, the condition of the moving object being inside R may be true *sometime* ($\exists t$) or *always* ($\forall t$) within $[t_1, t_2]$. Secondly, an uncertain moving object, in addition completely failing to be inside R, may either *possibly* or *definitely* satisfy the spatial aspect of the condition at a particular time-instant $t \in [t_1, t_2]$. In simpler terms, if some PMC(Tr) is inside R at t, there is a possibility that it has been the actual motion of the object – however, this need not be the case as there may have been another PMC(Tr) that the object has taken along its motion. Let VTr denote the bounding volume of (the union of) all the possible motion curves for a given trajectory Tr – i.e., the sequence of sheared cylinders (cf. Figure 3.5). Formally, the concept of *possibly* can be specified as $\exists PMC(Tr) \subset VTr$ and the one of *definitely* can be specified as $\forall PMC(Tr) \subset VTr$.

Given the two domains of quantification – spatial and temporal – with two quantifiers each, we have a total of $2^2 \cdot 2! = 8$ operators, and their combinations yield the following variants of the spatio-temporal range query for uncertain trajectories:

- Q_R^{PS}: *Possibly_Sometime_Inside(T,R,t_1,t_2)* $(\exists PMC(TR))(\exists t)Inside(R,PMC(TR),t)$
  Semantics: *true* iff there exists a possible motion curve $PMC(Tr)$ and there exists a time $t \in [t_1, t_2]$ such that $PMC(Tr)$ at the time t, is inside the region R.
- Q_R^{PA}: *Possibly_Always_Inside(Tr,R,t_1,t_2)*
  $(\exists PMC(Tr))(\forall t)Inside(R,PMC(Tr),t)$
  Semantics: *true* iff there exists a possible motion curve $PMC(Tr)$ of the trajectory T which is inside the region R for every t in $[t_1, t_2]$.
  Intuitively, this predicate captures the fact that the object may take (at least one) *specific* possible route, which is entirely contained within the region R, during

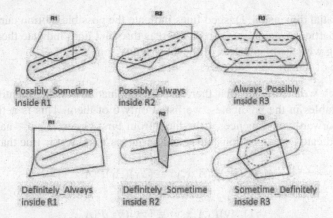

Fig. 3.9 Illustration of the Predicates Capturing the Different Quantifiers of Spatial and Temporal Domains

the whole query time interval.

- Q_R^{AP}: *Always_Possibly_Inside(Tr,R,t_1,t_2)*
  $(\forall t)(\exists PMC(Tr))Inside(R,PMC(Tr),t)$
  Semantics: *true* iff for every time value $t \in [t_1,t_2]$, there exists some (not necessarily unique) $PMC(Tr)$ inside (or on the boundary of) the region R at t.
- Q_R^{AD}: *Always_Definitely_Inside(Tr,R,t_b,t_e)*
  $(\forall t)(\forall PMC(Tr))Inside(R,PMC(Tr),t)$
  Semantics: *true* iff at every time $t \in [t_1,t_2]$, every possible motion curve $PMC(Tr)$ of the trajectory Tr, is in the region R. Thus, no matter which possible motion curve the object takes, it is guaranteed to be within the query region R throughout the entire interval $[t_1,t_2]$.
- Q_R^{DS}: *Definitely_Sometime_Inside(Tr,R,t_b,t_e)*
  $(\forall PMC(Tr))(\exists t)Inside(R,PMC(Tr),t)$
  Semantics: *true* iff for every possible motion curve $PMC(Tr)$ of the trajectory Tr, there exists some time $t \in [t_1,t_2]$ in which the particular motion curve is inside the region R. Intuitively, no matter which possible motion curve within the uncertainty zone is taken by the moving object, it will intersect the region at some time t between t_b and t_e. However, the time of the intersection may be different for different possible motion curves.
- Q_R^{SD}: *Sometime_Definitely_Inside(Tr,R,t_b,t_e)*
  $(\exists t)(\forall PMC(Tr))Inside(R,PMC(Tr),t)$
  Semantics: *true* iff there exists a time point $t \in [t_1,t_2]$ at which every possible route $PMC(Tr)$ of the trajectory Tr is inside the region R. In other words, no matter which possible motion curve is taken by the moving object, at the specific time t the object will be inside the query region.

Figure 3.9 illustrates the intuition behind plausible scenarios for the predicates specifying the properties of an uncertain trajectory with respect to a range query, projected in the spatial dimension. Dashed lines indicate the possible motion curve(s) due to which a particular predicate is true, whereas the solid lines indicate the expected routes, along with the boundaries of the uncertainty zone.

A couple of remarks are in order:

1. Although we mentioned that there are 8 combinations of the quantifiers over the variables in the predicates, we listed only 6 of them. This is actually a straightforward consequence of the facts from First Order Logic – namely, for any predicate P, given a constant A and variables x and y, it is true that:

$$(\exists x)(\exists y)P(A,x,y) \equiv (\exists y)(\exists x)P(A,x,y)$$

and

$$(\forall x)(\forall y)P(A,x,y) \equiv (\forall y)(\forall x)P(A,x,y)$$

This is regardless of the domain of (interpretation of) the variables and the semantics of the predicate P. Hence, we have that *Possibly_Sometime_Inside* is equivalent to *Sometime_Possible_Inside*; and *Definitely_Always_Inside* is equivalent to *Always_Definitely_Inside*. Hence, in effect we have 6 different predicates.

2. Similarly, the formula:

$$(\exists x)(\forall y)P(A,x,y) \Rightarrow (\forall y)(\exists x)P(A,x,y)$$

is a tautology. In effect, this means that the predicate *Possibly_Always_Inside* is stronger than *Always_Possibly_Inside*, in the sense that whenever *Possibly_Always_Inside* is true, *Always_Possibly_Inside* is guaranteed to be true.

We observe that the converse need not be true. As illustrated in Figure 3.9, the predicate *Always_Possibly_Inside* may be satisfied due to two or more possible motion curves, none of which satisfies *Possibly_Always_Inside* by itself. When the region R is convex, however, those two predicates are equivalent (cf. [149]).

3. For the same reason as above, we conclude that *Sometime_Definitely_Inside* is stronger than *Definitely_Sometime_Inside*, however, the above two predicates are not equivalent when the region R is convex. In Figure 3.9 this is shown by R_2 satisfying *Definitely_Sometime_Inside*, however, since it does not contain the entire uncertainty disk at any time-instant, it cannot satisfy *Sometime_Definitely_Inside*.

The algorithms for processing the respective predicates involve techniques from Computational Geometry (CG) (Red-Blue Intersection; Minkowski Sum/Difference) and their detailed presentation is beyond the level of detail for this Chapter. Their detailed implementation, along with complexity analysis, is available in [149]. We note that in the global context of query processing, [149] focused on the refinement stage.

In a sense, the predicates described above correspond to the, so called, "MAY" and "MUST" cases for range queries over uncertain trajectories (cf. [104, 103, 162])

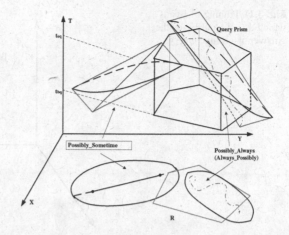

Fig. 3.10 Satisfying the Possibly_Sometimes and Possibly_Always Predicates for Beads

and, more specifically, the "Inside" property is discussed as a predicate in the generic query interface discussed in [88].

3.5.1.3 Continuous Range Queries for Beads/Necklaces

As demonstrated in [145], capturing qualitative relationships between a range query and uncertain trajectories whose uncertainty model is the one of space-time prisms (equivalently, beads) can be done using the same logical formalizations from [149]. Not only the same predicates are applicable, but also the relationships among them in terms of *Possibly Always Inside* being stronger than *Always_Possibly_Inside*, and *Sometime_Definitely_Inside* being stronger than *Definitely_Sometime_Inside* are valid.

The main difference is that the bead as a spatio-temporal structure yields a bit more complicated refinement algorithms than the ones used in [149] for sheared cylinders. As an illustrating example, Figure 3.10 shows Possible Motion Curves that cause the two predicates with existential quantifier over the spatial domain (*"Possibly"*) to be true.

Although the model of beads is more complicated for the refinement stage, it opens a room for improving the overall query processing when it comes to the *pruning* phase. Namely, one can utilize vertical cylinders surrounding a particular bead to eliminate a subset of the candidate trajectories from the answer-set more efficiently. As shown in Figure 3.11, the vertical cylinder surrounding the bead B_i does not intersect the query region R, hence, there is no need for detailed verification of any predicate capturing the uncertain range query with respect to B_i. The benefits of two pruning strategies are discussed in more details in [145].

As an illustrative example of pruning phase, below we show the steps of the algorithm for processing *Possibly_Sometime_Inside* predicate. Let El_i denote the ellipse which is the (X,Y) projection of the bead B_i, and let F_i^l ($= (x_i, y_i)$) and F_i^u

Fig. 3.11 Pruning of Beads
which do not Qualify for the
Answer of a Range Query

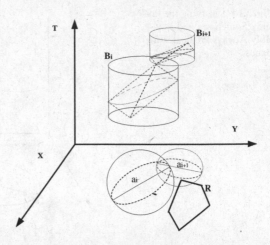

$(= (x_{i+1}, y_{i+1}))$ denote the 2D projections of its *lower* and *upper* foci in temporal sense – i.e., F_i^l occurs at time t_i and F_i^u at time t_{i+1}. For complexity analysis, assume that the region R has m edges/vertices, and an one-time pre-processing cost of $O(m)$ has been performed to determine the angles in-between its consecutive vertices with respect to a given point in R's interior [107]. The refinement algorithm can be specified as follows:

1. **If** $(t_i \in [t_1, t_1] \land F_i^l \in R) \lor (t_{i+1} \in [t_1, t_2] \land F_i^u \in R)$
2.   **return true**
3.   **else if** $(El_i \cap R \neq \emptyset)$
4.   **return true**
7. **return false**

Each of the disjuncts in line 1. can be verified in $O(\log m)$ due to the convexity of R (after the one-time pre-processing cost of $O(m)$) [107]. Similarly, by splitting the ellipse in monotone pieces (e.g., with respect to the major axis), one can check its intersection with R in $O(\log m)$, which is the upper bound on the time-complexity of the algorithm.

We note that many of the works on formalizing the predicates that capture different types of spatio-temporal range queries are geared towards extending the querying capabilities of MODs. Consider, for example the following query:

$\mathbf{Q}_R^U$: *"Retrieve all the objects which are possibly within a region R, always between the earliest[5] time when the object A arrives at locations L_1 and the latest time when it arrives at location L_2"*.

If the corresponding predicates are available, this query can be specified in SQL as:

```
WITH Earliest(times) AS
   SELECT When_At(trajectory,L_1)
   FROM MOD
```

---

[5] Observe that a given object may pass through a given point along its route more than once

```
    WHERE oid = A
WITH Latest(times) AS
    SELECT When_At(trajectory,L_2)
    FROM MOD
    WHERE oid = A
SELECT M1.oid
FROM MOD as M1
WHERE
Possibly_Always_Inside(M1.trajectory,R,
                            MIN(Earliest.times),
                            MAX(Latest.times))
```

3.5.1.4 Uncertain Range Queries on Road Networks

When the motion of a given object is constrained to an existing road network, one of the sources of its location uncertainty is due to the fact that the objects speed may vary between some v_{min} and v_{max} along a given segment – which we described in Section 3.4.3. However, there is another source of the uncertainty of such motion – namely, the low sampling-rate of the on-board GPS devices e.g., due to unavailability of satellite coverage in dense downtown areas. The main consequence of this is that the distance between two consecutive sampled positions can be large: e.g., it can be over 1.3km when sampling every 2 minutes, even if a vehicle is moving at the speed as low as 40km/h. The additional uncertainty is reflected in the fact that there may be many possible paths connecting the two consecutively sampled positions, which satisfy the temporal constraints of the actual consecutive samples. The problem is even more severe for vehicles travelling with higher speeds, as there may be several intersections between two consecutive samples.

Fig. 3.12 Uncertainty on Road Networks Due to Low Location-Sampling Frequency (object may take different routes between intersections)

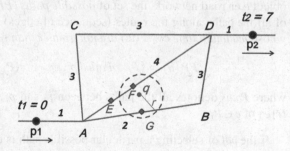

As an example, consider the scenario depicted in Figure 3.12. It shows two consecutive location-samples: L_1 at $t_1 = 0$, and L_2 at $t_2 = 7$. There are three possible routes between vertices (intersections) A and D: — $(\overline{AC}, \overline{CD})$ with travel time $4 + 2 = 6$ time units; — $\overline{AD}$ with travel time of 4; — and $(\overline{AB}, \overline{BD})$, with a travel time of $2 + 3 = 5$ time units. Given the information about minimum travel time cost – e.g.,

1 time unit between L_1 and A, as well as between D and L_2, is 1 time unit, consider the following query:

Q_R:*Retrieve all the moving objects that are within distance r from the location Q between t' = 3.5 and t" = 4.*

Clearly, it is impossible that the object has travelled along the route $(\overline{AC}, \overline{CD})$ because with the maximum speed at each segment, the earliest time of arrival at the location L_2 would be 8. This leaves only two *possible paths*: $\overline{AD}$ and $(\overline{AB}, \overline{BD})$. Following are the main observations regarding these plausible routes:

- If object a travelled along $\overline{AD}$ with the maximum speed, it will definitely be inside the spatio-temporal cylinder (based at the 2D disk centered at q and with radius r) between t = 3.5 and t = 4. However, now the question becomes, *what if the moving object did not travel using the maximum speed?* What is the probability of a satisfying Q_R, given some *pdf* of its speed?
- If object a travelled along $(\overline{AB}, \overline{BD})$ using the maximum speed, it will *not* qualify as an answer to Q_R. However, if the moving object uses smaller speed, then there may be a possibility of it entering the spatio-temporal query cylinder sometime during the time-interval of interest. Namely, the object can be at the location L_Q along the segment $\overline{BD}$ at any time during $t = 3.3$ and $t = 3.7$. Now the question again becomes, given the *pdf* of its speed, what is the probability of a satisfying Q_R. As an additional observation, we note that the object a can be anywhere within a particular segment at a given time-instant – as illustrated in Figure 3.12 for the time $t = 3.7$.

The models for uncertain trajectories on road networks in terms of possible locations at a given time-instant have been considered in [5, 45, 83]. The combination of the effects of choosing possible path together with the location uncertainty along a particular one has been formalized in [178].

Definition 3.5. Given two trajectory samples (t_i, p_i) and (t_{i+1}, p_{i+1}) of a moving object a on road network, the set of *possible paths (PP$_i$)* between t_i and t_{i+1} consists of all the paths along the routes (sequence of edges) that connect p_i and p_{i+1}, and whose *minimum time costs* (tc) *are not greater than* $t_{i+1} - t_i$, i.e.,

$$PP_i(a) = \{P_j \in Paths(p_i, p_{i+1}) | tc(P_j) \le t_{i+1} - t_i\},$$

where *Paths* denotes all the paths between p_i and p_{i+1}, $tc(P_j)$ is the sum of all the $tc(e)$ of $e \in P_j$.

If the *pdf* of selecting a particular possible path is uniform, then:

$$Pr_{i,j}(a) = Pr[PP_i(a) = P_j] = \frac{1}{|PP_i|}$$

where $|PP_i|$ denotes the cardinality of the set of all the possible paths PP_i. As another example, if the probability of a particular path being taken by an object a is inversely proportional to time-cost of that path, then:

$$Pr_{i,j}(a) = Pr[PP_i(a) = P_j] = \frac{1/tc(P_j)}{\sum_{P_x \in PP_i(a)} 1/tc(P_x)}$$

Even if a particular path P_j is considered, the location of the moving object at a given time-instant $t \in (t_i, t_{i+1})$ need not be crisp (i.e., certain) because the speed along P_j may fluctuate. However, the set of possible locations can be restricted as follows:

Definition 3.6. Given a path $P_j \in PP_i(a)$, the *Possible Locations* of a given moving object a with respect to P_j at $t \in [t_i, t_{i+1}]$ is the set of all the positions p along P_j from which a can reach p_i (respectively, p_{i+1}) within time period $t - t_i$ (respectively, $t_{i+1} - t$) i.e.,

$$PL_{i,j}(t) = \left\{ p \in P_j \left| \begin{array}{l} tc_{P_j}(p_i, p) \leq t - t_i \\ tc_{P_j}(p, p_{i+1}) \leq t_{i+1} - t \end{array} \right. \right\} \tag{3.5}$$

As an example, in the case of a uniform *pdf*, the probability that the object a is between positions p_A and p_B along a possible path P_j, whose network-distance is $d(p_A, p_B)$, is:

$$Pr[p_a(t) \in [p_A, p_B]] = Pr_{i,j}(a) \cdot \frac{d(p_A, p_B)}{\overline{PL_{i,j}(t)}} \tag{3.6}$$

where $\overline{PL_{i,j}(t)}$ denotes the the network-length of $PL_{i,j}(t)$. Formula 3.6 illustrates the joint consideration of the probability that a particular path P_j is being selected from among the possible ones, together with the probability of the object being somewhere along the segment $\overline{p_A, p_B}$ at a given time-instant t [178].

Clearly, given an existing road-map along with the *(location,time)* samples, a methodology is needed to construct all the possible trajectories that satisfy the temporal constraints of the samplings. In addition, one needs to determine the *pdf*s of the location uncertainty along different possible paths. Algorithmic solutions for two types of probabilistic range queries: *snapshot* (instantaneous) and *continuous*, are presented in [178], along with a novel indexing structure – Uncertain Trajectory Hierarchy (UTH), used to index the road network, object movement and trajectories in a hierarchical style and to improve the overall efficiency of the query processing.

3.5.2 Nearest-Neighbor Queries for Uncertain Trajectories

We now present some of the techniques that have addressed variants of the problem of efficient management of Nearest-Neighbor (NN) queries for uncertain trajectories. Before we proceed with the details, we note that an assumption commonly used in the literature (e.g., [21, 142]) is that the locations of the uncertain objects are *independent* random variables.

The basic form of spatio-temporal range query is:

$\mathbf{Q}_{NN}$: *Retrieve the nearest neighbors of the trajectory Tr_q^u between t_1 and t_2.*

where Tr_q^u denotes the uncertain querying trajectory.

3.5.2.1 NN Query for Cone Uncertainty Model

Recall Figure 3.8 used to explain the intuition behind spatio-temporal range query processing for cone-like model of uncertainty. If we take a horizontal "slice" at a particular time-instant, we will obtain all the spatial locations of the objects at that time-instant.

Fig. 3.13 Relationship A-
mong Uncertain Moving
Objects when the Querying
Object is Crisp

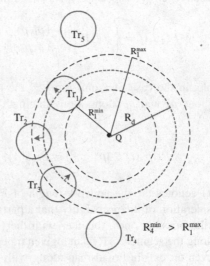

Consider a scenario in which that we are given a query object o_Q whose location at a particular time instant is *crisp*, i.e., a 2D point **Q**, with no uncertainty associated with it, and assume that the possible locations of the other objects are disks with radius r (cf. Figure 3.13).

An important observation was made in [21], which can be used to effectively prune all the objects that cannot qualify to have a non-zero probability of being a nearest neighbor to o_Q. Namely, the distance between **Q** and the *most distant point of the closest disk* (e.g., R_{max} in Figure 3.13), is an *upper bound* on the distance that *any possible nearest neighbor* of Tr_q can have. Consequently, any object o_i (a snapshot of a trajectory Tr_i) whose closest possible distance to **Q**, denoted with R_i^{min}, is larger than R_{max}, has a zero probability of being a nearest neighbor to Tr_q and can therefore be safely pruned. As can be seen from Figure 3.13, $R_4^{min} > R_1^{max}$ and similarly $R_5^{min} > R_1^{max}$, which means that Tr_4^u and Tr_5^u have zero probability of being a nearest neighbor of Tr_q.

Once the trajectories that do not qualify to be in the answer set of an NN query have been pruned, the next task is to evaluate the probability of a given trajectory Tr_i^u being within a given distance R_d from **Q**:

$$P_{i,Q}^{WD}(R_d) = \int\int_A pdf_i(x,y)\,dx\,dy \qquad (3.7)$$

where A, the integration bound, denotes the area of the intersection of the disk with radius R_d centered at $\mathbf{Q}$ and the uncertainty disk of Tr_i, with a corresponding $pdf_i(x,y)$.

Then, in order to calculate the probability that the trajectory of a given object, Tr_j^u, is a nearest neighbor of the crisp querying object Tr_q at a given time instant, one needs to consider:

1. The probability of Tr_j being within distance $\leq R_d$ from Tr_q; combined with:
2. The probability that *every other* object Tr_i $(i \neq j)$ is at a distance greater than R_d from the location $\mathbf{Q}$ of Tr_q; and
3. The fact that the distributions of the objects are assumed to be independent from each other.

Using these observations, the generic formula for the nearest-neighbor probability (cf. [21]) is:

$$P_{j,Q}^{NN} = \int_0^\infty pdf_{j,Q}^{WD}(R_d) \cdot \prod_{i \neq j}(1 - P_{i,Q}^{WD}(R_d))\,dR_d \qquad (3.8)$$

As pointed out in [21], the boundaries of the integration need not be 0 and ∞ because the effective boundary of the region for which an object *can qualify* to be a nearest neighbor of Tr_q is the ring centered at $\mathbf{Q}$ with radii R_{min} and R_{max}. More specifically, $pdf_{j,Q}^{WD}(R_d)$ is 0 for any $R_d < R_j^{min}$ and $1 - P_{i,Q}^{WD}(R_d)$ is 1 for $R_d < R_i^{min}$.

By sorting the objects that have a non-zero probability of being nearest neighbors according to the minimal distances of their boundaries from $\mathbf{Q}$, one can break the evaluation of the integral from Equation 3.8 into subintervals corresponding to each R_{min_i} and the computation of the $P_{j,Q}^{NN}$ can be performed in a more efficient manner, based on the sorted distances and the corresponding intervals [21]. The importance of this observation is in the fact that the the integrals (cf. Equation 3.8) are likely to be computed numerically. For a uniform *pdf* of the location uncertainty, the objects can be sorted according to the distances of their expected locations from the querying object.

3.5.2.2 NN Query for Sheared Cylinders – Continuity and Time-Parameterization

While the methodology explained above is sound for evaluating a snapshot (i.e., instantaneous) NN queries, an important property of the NN queries in MOD settings is that their answer over the time-interval of interest needs to be parameterized [139]. In other words, as the querying object itself, as well as the other objects are continuously moving, the nearest neighbors will change over time. To illustrate this feature, assume that we have a MOD with four trajectory-segments
$Tr_1 = \{(120, 60, 10), (220, 300, 20)\}$
$Tr_2 = \{(310, 100, 10), (190, 260, 20)\}$

Fig. 3.14 NN Queries for
Uncertain Trajectories: The
Continuity Aspect of the
Answer

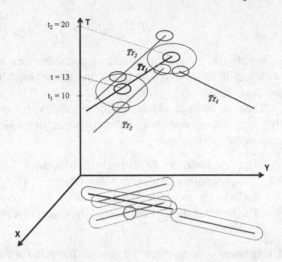

$Tr_3 = \{(150, 100, 10), (30, 260, 20)\}$
$Tr_4 = \{(370, 570, 10), (270, 330, 20)\}$
corresponding to the scenario in Figure 3.14 without the uncertainty component, and consider the following query:

Q_NN: Retrieve the nearest neighbor of the trajectory Tr_1 between $t_1 = 10$ and $t_2 = 20$.

Using the existing approaches [105, 139]), the answer A_{Q_NN} to the query is the set $\{[Tr_3, (10, 15)], [Tr_2, (15, 20)]\}$, meaning that during the first 5 seconds of the time interval of interest, i.e., between $t_1 = 10$ and $t = 15$, the nearest neighbor of Tr_1 is the trajectory Tr_3 and during the last 5 seconds, i.e., between $t = 15$ and $t_2 = 20$, the nearest neighbor of Tr_1 is the trajectory Tr_2.

However, if we take the uncertainty into consideration (cf. Figure 3.14), assuming that at every time-instant, an object can be anywhere within a disk with a 30 meter radius, we observer, for example that at time $t = 13(< 15)$, both Tr_3 and Tr_2 have a *non-zero probability* of being the NN-trajectory to Tr_1. However, that is not the case for Tr_4 which, at $t = 13$, cannot possibly be the nearest neighbor to Tr_1. Moreover, at $t_2 = 20$, we note that Tr_4—which was not part of the answer A_{Q_NN} for crisp trajectories—also has a non-zero probability of being the NN-trajectory to Tr_1. Hence, now some new important aspects emerge:

- *Syntax and Semantics of the Answer*: how can we capture the time-parameterized nature of the answer in a compact manner?
- *Ranking:* how can we establish the rank of a given trajectory's probability (e.g., highest or lowest) of being a nearest neighbor [133] at a particular time instant?
- *Continuity:* how can we efficiently detect the changes to the continuous ranking of the objects that qualify to be nearest neighbors (with non-zero probability) throughout the time-interval of interest?

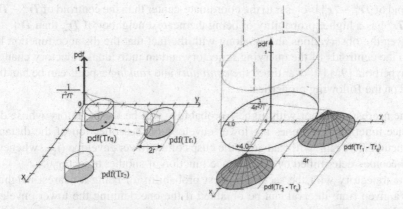

Fig. 3.15 Left: Uncountably many integrations needed to evaluate Within_Distance probability in "native" space; Right: After convolution, the querying trajectory is fixed and crisp

The first source of complication comes when the querying object is no longer crisp – how is the probability of being within distance evaluated in such cases? As shown in the left portion of Figure 3.15, we need to take infinitely many integrations over the entire disk of possible locations for the uncertain trajectory. However, since the relevant part for determining the nearest neighbor status between a given trajectory and the querying trajectory is their *distance*, it was observed in [148] that one may actually focus on a random variable specifying the difference between the two random variables: – one corresponding to a particular trajectory; – one corresponding to the querying trajectory. It is a consequence of the laws of probability theory that the difference-variable will have a *pdf* which is a *convolution* of the *pdf*s of the original variables and, in addition, as demonstrated in [148] – if the original *pdf*s have circular symmetry, so will their convolution. What is enabled by this observation is that one can "snap" the querying trajectory to the origin of the respective spatial coordinate system, and calculate the $P_{i,q}^{WD}$ using the results from [21], except the non-querying trajectories will be transformed by: – translation of the expected location; – modification of their location *pdf*. An illustration of this is provided in the right portion of Figure 3.15 – in effect, reducing an extra-level of an outer integration.

Most importantly, though, since the transformation described above is applicable to every time-instant, one can tackle the continuity aspect by using the, so called, *difference trajectories*. Specifically, instead of considering the original expected trajectories in the MOD to evaluate the expected distance from the querying trajectory, one can assume that the querying trajectory is "snap"-ed to the origin, and consider the modified trajectories to evaluate the change of the mutual distance. The main consequence of this, which is enabling the design of the efficient algorithm for calculating the answer to the continuous NN query for uncertain trajectories (again, assuming the location *pdf* has circular symmetry around the centroid) is that if the centroid of $Tr_i^u - Tr_q^u$ is closer to the coordinate-center than the centroid of $Tr_j^u - Tr_q^u$, then Tr_i^u has a higher probability of being the nearest neighbor of Tr_q^u than Tr_j^u.

Given the observations above, along with the fact that the distance function between the centroids of the querying trajectory and an individual trajectory changes as a hyperbola [9, 119] over time[6] the *continuity* and *ranking* aspects can be handled based on the following properties:

- The nearest neighbor with highest probability will be the trajectory whose distance function determines the lower envelope of the collection of the distance function. The rank will change in the cusps of the lower envelope (i.e., whenever it becomes determined by the distance function of another trajectory).
- The trajectory with the second-highest probability of being a nearest neighbor in a given time-interval can be obtained if the one defining the lower envelope in that time-interval is removed (and recursively for the k-th highest probability ($k \geq 2$).
- Regardless of the particular *pdf*, for as long as the uncertainty zone of the objects' locations is bounded by a circle with radius r, every trajectory whose distance function is further than $4 \times r$ from the lower envelope can be pruned from consideration for a nearest neighbor with non-zero probability.

Fig. 3.16 Time-parameterization of the Answer to a Continuous Nearest Neighbor Query for Uncertain Trajectories (sheared cylinder model)

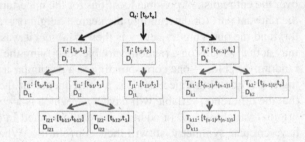

When it comes to the structure of the answer that is to be presented to the user [148] postulates that one compact structure can be obtained by splitting the time-interval of interest, say $[t_b, t_e]$, into sub-intervals $[t_b, t_1], [t_1, t_2], \ldots, [t_{n-1}, t_e]$ so that the trajectory that has the highest probability of being the nearest neighbor of Tr_q^u in each sub-interval is unique.

Subsequently, each such sub-interval can be further split into its own sub-intervals – e.g., $[t_{i-1}, t_i]$ is split into $[t_{i-1}, t_{(i-1),1}], [t_{(i-1),1}, t_{(i-1),2}], \ldots, [t_{(i-1),(k-1)}, t_i]$. To each of this sub-intervals, again a unique trajectory is matched – representing the trajectory which would have been the actual highest-probability nearest neighbor of Tr_q^u, if the MOD did not contain Tr_{i-1}^u. Towards that, a tree-structure called IPAC-NN tree (Intervals of Possible Answers to Continuous-NN) was introduced, shown in Figure 3.16 with the following properties:

---

[6] Since squaring each distance function will not disturb the relative ordering, one may readily work with the corresponding parabolae.

- The root of the tree is node labelled with the description of the parameters of interest for the specification of the query, e.g., Tr_q^u, along with $[t_b, t_e]$.
- The root has one child for each sub-interval of $[t_b, t_e]$, throughout which there is a unique uncertain trajectory Tr_i^u having the highest probability of being the nearest neighbor to Tr_q^u. Each child of the root is labelled with the corresponding trajectory (e.g., Tr_i^u) and the time-interval of its validity as the highest probability nearest neighbor (e.g., t_{i-1}, t_i) in Figure 3.16).
- After obtaining the respective labels from the respective parent-node, each child-node checks whether if it is removed from the MOD, there could still be some object with nonzero probability of being the nearest neighbor of Tr_q^u in the time sub-interval of its label.

  - If so, then it is an internal node, and each internal node follows the principle of splitting its own (sub)interval like it has been done in the root, and uses the same labelling for its children.
  - If not, then that node is a leaf-node.

The construction of the IPAC-NN tree is based on the algorithm for constructing the collection of lower envelopes of the distance functions.

3.5.2.3 NN Query for Beads Uncertainty Model

Recall that at the heart of the space-time prisms (beads) is the assumption that the motion of the objects is constrained by some maximal velocity v_{max}, and the objects can take any speed $v \in [v_{min}, v_{max}]$ in-between two consecutive *(location,time)* updates.

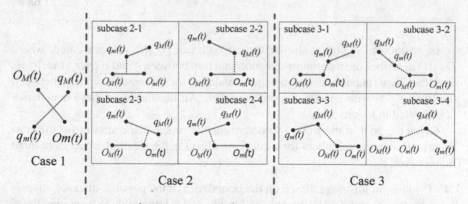

Fig. 3.17 Different Cases of Evaluating the Distance Between Two Objects with Uncertain Velocities (cf. [67], with permission)

The assumption for uncertain speed and crisp update points clearly affects how the minimum possible distance between two objects can vary in-between updates.

To capture the different variations, in [67] three basic cases of the minimum distance were identified (cf. Figure 3.17):

- The two objects are moving along paths that intersect.
- The two objects are on the expected segments that do not intersect, and the minimum distance is based on a perpendicular from a point on one segment to a point on the other.
- The two objects are on expected segments that do not intersect, however, their minimum distance happens when each of the two objects is located in some of the end-points of the expected segment of motion.

Based on the three cases for instantaneous distance values, when it comes to monitoring the distance between a given object and the querying object over a time-interval, the so called *function-switching time points* are determined. The key property is that in-between two consecutive function-switching time points the function describing the variation of the minimum-distance between the querying object q and a moving object o (denoted $d_{o,q}(t)$ in [67]) is one and the same function of time.

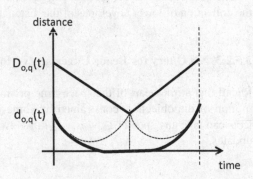

Fig. 3.18 Boundaries on the Possible Distances Between Two Objects (adapted from [67], with permission)

In addition to $d_{o,q}(t)$, a similar formalization of $D_{o,q}(t)$ was presented, where $D_{o,q}(t)$ describes the maximum-distance function between q and o over time. In effect, these two functions determine the boundaries for the possible distance between the two objects with uncertain speeds, q and o. An illustration of these boundaries is presented in Figure 3.18.

Given the goal of the work – to determine the probabilistic answer for the continuous K- nearest neighbors for a querying object q, the solution proceeds in three main stages:

1. *Pruning*: in this stage, based on the boundaries of the possible-distance zones – i.e., $D_{o,q}(t)$ and $d_{o,q}(t)$, the objects for which it is impossible to be among the K closest ones to q during the time-interval of interest are eliminated from further consideration.
2. *Candidate-distilling*: during this stage, sub-intervals are identified, during which the set of possible K nearest neighbors consists of same objects. To determine the time-instants during which the change occurs, one needs to determine the

time t_c when $d_{o_i,q}(t_c = D_{o_j,q}(t_c)$ – i.e., the minimum-distance of the object o_i becomes equal to the maximum-distance of the object o_j with respect to the querying object q. In this case, if o_j was among the $PKNN(q)$ (i.e., possibly a-mong the K nearest neighbors) before t_c, then it will be substituted by o_i at/after t_c.

3. *Ranking*: in this phase, a confidence value, based on a reasonable probability-model is determined for each object among the candidates.

In [68], an indexing structure was proposed – TPR^e tree – which can be applied to index trajectories with uncertain velocities, thereby avoiding unnecessary I/O operations from the secondary storage. Extending the paradigm from [67], scalable efficient techniques were presented to process probabilistic variants of the $K - NN$ query, along with a variant of the range query – a moving range (i.e., within a given distance from a moving object).

3.5.2.4 NN Query for Uncertain Trajectories on Road Networks

As discussed in Section 3.4.3, assuming that the periodic *(location,time)* samples do not contain any errors, the main source of uncertainty for the objects moving along road networks is the fluctuation of the speed between some minimal value v_{min} and maximal value v_{max}. Typically, in these settings, v_{max} corresponds to some speed-limits dictated for a particular type of road segment (e.g., highway portion vs. street in downtown area).

When it comes to processing NN query on road networks, the key aspect is the *distance function*. Contrary to the motion in a free 2D space, where the distance at a particular time-instant is the L_2 – Euclidian distance, as we mentioned in Section 3.4 in these settings, the distance can be evaluated only based on the existing network, i.e., the underlying graph representing it [67, 105, 129, 168]. Hence, the typical strategies rely on either the shortest path distance, or shortest travel-time distance.

A recent approach for tackling Continuous $K - NN$ queries for objects moving along segments of road networks with Uncertain speeds (CUkNN) has been presented in [91]. In a similar spirit to [67], for a given object o and a querying object q, two bounding distance-functions are presented:

1. $MaxD_{q,o,}(t)$ determining how the maximum distance between o and q varies over time.
2. $MinD_{q,o,}(t)$ determining how the minimum distance between o and q varies over time.

When calculating the distance functions, given the bounds v_{min} and v_{max}, for each of o and q at a given time-instant, the closest possible location and the furthest possible location from a vertex (e.g., an intersection in the graph-based network model) along the direction of movement in the current segment are obtained. The shortest path distance between the vertices of the graph incident to the edges along which o and q travel, together with the bounds for closest/furthest possible locations, is used to calculate the total value of $MinD_{q,o,}(t)$ and $MaxD_{q,o,}(t)$. The crux is that these two functions – which were hyperbolae for the motion in free 2D space (or, equivalently, parabolae) – now correspond to *line segments* in the *(distance,time)* space.

The typical CUkNN query processing proceeds with the standard three stages: *pruning* – eliminating the objects with zero probability of being one of the K nearest neighbors of q; *refinement* – where the possible candidates are ranked; and *probability evaluation* – the last phase where the actual probabilities for the objects from the refinement phase are calculated. We close this part with a note that a methodology for processing NN query in the settings in which a model based on location-uncertainty is coupled with a network distance function based on shortest travel-time has recently been presented in [147]

3.5.3 Potpourri: Some Miscellaneous Queries/Predicates for Uncertain Trajectories

The range and nearest neighbor queries have been identified as important categories even in traditional databases settings. However, there are certain predicates that are topological in nature [33, 76] which have recently been considered in the context of uncertain spatio-temporal data.

Given that the beads (or, space-time prisms) can be described by polynomial constraints (cf. Section 3.4), various queries that are well-suited for constrained database paradigm can be explored. For example, one can envision predicates like *inPrism(r,p,q,v)* specifying that the point r is inside the space-time prism defined by points p and q, with a maximum speed v, where p is preceding q in time.

An example query that is of interest to geographers [61] is the, so called, *alibi-query*. Given two space-time prisms, representing the uncertain motions of two individuals, the alibi-query asks whether those objects had a chance to meet – essentially, whether their corresponding space-time prisms intersect. It was observed that relying on the quantifier-elimination approaches for first order theories to process the alibi-query was computationally cost-prohibitive, and an analytic solution to this problem was presented in [82].

The *inside*-ness (cf. Section 3.5.1.2) can also be viewed as a topological property describing a possible relationship between an uncertain trajectory and a spatio-temporal range corresponding to the query-prism. This is but an example of the perspective taken in [94], where different topological predicates for uncertain trajectories under the, so called, *pendant* model are discussed. The pendant model is, in some sense, equivalent to the beads for moving points, however, the formalization in [94] addresses both uncertain moving points (*unmpoint*) and uncertain moving regions (*unmregion*). Extending the work in [34] presenting the *STP* framework for formalizing Spatio-Temporal Predicates, a collection of Spatio-Temporal Uncertain Predicates (*SUTPs*) is presented, based on the pendant model. Formally, a SUTP is a boolean expression containing:

- *Topological predicates*: disjoint, meet, overlap, covers, coveredBy, equal, inside, contains.
- *Traditional logic operators*: ¬, ∨, ∧, ∃, ∀.
- *Set expressions*: ∪, ∩, ∈, ⊂.
- *Model-operations*: e.g., at_instance; temp_select (along with the distance function *dist*).

An interesting geometric approach towards the scalability aspect of the probabilistic NN query processing is presented in [4]. Namely, in a similar spirit to [22], the work introduces the concept of Probabilistic Voronoi Diagram that can be used to prune the objects that do not qualify to be in the answer set of the NN-query – i.e., ones whose Voronoi cells are not adjacent to the Voronoi cell of the querying object.

3.6 Summary

In this Chapter, we presented an overview of the research trends addressing problems related to modelling and querying of uncertain trajectories data. After a selected historic overview of the treatment of uncertainty in a few disciplines which, in one way or another, have impacted the evolution of the thought related to the theme of this Chapter, we gave a more focused overview of the uncertainty in the fields of spatial and temporal databases. In the last two sections, we gave detailed discussions related to different models of trajectories that capture the uncertain data, and the issues that arise when processing queries over such data.

We now bring a few observations regarding the role and impact of trajectories uncertainty in a broader context and application settings/requirements[7]:

- In order to reduce the communication and bandwidth consumption, moving objects may have a "contract" with the MOD server based, e.g., on the distance-based dead-reckoning policy [161, 162]. In these settings, each object will periodically transmit an update of the form *(location, time, velocity)* and will not transmit any other location update, for as long as the deviation between the actual location (as observed e.g., by the on-board GPS device) and the expected location in the MOD does not exceed certain threshold. Clearly, in-between updates, the MOD server cannot have any certain knowledge about the objects whereabouts. In the case that the objects are moving along a road-network, clearly, the location uncertainty can be reduced [45].
- In order to reduce the storage requirements for the large quantities of *(location,time)* data, sometimes MOD servers may apply data-reduction techniques [15]. Clearly, reducing the size of the data points will ultimately introduce an uncertainty, although the size vs. imprecision trade-offs can be managed (e.g., one can

---

[7] Note that Chapter 1 and Chapter 4 address in detail the topics of trajectory data reduction and privacy, respectively.

guarantee a particular error-bound). Clearly, this will affect the (im)precision in the answers to the spatio-temporal queries in such MOD [15]. To couple the management of the transmission cost and storage costs, sometimes one may delegate (part of) the responsibility of spatio-temporal data reduction to the moving objects themselves, asking them to periodically transmit a historic data of their trip after applying reduction techniques [87, 144]

- In wireless sensor networks scenarios, in the case that some imprecision in time/space is acceptable, some reduction in the *(location,time)* data-items locally generated by the tracking sensors can be spared from transmission to a dedicated sink. This can significantly contribute towards saving the scarce energy resources of the nodes [146, 165], and prolong the network lifetime.
- Another application domain in which the uncertainty arises as a requirement is the location-privacy [25, 101, 166]. One of the most popular techniques – spatial cloaking – actually blurs the user's exact location into a cloaked area, satisfying some "quality threshold" – e.g., the available location information contains an uncertainty disk with area larger than the desired threshold.

We believe that the spatio-temporal uncertainty will keep on playing an important role in many application domains in the future. One challenge is coming up with a unified model of location and time uncertainties, along with corresponding query constructs and processing strategies. Many aspects of trajectory data clustering [47, 75] and warehousing [153] will inevitably require a formal treatment of uncertainty. Similarly, many applications that rely on maintaining spatio-temporal variograms [96, 151] will need to incorporate some type of uncertainty. One field that could potentially benefit from proper adaptive use of uncertainty is visualization of mobile data, both in large center displays, as well as limited resolution displays on board moving vehicles [95].

Acknowledgments: Many thanks to Reynold Cheng, Ralf Guting, Yuan-Ko Huang, Bart Kuijpers, Ralph Lange, Markus Schneider, Egemen Tanin, Yufei Tao and Stephan Winter (along with their collaborators) for generously permitting the use of some of the figures appearing in their works throughout this Chapter.

Special thanks to Hans-Peter Kriegel for his constructive suggestions.
As stated earlier, this Chapter was written with an intention of providing an overview of the topic of uncertainty in location trajectories in a manner that will balance the breadth of the coverage vs. depth of exposing the main issues in a few selected problems and solutions, as well as a tutorial-like source for both non-experts and researchers in the field. Any omissions, with sincere apologies, are sole responsibility of the author.

References

1. Abiteboul, S., Kanellakis, P.C., Grahne, G.: On the representation and querying of sets of possible worlds. In: SIGMOD Conference, pp. 34–48 (1987)
2. Aggarwal, C.C., Yu, P.S.: A survey of uncertain data algorithms and applications. IEEE Trans. Knowl. Data Eng. 21(5), 609–623 (2009)
3. Alchourrón, C., Gärdenfors, P., Makinson, D.: On the logic of theory change. Journal of Symbolic Logic 50, 510–530 (1985)
4. Ali, M.E., Tanin, E., Zhang, R., Kotagiri, R.: Probabilistic voronoi diagrams for processing moving nearest neighbor queries (2011). (personal communication, manuscript under revision)
5. de Almeida, V.T., Güting, R.H.: Supporting uncertainty in moving objects in network databases. In: GIS, pp. 31–40 (2005)
6. Antova, L., Koch, C., Olteanu, D.: $10^{(10^6)}$ worlds and beyond: efficient representation and processing of incomplete information. VLDB Journal 18(5), 1021–1040 (2009)
7. Apt, K., Blair, H.: Arithmetic classification of perfect models of stratified programs. In: R. Kowalski, K. Bowen (eds.) Proc. 5th International Conference and Symposium on Logic Programming, pp. 765–779. Seattle, Washington (August 15-19, 1988)
8. Beckmann, N., Kriegel, H., Schneider, R., Seeger, B.: The r* tree: An efficient and robust access method for points and rectangles. In: ACM SIGMOD (1990)
9. Benetis, R., Jensen, C.S., Karciauskas, G., Saltenis, S.: Nearest and reverse nearest neighbor queries for moving objects. VLDB Journal 15(3), 229–249 (2006)
10. Bernecker, T., Emrich, T., Kriegel, H.P., Mamoulis, N., Renz, M., Züfle, A.: A novel probabilistic pruning approach to speed up similarity queries in uncertain databases. In: ICDE, pp. 339–350 (2011)
11. Bernecker, T., Kriegel, H.P., Mamoulis, N., Renz, M., Züfle, A.: Scalable probabilistic similarity ranking in uncertain databases. IEEE Trans. Knowl. Data Eng. 22(9), 1234–1246 (2010)
12. Biazzo, V., Giugno, R., Lukasiewicz, T., Subrahmanian, V.S.: Temporal probabilistic object bases. IEEE Trans. Knowl. Data Eng. 15(4), 921–939 (2003)
13. Buh, I.: Epistemic Logic in the Middle Ages. Routledge (1993)
14. Cai, Y., Hua, K., Cao, G.: Processing range-monitoring queries on heterogeneous mobile objects. In: International conference on Mobile Data Management (MDM) (2004)
15. Cao, H., Wolfson, O., Trajcevski, G.: Spatio-temporal data reduction with deterministic error bounds. VLDB Journal 15(3) (2006)
16. Cavallo, R., Pittarelli, M.: The theory of probabilistic databases. In: VLDB, pp. 71–81 (1987)
17. Chellas, B.: Modal Logic: An Introduction. Cambridge University Press (1980)
18. Chen, J., Cheng, R., Mokbel, M.F., Chow, C.Y.: Scalable processing of snapshot and continuous nearest-neighbor queries over one-dimensional uncertain data. VLDB Journal 18(5), 1219–1240 (2009)
19. Chen, T., Schneider, M.: Data structures and intersection algorithms for 3d spatial data types. In: GIS (2009)
20. Cheng, R., Chen, J., Mokbel, M.F., Chow, C.Y.: Probabilistic verifiers: Evaluating constrained nearest-neighbor queries over uncertain data. In: ICDE (2008)
21. Cheng, R., Kalashnikov, D.V., Prabhakar, S.: Querying imprecise data in moving objects environments. IEEE-Trans. Knowl. Data Eng. 16(9) (2004)
22. Cheng, R., Xie, X., Yiu, M.L., Chen, J., Sun, L.: Uv-diagram: A voronoi diagram for uncertain data. In: ICDE, pp. 796–807 (2010)
23. Chomicki, J.: Temporal query languages: A survey. In: ICTL, pp. 506–534 (1994)
24. Chon, H.D., Agrawal, D., Abbadi, A.E.: Range and knn query processing for moving objects in grid model. Mobile Networks and Applications 8 (2003)
25. Chow, C.Y., Mokbel, M.F., Liu, X.: Spatial cloaking for anonymous location-based services in mobile peer-to-peer environments. GeoInformatica 15(2), 351–380 (2011)

26. Civilis, A., Jensen, C.S., Pakalnis, S.: Techniques for efficient road-network-based tracking of moving objects. IEEE Trans. Knowl. Data Eng. **17**(5) (2005)
27. of Defense, U.S.D.: Navstar gps: Global positioning system standard (2008)
28. Demiryurek, U., Pan, B., Kashani, F.B., Shahabi, C.: Towards modeling the traffic data on road networks. In: GIS-IWCTS (2009)
29. Detlovs, V., Podnieks, K.: Introduction to Mathematical Logic. University of Latvia (2011). Hyper-textbook for students
30. Ding, Z.: Utr-tree: An index structure for the full uncertain trajectories of network-constrained moving objects. In: MDM, pp. 33–40 (2008)
31. Ding, Z., Güting, R.H.: Uncertainty management for network constrained moving objects. In: DEXA, pp. 411–421 (2004)
32. Egenhofer, M., Franzosa, R.: Point set topological relations. International Journal of Geographical Information Systems **5** (1991)
33. Egenhofer, M.J., Dube, M.P.: Topological relations from metric refinements. In: GIS (2009)
34. Erwig, M., Schneider, M.: Spatio-temporal predicates. IEEE Trans. Knowl. Data Eng. **14**(4), 881–901 (2002)
35. Etzion, O., Jajodia, S., (eds.), S.S.: Temporal Databases: Research and Practice. Springer, LNCS (1998)
36. Fagin, R., Ullman, J., Vardi, M.: On the semantics of updates in databases. In: Proc. ACM PODS, pp. 352–365 (1983)
37. Gedik, B., Liu, L.: Mobieyes: A distributed location monitoring service using moving location queries. IEEE Transactions on Mobile Computing **5**(10) (2006)
38. George, B., Kim, S., Shekhar, S.: Spatio-temporal network databases and routing algorithms: A summary of results. In: SSTD (2007)
39. Ghica, O., Trajcevski, G., Zhou, F., Tamassia, R., Scheuermann, P.: Selecting tracking principals with epoch-awareness. In: Proceedings of the 18th ACM SIGSPATIAL GIS Conference, pp. 222–231 (2010)
40. Ginsberg, M.L., Smith, D.E.: Reasoning about action i: A possible worlds approach. Artif. Intell. **35**(2), 165–195 (1988)
41. Goguen, J.: The logic of inexact concepts. Synthese **19**, 325–373 (1969)
42. Goldblatt, R.: Mathematical modal logic: A history of its evolution (2006). Http://homepages.mcs.vuw.ac.nz/ rob/papers/modalhist.pdf
43. Goodchild, M.F., Zhang, J., Kyriakidis, P.C.: Discriminant models of uncertainty in nominal fields. T. GIS **13**(1), 7–23 (2009)
44. von Gottfried Wilhelm Leibniz, F.: Theodicy: Essays on the Goodness of God, the Freedom of Man, and the Origin of Evil. Open Court (1998)
45. Gowrisankar, N., Nittel, S.: Reducing uncertainty in location prediction of moving objects in road networks. In: GIScience (2002)
46. Grant, J., Horty, J., Lobo, J., Minker, J.: View updates in stratified disjunctive deductive databases. Journal of Automated Reasoning **11**, 249–267 (1993)
47. Gudmundsson, J., van Kreveld, M.J.: Computing longest duration flocks in trajectory data. In: GIS (2006)
48. Güting, R.H.: Gral: An extensible relational database system for geometric applications. In: VLDB (1989)
49. Güting, R.H.: An introduction to spatial database systems. VLDB J. **3**(4), 357–399 (1994)
50. Güting, R.H., de Almeida, V.T., Ding, Z.: Modeling and querying moving objects in networks. VLDB Journal **15**(2) (2006)
51. Güting, R.H., Behr, T., Düntgen, C.: Secondo: A platform for moving objects database research and for publishing and integrating research implementations. IEEE Data Eng. Bull. **33**(2), 56–63 (2010)
52. Güting, R.H., Böhlen, M.H., Erwig, M., Jensen, C.S., Lorentzos, N., Nardeli, E., Schneider, M., Viqueira, J.R.R.: Spatio-temporal models and languages: An approach based on data types. In: Spatio-Temporal Databases – the Chorochronos Approach (2003)

53. Güting, R.H., Böhlen, M.H., Erwig, M., Jensen, C.S., Lorentzos, N., Schneider, M., Vazir-giannis, M.: A foundation for representing and queirying moving objects. ACM TODS (2000)

54. Güting, R.H., Böhlen, M.H., Erwig, M., Jensen, C.S., Lorentzos, N.A., Schneider, M., Vazir-giannis, M.: A foundation for representing and querying moving objects. ACM Trans. Database Syst. **25**(1) (2000)

55. Güting, R.H., Schneider, M.: Realm-based spatial data types: The rose algebra. VLDB J. **4**(2), 243–286 (1995)

56. Güting, R.H., Schneider, M.: Moving Objects Databases. Morgan Kaufmann (2005)

57. Guttman, A.: R-trees: A dynamic index structure for spatial searching. In: SIGMOD Conference, pp. 47–57 (1984)

58. Haas, P.J., Suciu, D.: Special issue on uncertain and probabilistic databases. VLDB Journal **18**(5), 987–988 (2009)

59. Hadjielefteriou, M., Kollios, G., Bakalov, P., Tsotras, V.: Complex spatio-temporal pattern queries. In: VLDB (2005)

60. Hadjieleftheriou, M., Kollios, G., Tsotras, V.J., Gunopulos, D.: Efficient indexing of spa-tiotemporal objects. In: EDBT (2002)

61. Hägerstrand, T.: What about people in regional science? Papers of the Regional Science Association **24**, 7–21 (1970)

62. He, G., Hou, J.C.: Tracking targets with quality in wireless sensor networks. In: 13th IEEE International Conference on Network Protocols (ICNP) (2005)

63. Herrick, P.: The Many Worlds of Logic. Oxford University Press (1999)

64. Hintikka, J.: Knowledge and Belief: An Introduction to the Logic of the Two Notions. Cornell University Press (1962)

65. Hjaltason, G.R., Samet, H.: Distance browsing in spatial databases. ACM Trans. Database Syst. **24**(2), 265–318 (1999)

66. Hornsby, K., Egenhofer, M.J.: Modeling moving objects over multiple granularities. Ann. Math. Artif. Intell. **36**(1-2), 177–194 (2002)

67. Huang, Y.K., Chen, C.C., Lee, C.: Continuous k -nearest neighbor query for moving objects with uncertain velocity. GeoInformatica **13**(1) (2009)

68. Huang, Y.K., Lee, C.: Efficient evaluation of continuous spatio-temporal queries on moving objects with uncertain velocity. GeoInformatica **14**(2), 163–200 (2010)

69. Ilarri, S., Mena, E., Illarramendi, A.: Location-dependent query processing: Where we are and where we are heading. ACM Comput. Surv. **42**(3) (2010)

70. online issue, F.: (2006). Http://www.forbes.com/home/digitalentertainment/2006/ 04/13/google-aol-yahoo-cx_rr_0417maps.html

71. Iwerks, G.S., Samet, H., Smith, K.P.: Maintenance of k-nn and spatial join queries on con-tinuously moving points. ACM Trans. Database Syst. **31**(2) (2006)

72. Jensen, C.S., Snodgrass, R.T.: Temporal data management. IEEE Trans. Knowl. Data Eng. **11**(1), 36–44 (1999)

73. Jeong, J., Guo, S., He, T., Du, D.: Apl: Autonomous passive localization for wireless sensors deployed in road networks. In: INFOCOM (2008)

74. Jeung, H., Liu, Q., Shen, H.T., Zhou, X.: A hybrid prediction model for moving objects. In: ICDE, pp. 70–79 (2008)

75. Jeung, H., Yiu, M.L., Zhou, X., Jensen, C.S., Shen, H.T.: Discovery of convoys in trajectory databases. PVLDB **1**(1) (2008)

76. Kainz, W., Egenhofer, M., Greasley, I.: Modeling spatial relations and operations with par-tially ordered sets. International Journal of Geographical Information Systems **7**(3) (1993)

77. Kanjilal, V., Liu, H., Schneider, M.: Plateau regions: An implementation concept for fuzzy regions in spatial databases and gis. In: IPMU, pp. 624–633 (2010)

78. Keogh, E.J.: A decade of progress in indexing and mining large time series databases. In: VLDB (2006)

79. Keogh, E.J., Chakrabarti, K., Mehrotra, S., Pazzani, M.J.: Locally adaptive dimensionality reduction for indexing large time series databases. In: SIGMOD Conference (2001)

80. Koubarakis, M., Sellis, T., Frank, A., Grumbach, S., Güting, R., Jensen, C., Lorentzos, N., Manolopoulos, Y., Nardelli, E., Pernici, B., Scheck, H.J., Scholl, M., Theodoulidis, B., Tryfona, N. (eds.): Spatio-Temporal Databases – the CHOROCHRONOS Approach. Springer-Verlag (2003)
81. Kripke, S.: Semantical considerations on modal logic. Acta Philosophica Fennica **16** (1963)
82. Kuijpers, B., Grimson, R., Othman, W.: An analytic solution to the alibi query in the space-time prisms model for moving object data. International Journal of Geographical Information Science **25**(2), 293–322 (2011)
83. Kuijpers, B., Miller, H.J., Neutens, T., Othman, W.: Anchor uncertainty and space-time prisms on road networks. International Journal of Geographical Information Science **24**(8), 1223–1248 (2010)
84. Kuijpers, B., Moelans, B., Othman, W., Vaisman, A.A.: Analyzing trajectories using uncertainty and background information. In: SSTD, pp. 135–152 (2009)
85. Kuijpers, B., Othman, W.: Modelling uncertainty on road networks via space-time prisms. Int.l Journal on GIS **23**(9) (2009)
86. Kuijpers, B., Othman, W.: Trajectory databases: data models, uncertainty and complete query languages. Journal of Computer and System Sciences (2009). Doi:10.1016/j.jcss.2009.10.002
87. Lange, R., Farrell, T., Dürr, F., Rothermel, K.: Remote real-time trajectory simplification. In: PerCom (2009)
88. Lange, R., Weinschrott, H., Geiger, L., Blessing, A., Dürr, F., Rothermel, K., Schütze, H.: On a generic uncertainty model for position information. In: QuaCon, pp. 76–87 (2009)
89. Lee, D.L., Zhu, M., Hu, H.: When location-based services meet databases. Mobile Information Systems **1**(2), 81–90 (2005)
90. Lema, J.A., Forlizzi, L., Güting, R.H., Nardelli, E., Schneider, M.: Algorithms for moving objects databases. Computing Journal **46**(6) (2003)
91. Li, G., Li, Y., Shu, L., Fan, P.: Cknn query processing over moving objects with uncertain speeds in road networks. In: APWeb, pp. 65–76 (2011)
92. Lian, X., 0002, L.C.: Probabilistic ranked queries in uncertain databases. In: EDBT, pp. 511–522 (2008)
93. Lin, D., Cui, B., Yang, D.: Optimizing moving queries over moving object data streams. In: DASFAA, pp. 563–575 (2007)
94. Liu, H., Schneider, M.: Querying moving objects with uncertainty in spatio-temporal databases. In: DASFAA (1), pp. 357–371 (2011)
95. Lynch, J.D., Chen, X., Hui, R.B.: A multimedia approach to visualize and interact with large scale mobile lidar data. In: ACM Multimedia, pp. 1689–1692 (2010)
96. Ma, C.: Spatio-temporal variograms and covariance models. Advances in Applied Probability **37**(3), 706–725 (2005)
97. Mamoulis, N., Cao, H., Kollios, G., Hadjieleftheriou, M., Tao, Y., Cheung, D.W.: Mining, indexing, and querying historical spatiotemporal data. In: International Conference on Knowledge Discovery and Data Mining, pp. 236–245 (2004)
98. Minker, J., Lobo, J., Rajasekar, A.: Circumscription and disjunctive logic programming. In: V. Lifschitz (ed.) Artificial Intelligence and Mathematical Theory of Computation, pp. 281–304. Academic Press (1991)
99. Mokbel, M.F., Aref, W.G.: SOLE: Scalable on-line execution of continuous queries on spatio-temporal data streams. VLDB Journal **17**(5), 971–985 (2008)
100. Mokbel, M.F., Aref, W.G.: SOLE: scalable on-line execution of continuous queries on spatio-temporal data streams. VLDB Journal **17**(5), 971–995 (2008)
101. Mokbel, M.F., Chow, C.Y., Aref, W.G.: The new casper: Query processing for location services without compromising privacy. In: VLDB (2006)
102. Mokbel, M.F., Xing, X., Hammad, M., Aref, W.G.: Continuous query processing of spatio-temporal data streams in place. The GeoInformatica Journal **9**(4) (2005)
103. Mokhtar, H., Su, J.: Questo: A query language for uncertain and exact spatio-temporal objects. In: ADBIS, pp. 184–198 (2008)

104. Mokhtar, H., Su, J., Ibarra, O.: On moving objects queries. In: PODS (2002)
105. Mouratidis, K., Yiu, M.L., Papadias, D., Mamoulis, N.: Continuous nearest neighbor moni-
     toring in road networks. In: VLDB, pp. 43–54 (2006)
106. du Mouza, C., Rigaux, R.: Multi-scale classification of moving objects trajectories. In: SS-
     DBM (2004)
107. O'Rourke, J.: Computational Geometry in C. Cambridge University Press (2000)
108. Othman, W.: Uncertainty management in trajectory databases. Phd thesis, Hasselt University
     (Belgium) (2009)
109. Pasquale, A.D., Forlizzi, L., Jensen, C.S., Manolopoulos, Y., Nardelli, E., Pfoser, D., Proietti,
     G., Saltenis, S., Theodoridis, Y., Tzouramanis, T.: Access methods and query processing
     techniques. In: Spatio-Temporal Databases: the Chorochronos Approach (2003)
110. Pattem, S., Poduri, S., Krishnamachari, B.: Energy-quality tradeoffs for target tracking in
     wireless sensor networks. In: IPSN (2003)
111. Pei, J., Hua, M., Tao, Y., Lin, X.: Query answering techniques on uncertain and probabilistic
     data: tutorial summary. In: ACM SIGMOD (2008)
112. Pelanis, M., Saltenis, S., Jensen, C.S.: Indexing the past, present, and anticipated future po-
     sitions of moving objects. ACM Trans. Database Syst. **31**(1) (2006)
113. Pfoser, D., Jensen, C.S.: Capturing the uncertainty of moving objects representation. In: SSD
     (1999)
114. Pfoser, D., Tryfona, N.: Capturing fuzziness and uncertainty of spatiotemporal objects. In:
     ADBIS, pp. 112–126 (2001)
115. Pfoser, D., Tryfona, N., Jensen, C.S.: Indeterminacy and spatiotemporal data: Basic defini-
     tions and case study. GeoInformatica **9**(3) (2005)
116. Popivanov, I., Miller, R.J.: Similarity search over time-series data using wavelets. In: ICDE
     (2002)
117. Prabhakar, S., Xia, Y., Khalashnikov, D., Aref, W., Hambrusch, S.: Query indexing and ve-
     locity constrained indexing: Scalable techniques for continuous queries on moving objects.
     IEEE - TRANS. KNOWL. DATA ENG. **51**(10) (2002)
118. Qi, Y., Jain, R., Singh, S., Prabhakar, S.: Threshold query optimization for uncertain data. In:
     SIGMOD Conference, pp. 315–326 (2010)
119. Raptopoulou, K., Papadopoulos, A., Manolopoulos, Y.: Fast nearest-neighbor query process-
     ing in moving-object databases. GeoInformatica **7**(2), 113–137 (2003)
120. Renz, M., Cheng, R., Kriegel, H.P., Züfle, A., Berneckcr, T.: Similarity search and mining in
     uncertain databases (tutorial). PVLDB **3**(2), 1653–1654 (2010)
121. Reynolds, A., Ioacono, G.L.: On the simulation of particle trajectories in turbulent flows.
     Physics of Fluids **16**(12) (2004)
122. Rigaux, P., Scholl, M., Voisard, A.: Introduction to Spatial Databases: Applications to GIS.
     Morgan Kauffmann (2000)
123. Roberts, F.: Tolerance geometry. Notre Dame Journal of Formal Logic **14**(1), 68–76 (1973)
124. Roussopoulos, N., Kelley, S., Vincent, F.: Nearest neighbor queries. In: SIGMOD Confer-
     ence, pp. 71–79 (1995)
125. Sarma, A.D., Benjelloun, O., Halevy, A.Y., Nabar, S.U., Widom, J.: Representing uncertain
     data: models, properties, and algorithms. VLDB Journal **18**(5), 989–1019 (2009)
126. Schiller, J., (editors), A.V.: Location-Based Services. Morgan Kaufmann (2004)
127. Schneider, M.: Uncertainty management for spatial data in databases: Fuzzy spatial data
     types. In: SSD, pp. 330–351 (1999)
128. Schneider, M.: Fuzzy spatial data types for spatial uncertainty management in databases. In:
     Handbook of Research on Fuzzy Information Processing in Databases, pp. 490–515 (2008)
129. Shahabi, C., Kolahdouzan, M.R., Sharifzadeh, M.: A road network embedding technique for
     k-nearest neighbor search in moving object databases. GeoInformatica **7**(3), 255–273 (2003)
130. Shekhar, S., Chawla, S.: Spatial Databases: A Tour. Prentice Hall (2003)
131. Sistla, A.P., Wolfson, O., Chamberlain, S., Dao, S.: Modeling and querying moving objects.
     In: 13th Int'l Conf. on Data Engineering (ICDE) (1997)
132. Snodgrass, R.T., Böhlen, M.H., Jensen, C.S., Steiner, A.: Transitioning temporal support in
     tsql2 to sql3. In: Temporal Databases, Dagstuhl, pp. 150–194 (1997)

133. Soliman, M.A., Ilyas, I.F., Chang, K.C.C.: Top-k query processing in uncertain databases. In: ICDE (2007)
134. Suciu, D., Dalvi, N.N.: Foundations of probabilistic answers to queries. In: ACM SIGMOD (2005). Tutorial
135. Sundararaman, B., Buy, U., Kshemkalyani, A.D.: Clock synchronization for wireless sensor networks: a survey. Ad Hoc Networks **3**(3), 281–323 (2005)
136. Szewczyk, R., Mainwaring, A.M., Polastre, J., Anderson, J., Culler, D.E.: An analysis of a large scale habitat monitoring application. In: SenSys, pp. 214–226 (2004)
137. Tanin, E., Chen, S., Tatemura, J., Hsiung, W.P.: Monitoring moving objects using low frequency snapshots in sensor networks. In: MDM (2008)
138. Tao, Y., Faloutsos, C., Papadias, D., 0002, B.L.: Prediction and indexing of moving objects with unknown motion patterns. In: SIGMOD Conference, pp. 611–622 (2004)
139. Tao, Y., Papadias, D.: Spatial queries in dynamic environments. ACM Trans. Database Syst. **28**(2) (2003)
140. Tao, Y., Papadias, D., Shen, Q.: Continuous nearest neighbor search. In: VLDB (2002)
141. Tao, Y., Papadias, D., Sun, J.: The tpr∗-tree: An optimized spatio-temporal access method for predictive queries. In: VLDB (2003)
142. Tao, Y., Xiao, X., Cheng, R.: Range search on multidimensional uncertain data. ACM Trans. Database Syst. **32**(3), 15 (2007)
143. Tøssebro, E., Nygård, M.: Uncertainty in spatiotemporal databases. In: ADVIS, pp. 43–53 (2002)
144. Trajcevski, G., Cao, H., Scheuermann, P., Wolfson, O., Vaccaro, D.: On-line data reduction and the quality of history in moving objects databases. In: MobiDE, pp. 19–26 (2006)
145. Trajcevski, G., Choudhary, A., Wolfson, O., Li, Y., Li, G.: Uncertain range queries for necklaces. In: MDM (2010)
146. Trajcevski, G., Ghica, O.C., Scheuermann, P.: Tracking-based trajectory data reduction in wireless sensor networks. In: SUTC/UMC, pp. 99–106 (2010)
147. Trajcevski, G., Tamassia, R., Cruz, I., Scheuermann, P., Hartglass, D., Zamierowski, C.: Ranking continuous nearest neighbors for uncertain trajectories. VLDB Journal (2011). (accepted, to appear)
148. Trajcevski, G., Tamassia, R., Ding, H., Scheuermann, P., Cruz, I.F.: Continuous probabilistic nearest-neighbor queries for uncertain trajectories. In: EDBT (2009)
149. Trajcevski, G., Wolfson, O., Hinrichs, K., Chamberlain, S.: Managing uncertainty in moving objects databases. ACM Trans. Database Syst. **29**(3) (2004)
150. Trajcevski, G., Wolfson, O., Zhang, F., Chamberlain, S.: The geometry of uncertainty in moving objects databases. In: International Conference on Extending Database Technology (EDBT) (2002)
151. Umer, M., Kulik, L., Tanin, E.: Spatial interpolation in wireless sensor networks: localized algorithms for variogram modeling and kriging. GeoInformatica **14**(1), 101–134 (2010)
152. UNISYS: 2009 hurrican/tropical data for atlantic. http://weather.unisys.com/hurricane/atlantic/2009/index.htm (2009)
153. Vaisman, A.A., Zimányi, E.: What is spatio-temporal data warehousing? In: DaWaK, pp. 9–23 (2009)
154. Vazirgiannis, M., Wolfson, O.: A spatiotemporal model and language for moving objects on road networks. In: SSTD (2001)
155. Vlachos, M., Hadjielefteriou, M., Gunopulos, D., Keogh, E.: Indexing multidimensional time-series. VLDB Journal **15**(1) (2006)
156. Weibel, R.: Generalization of spatial data: Principles and selected algorithms. In: Algorithmic Foundations of Geographic Information Systems. LNCS Springer Verlag (1996)
157. Wilke, G., Frank, A.F.: Tolerance geometry – euclid's first postulate for points and lines with extension. In: Proceedings of the 18th ACM SIGSPATIAL GIS Conference, pp. 162–171 (2010)
158. Winslett, M.: Reasoning about action using a possible model approach. In: Proc. of the Seventh National Conference on Artificial Intelligence, pp. 89–93 (1988)

159. Winter, S., Yin, Z.C.: Directed movements in probabilistic time geography. International Journal of Geographical Information Science **24**(9) (2010)
160. Winter, S., Yin, Z.C.: The elements of probabilistic time geography. GeoInformatica **15**(3), 417–434 (2011)
161. Wolfson, O., Chamberlain, S., Dao, S., Jiang, L., Mendez, G.: Cost and imprecision in modeling the position of moving objects. In: 14 Int'l Conf. on Data Engineering (ICDE) (1998)
162. Wolfson, O., Sistla, A.P., Chamberlain, S., Yesha, Y.: Updating and querying databases that track mobile units. Distributed and Parallel Databases **7** (1999)
163. Wolfson, O., Xu, B., Chamberlain, S., Jiang, L.: Moving objects databases: Issues and solutions. In: SSDB (1999)
164. Xing, X., Mokbel, M.F., Aref, W.G., Hambrusch, S.E., Prabhakar, S.: Scalable spatiotemporal continuous query processing for location-aware services. In: International Conference on Scientific and Statistical Database Management (SSDBM) (2004)
165. Xu, Y., Lee, W.C.: Compressing moving object trajectory in wireless sensor networks. IJDSN **3**(2) (2007)
166. Yang, J., Hu, M.: Trajpattern: Mining sequential patterns from imprecise trajectories of mobile objects. In: International Conference on Extending Database Technology, pp. 664–681 (2006)
167. Yi, K., Li, F., Kollios, G., Srivastava, D.: Efficient processing of top-k queries in uncertain databases. In: ICDE, pp. 1406–1408 (2008)
168. Yiu, M.L., Mamoulis, N., Papadias, D.: Aggregate nearest neighbor queries in road networks. IEEE Trans. Knowl. Data Eng. **17**(6) (2005)
169. Yoon, H., Shahabi, C.: Robust time-referenced segmentation of moving object trajectories. In: ICDM (2008)
170. Yu, B., Kim, S.H., Alkobaisi, S., Bae, W.D., Bailey, T.: The tornado model: Uncertainty model for continuously changing data. In: DASFAA, pp. 624–636 (2007)
171. Yu, X., Mehrotra, S.: Capturing uncertainty in spatial queries over imprecise data. In: DEXA, pp. 192–201 (2003)
172. Yu, X., Pu, K.Q., Koudas, N.: Monitoring k-nearest neighbor queries over moving objects. In: ICDE, pp. 631–642 (2005)
173. Zadeh, L.A.: Commonsense knowledge representation based on fuzzy logic. IEEE Computer **16**(10), 61–65 (1983)
174. Zadeh, L.A.: Granular computing - computing with uncertain, imprecise and partially true data. In: Int.l Symp. on Spatial Data Quality (2007)
175. Zhang, Q., Sobelman, G.E., He, T.: Gradient-based target localization in robotic sensor networks. Pervasive and Mobile Computing **5**(1) (2009)
176. Zhang, Y., Cheng, R., Chen, J.: Evaluating continuous probabilistic queries over imprecise sensor data. In: DASFAA (1), pp. 535–549 (2010)
177. Zhao, Y., Aggarwal, C.C., Yu, P.S.: On wavelet decomposition of uncertain time series data sets. In: CIKM, pp. 129–138 (2010)
178. Zheng, K., Trajcevski, G., Zhou, X., Scheuermann, P.: Probabilistic range queries for uncertain trajectories on road networks. In: EDBT, pp. 283–294 (2011)
179. Zhong, Z., Zhu, T., Wang, D., He, T.: Tracking with unreliable node sequences. In: INFOCOM, pp. 1215–1223 (2009)
180. Zhu, X., Sarkar, R., Gao, J., Mitchell, J.S.B.: Light-weight contour tracking in wireless sensor networks. In: INFOCOM, pp. 1175–1183 (2008)

Chapter 4
Privacy of Spatial Trajectories

Chi-Yin Chow and Mohemad F. Mokbel

Abstract The ubiquity of mobile devices with global positioning functionality (e.g., GPS and Assisted GPS) and Internet connectivity (e.g., 3G and Wi-Fi) has resulted in widespread development of location-based services (LBS). Typical examples of LBS include local business search, e-marketing, social networking, and automotive traffic monitoring. Although LBS provide valuable services for mobile users, revealing their private locations to potentially untrusted LBS service providers pose privacy concerns. In general, there are two types of LBS, namely, snapshot and continuous LBS. For snapshot LBS, a mobile user only needs to report its current location to a service provider once to get its desired information. On the other hand, a mobile user has to report its location to a service provider in a periodic or on-demand manner to obtain its desired continuous LBS. Protecting user location privacy for continuous LBS is more challenging than snapshot LBS because adversaries may use the spatial and temporal correlations in the user's a sequence of location samples to infer the user's location information with a higher degree of certainty. Such user spatial trajectories are also very important for many applications, e.g., business analysis, city planning, and intelligent transportation. However, publishing original spatial trajectories to the public or a third party for data analysis could pose serious privacy concerns. Privacy protection in continuous LBS and trajectory data publication has increasingly drawn attention from the research community and industry. In this chapter, we describe the state-of-the-art privacy-preserving techniques for continuous LBS and trajectory publication.

Chi-Yin Chow
City University of Hong Kong, China

Mohemad F. Mokbel
University of Minnesota - Twin Cities, USA

4.1 Introduction

With the advanced location-detection technologies, e.g., global positioning system (GPS), cellular networks, Wi-Fi, and radio frequency identification (RFID), location-based services (LBS) have become ubiquitous [6, 30, 42]. Examples of LBS include local business search (e.g., searching for restaurants within a user-specified range distance from a user), e-marketing (e.g., sending e-coupons to nearby potential customers), social networking (e.g., a group of friends sharing their geo-tagged messages), automotive traffic monitoring (e.g., inferring traffic congestion from the position and speed information periodically reported from probe vehicles), and route finder applications (e.g., finding a route with the shortest driving time between two locations). There are two types of LBS, namely, *snapshot* and *continuous* LBS. For snapshot LBS, a mobile user only needs to report its current location to a service provider once to get its desired information. On the other hand, a mobile user has to report its location to a service provider in a periodic or on-demand manner to obtain its desired continuous LBS.

Although LBS provide many valuable and important services for end users, revealing personal location data to potentially untrustworthy service providers could pose privacy concerns. Two surveys reported in July 2010 found that more than half (55%) of LBS users show concern about their loss of location privacy [57] and 50% of U.S. residents who have a profile on a social networking site are concerned about their privacy [39]. The results of these surveys confirm that location privacy is one of the key obstacles for the success of location-dependent services. In fact, there are many real-life scenarios where perpetrators abuse location-detection technologies to gain access to private location information about victims [13, 15, 54, 55].

Privacy in continuous LBS is more challenging than snapshot LBS because adversaries could use the spatial and temporal correlations in the user's location samples to infer the user's private location information. Such user spatial trajectories are also very important for many real-life applications, e.g., business analysis, city planning, and intelligent transportation. However, publishing original spatial trajectories to the public or a third party for data analysis could pose serious privacy concerns. Privacy protection in continuous LBS and trajectory data publication has increasingly drawn attention from the industry and academia. In this chapter, we describe existing privacy-preserving techniques for continuous LBS and trajectory data publication.

The rest of this chapter is organized as follows. Section 4.2 presents the derivation of spatial trajectory privacy. Section 4.3 discusses the state-of-the-art privacy-preserving techniques for continuous LBS. Section 4.4 gives existing privacy protection techniques for trajectory publication. Finally, Section 4.5 concludes this chapter with research directions in privacy-preserving continuous LBS and trajectory data publication.

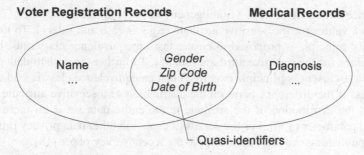

Fig. 4.1 Using quasi-identifiers to join two microdata sets.

4.2 The Derivation of Spatial Trajectory Privacy

This section gives the derivation of spatial trajectory privacy from data privacy and location privacy.

4.2.1 Data Privacy

Many agencies and other organizations often need to publish microdata, i.e., tables that contain unaggregated information about individuals, (e.g., medical, voter registration, census, and customer data) for many practical purposes such as demographic and public health research. In general, microdata are stored in a table where each row corresponds to one individual. In order to avoid the identification of records in microdata, known identifiers (e.g., name and social security number) must be removed. However, joining such *de-identified* microdata with other released microdata may still pose *data privacy* issues for individuals [49]. A study estimated that 87% of the population of the United States can be uniquely identified by using the collection of non-identity attributes, i.e., gender, date of birth, and 5-digit zip code [52]. In fact, those three attributes were used to link Massachusetts, USA voter registration records including name, gender, zip code and date of birth to *de-identified* medical data from Group Insurance Company including gender, zip code, date of birth and diagnosis to identify the medical records of the governor of Massachusetts in the medical data [52], as illustrated in Figure 4.1. Terminologically, attributes whose values taken together can potentially identify an individual record are referred to as *quasi-identifiers* and a set of records that have the same values for the quasi-identifiers in a released microdata is defined as an *equivalence class*.

Data privacy-preserving techniques are developed to anonymize microdata. Several data privacy principles are proposed to limit disclosure of anonymized microdata. For example, *k*-**anonymity** requires each record to be indistinguishable with other at least $k - 1$ records with respect to the quasi-identifier, i.e., each equivalence class contains at least k records [35, 49, 51, 52]. However, a *k*-anonymized equivalence class suffers from a homogeneity attack if all records in the class have less than k values for the sensitive attribute (e.g., disease and salary). To this end, *l*-diversity principle is proposed to ensure that an equivalence class must have at least l values for the sensitive attribute [38, 58]. To further strengthen data privacy protection, *t*-closeness principle is defined that an equivalence class is said to have t-closeness if the difference between the distribution of a sensitive attribute in this class and the distribution of the attribute in the entire data set is no more than a threshold parameter t [36]. For the details of these and other data privacy principles for data publishing, we refer the reader to the recent survey paper [18].

4.2.2 Location Privacy

In LBS, mobile users issue location-based queries to LBS service providers to obtain information based on their physical locations. LBS pose new challenges to traditional data privacy-preserving techniques due to two main reasons [41]. (1) These techniques preserve data privacy, but not the location-based queries issued by mobile users. (2) They ensure desired privacy guarantees for a snapshot of the database. In LBS, queries and data are continuously updated at high rates. Such highly dynamic behaviors need continuous maintenance of anonymized user and object sets.

Privacy-preserving techniques for LBS can be classified into three categories:

1. **False locations.** The basic idea of the techniques in this category is that the user sends either a fake location which is related to its actual location or its actual location with several fake locations to the LBS service provider in order to hide its location [28, 33, 62].
2. **Space transformation.** The techniques in this category transform the location information into another space where the spatial relationships among queries and data are encoded [20, 32].
3. **Spatial cloaking.** The main idea of spatial cloaking is to blur users' locations into cloaked spatial regions that are guaranteed to satisfy the k-anonymity [52] (i.e., the cloaked spatial region contains at least k users) and/or minimum region area privacy requirements [5, 14, 41] (i.e., the cloaked spatial region size is larger than a threshold) [2, 5, 7, 11, 12, 14, 19, 21, 22, 25, 31, 41, 64]. Spatial cloaking techniques have been extended to support road networks where a user's location is cloaked into a set of connected road segments so that the cloaked road segment set satisfies the privacy requirements of k-anonymity and/or minimum total road segment length [10, 34, 43, 56].

Anonymizing user locations is not the end of the story because database servers have to provide LBS based on anonymized user and/or object location information. Research efforts have also dedicated to dealing with privacy-preserving location-based queries, i.e., getting anonymous services from LBS service providers (e.g., [5, 20, 29, 31, 32, 41, 62]). These query processing frameworks can be divided into three main categories:

1. **Location obstruction.** The basic idea of location obstruction [62] is that a querying user first sends a query along with a fake location as an anchor to a database server. The database server keeps sending a list of nearest objects to the anchor to the user until the list of received objects satisfies the user's privacy and quality requirements.
2. **Space transformation.** This approach converts the original location of data and queries into another space. The space transformation maintains the spatial relationship among the data and query, in order to provide approximate query answers [20, 32] or exact query answers [20].
3. **Cloaked query area processing.** In this framework, a privacy-aware query processor is embedded into a database server to deal with the cloaked spatial region received either from a querying user [5, 29] or from a trusted third party [9, 31, 41]. For spatial cloaking in road networks, a query-aware algorithm is proposed to process privacy-aware location-based queries [3].

4.2.3 Trajectory Privacy

A spatial trajectory is a moving path or trace reported by a moving object in the geographical space. A spatial trajectory T is represented by a set of n time-ordered points, $T : p_1 \rightarrow p_2 \rightarrow \ldots \rightarrow p_n$, where each point p_i consists of a geospatial coordinate set (x_i, y_i) (which can be detected by a GPS-like device) and a timestamp t_i, i.e., $p_i = (x_i, y_i, t_i)$, where $1 \leq i \leq n$. Such spatial and temporal attributes of a spatial trajectory can be considered as powerful quasi-identifiers that can be linked to various other kinds of physical data objects [18, 44]. For example, a hospital releases a trajectory data set of its patients to a third-party research institute for analysis, as shown in Table 4.1. The released trajectory data set does not contain any explicit identifiers, such as patient name, but it contains a sensitive attribute (i.e., disease). Each record with a unique random ID, *RID*, corresponds to an individual, e.g., the record with $RID = 1$ means a patient visited locations $(1,5)$, $(6,7)$, $(8,10)$, and $(11,8)$ at timestamps 2, 4, 5, and 8, respectively. Suppose that an adversary knows that a patient of the hospital, Alice, visited locations $(1,5)$ and $(8,10)$ at timestamps 2 and 8, respectively. Since only the trajectory record with $RID = 1$ satisfies such spatial and temporal attributes, the adversary is 100% sure that Alice has HIV. This

Table 4.1 Patient trajectory data.

RID	Trajectory	Disease	...
1	$(1,5,2) \rightarrow (6,7,4) \rightarrow (8,10,5) \rightarrow (11,8,8)$	HIV	...
2	$(5,6,1) \rightarrow (3,7,2) \rightarrow (1,5,6) \rightarrow (7,8,7) \rightarrow (1,11,8) \rightarrow (6,5,10)$	Flu	...
3	$(4,7,2) \rightarrow (4,6,3) \rightarrow (5,1,6) \rightarrow (11,8,8) \rightarrow (5,8,9)$	Flu	...
4	$(10,3,5) \rightarrow (7,3,7) \rightarrow (4,6,10)$	HIV	...
5	$(7,6,3) \rightarrow (6,7,4) \rightarrow (6,10,6) \rightarrow (4,6,9)$	Fever	...

example shows that publishing *de-identified* user trajectory data can still cause serious privacy threats if the adversary has certain background knowledge.

In LBS, when a mobile user issues a continuous location-based query to a database server (e.g., "continuously inform me the traffic condition within 1 mile from my vehicle"), the user has to report its new location to the database server in a periodic or on-demand manner. Similarly, intelligent transportation systems require their users (e.g., probe vehicles) to periodically report their location and speed information to the system for analysis. Although such location-based queries and reports can be anonymized by replacing the identifiers of users with random identifiers, in order to achieve pseudonymity [47], the users may still suffer from privacy threats. This is because movements of whereabouts of users in public spaces can be openly observed by others through chance or engineered meetings [37]. In the worst case, if the starting location point of a trajectory is a home, an adversary uses reverse geocoding[1] [24] to translate a location point into a home address, and then uses a people-search-by-address engine (e.g., http://www.intelius.com and http://www.peoplefinders.com) to find the residents of the home address. Even though users generate a random identity for each of their location samples, multi-target tracking techniques (e.g., the multiple hypothesis tracking algorithm [48]) can be used to link anonymous location samples to construct target trajectories [26]. To this end, new techniques are developed to protect user spatial trajectory.

The key difference between continuous LBS and trajectory data publication with respect to challenges in privacy protection is twofold: (1) The scalability requirement of the privacy-preserving techniques for continuous LBS is much more important than that for trajectory data publication. This is because continuous LBS require the anonymization module to deal with a large number of real-time location updates at high rates while the anonymization process for trajectory data publication can be performed offline. (2) Global optimization can be applied to trajectory data publication because the anonymization process is able to analyze the entire (static) trajectory data to optimize its privacy protection or usability. However, global optimization is very difficult for continuous LBS, due to highly dynamic, uncertain user movements. Sections 4.3 and 4.4 present the state-of-the-art privacy-preserving techniques for continuous LBS and trajectory publication, respectively.

4.3 Protecting Trajectory Privacy in Location-based Services

In general, there are two categories of LBS based on whether they need a consistent user identity. A consistent user identity is not necessarily a user's actual identity or name because it can be an internal pseudonym.

---

[1] Reverse geocoding is the process of translating a human-readable address, such as a street address, from geographic coordinates, such as latitude and longitude.

- **Category-I LBS.** Some LBS require consistent user identities. For example, "Q1: let me know my friends' locations if they are within 2km from my location", "Q2: recommend 10 nearby restaurants to me based on my profile", and "Q3: continuously tell me the nearest shopping mall to my location". Q1 and Q2 require consistent user identities to let applications to find out their friends and profiles. Although Q3 does not need any consistent user identity, the query parameters (e.g., a particular shop name) can be considered as a virtual user identity that remains the same until the query expires.
- **Category-II LBS.** Other LBS do not require any consistent user identity, or even any user identity, such as "Q4: send e-coupons to users within 1km from my coffee shop".

In this section, we discuss seven privacy-preserving techniques for continuous LBS, including spatial cloaking, mix-zones, vehicular mix-zones, path confusion, path confusion with mobility prediction and data caching, Euler histogram-based on short IDs, and dummy trajectories, and indicate whether they support Category-I and/or II LBS from Sections 4.3.1 to 4.3.7, as summarized in Table 4.2.

Table 4.2 Privacy-preserving techniques for continuous LBS.

Techniques	Support Category-I LBS	Support Category-II LBS
Spatial cloaking	Yes	Yes
Mix-zones	No	Yes
Vehicular mix-zones	No	Yes
Path confusion	No	Yes
Path confusion with mobility prediction and data caching	No	Yes
Euler histogram-based on short IDs	No	Yes
Dummy trajectories	Yes	Yes

4.3.1 Spatial Cloaking

Mobile users have to reveal their locations to database servers in a periodic or on-demand manner to obtain continuous LBS. Simply applying a snapshot spatial cloaking technique (e.g., [2, 14, 19, 21, 22, 25, 31, 41, 64]) to each user location independently cannot ensure k-anonymity for a user spatial trajectory. Specifically, we present two techniques, namely, *trajectory tracing* [5] and *anonymity-set tracing* [8], that can reduce the protection of spatial cloaking.

- **Trajectory tracing attack.** In case that an attacker can save the cloaked spatial regions of a querying user U and capture the maximum movement speed

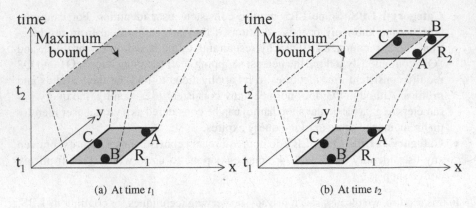

(a) At time t_1 (b) At time t_2

Fig. 4.2 Trajectory tracing attack.

max_{speed} of U based on U's historical movement patterns and/or other background information, e.g., the maximum legal driving speed and the maximum speed of U's vehicle, the attacker can use the trajectory tracing attack to reduce the effectiveness of spatial cloaking. Figure 4.2 depicts an example of the trajectory tracing attack, where the attacker collects a cloaked spatial region R_1 at time t_1 (represented by a solid rectangle in Figure 4.2a). The attacker cannot distinguish among the three users in R_1, i.e., A, B, and C. Figure 4.2b shows that the attacker collects another cloaked spatial region R_2 from U at time t_2. The attacker could use the most consecutive approach, where U moves at max_{speed} at any direction, to determine a *maximum bound* (represented by a dotted rectangle) that must contain U at time t_2. The maximum bound of R_1 at time t_2 can be determined by expanding each side of R_1 to a distance of $(t_2 - t_1) \times max_{speed}$. Since U must be inside R_2 and the maximum bound of R_1 at time t_2, the attacker knows that U is inside their overlapping area. Thus, the attacker knows that C is the querying user U.

- **Anonymity-set tracing attack.** An attacker could trace an anonymity set of a sequence of cloaked spatial regions of a continuous query to identify the query's sender [8]. Figure 4.3 gives an example of the anonymity-set tracing attack, where there are eight users A to H. The attacker collects two consecutive three-anonymous spatial regions at times t_1 and t_2, as depicted in Figures 4.3a and 4.3b, respectively. At time t_1, the probability of user A, C, or G being the query sender is 1/3. However, at time t_2, since only user A remains in the cloaked spatial region, the attacker knows that A is the query sender.

Patching and *delaying* techniques are proposed to prevent the trajectory tracing attack [5]. We describe these two techniques below.

- **Patching technique.** The patching technique is to combine the current cloaked spatial region and the maximum bound of the previous one such that the attacker can only know that the target user is inside the combined region. Figure 4.4a

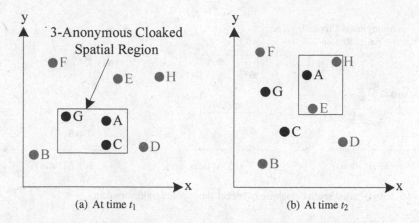

(a) At time t_1 (b) At time t_2

Fig. 4.3 Anonymity-set tracing attack.

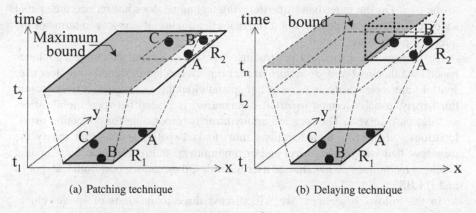

(a) Patching technique (b) Delaying technique

Fig. 4.4 Two techniques for preventing the trajectory tracing attack.

depicts an example for the patching technique in the running example, where the combination of the current cloaked spatial region at time t_2 (which is represented by a dotted rectangle) and the maximum bound of the cloaked spatial region released at time t_1 constitutes the user's cloaked spatial region R_2 (which is represented by a solid rectangle at time t_2).

- **Delaying technique.** The delaying technique is to suspend a location update until its cloaked spatial region is completely included in the maximum bound of the previous cloaked spatial region. As depicted in Figure 4.4b, the cloaked spatial region R_2 generated at time t_2 is suspended until R_2 fits into the maximum bound of the previous cloaked spatial region R_1 at time t_1. At time t_n, R_2 is reported to the database server.

In general, the patching technique generates larger cloaked spatial regions than the original ones, so it reduces the spatial accuracy of location updates that could

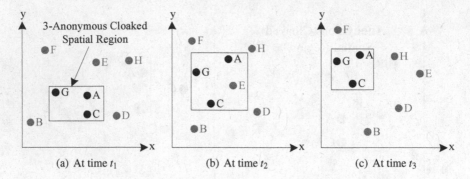

Fig. 4.5 Group-based spatial cloaking over real-time spatial trajectory data.

degrade the quality of services, in terms of the number of candidate answers returned to the user. On the other hand, the delaying technique does not reduce the spatial accuracy of location updates, but it degrades the quality of services in terms of the query response time.

To prevent the anonymity-set tracing attack, new spatial cloaking techniques based on either *real-time* or *historical* user trajectories are designed to protect user spatial trajectories. Similar to snapshot spatial cloaking techniques, a fully-trusted third party, usually termed *location anonymizer*, is placed between mobile users and database servers. The location anonymizer is responsible for collecting users' locations and blurring their locations into cloaked spatial regions that satisfy the user-specified k-anonymity level and/or minimum spatial region area. Since spatial cloaking techniques do not change user identities, they can support both Category I and II LBS.

In the following sections, we will discuss three main kinds of spatial cloaking techniques over user trajectories, namely, group-based, distortion-based, and prediction-based approaches, from Sections 4.3.1.1 to 4.3.1.3, respectively. All these techniques can prevent the anonymity-set tracing attack. The first two approaches are designed for real-time user trajectories, while the last one is for historical trajectory data.

4.3.1.1 Group-based Approach for Real-time Trajectory Data

The group-based algorithm is proposed to use real-time spatial trajectory data to protect trajectory privacy for continuous location-based queries [8]. The basic idea is that a querying user U forms a group with other $k-1$ nearby peers. Before the algorithm issues U's location-based query or reports U's location to a database server, it blurs U's location into a spatial region that contains all the group members as a cloaked spatial region. Figure 4.5 depicts an example of continuous spatial cloaking over real-time user locations. In this example, user A that issues a continuous location-based query at time t_1 requires its location to be k-anonymized, where

$k = 3$. At time t_1, a location anonymizer forms a group of users A, C, and G, so that A's cloaked spatial region contains all these group members, as represented by a rectangle in Figure 4.5a. The location anonymizer sends A's query with its cloaked spatial region to the database server. At later times t_2 and t_3, when A reports its new location to the location anonymizer, a new cloaked spatial region that contains the group members is formed, as shown in Figures 4.5b and 4.5c, respectively. The drawbacks of this approach are that users not issuing any query have to report their locations to the location anonymizer and the cloaked spatial region would become very large after a long time period. Such a large cloaked spatial region may incur high computational overhead at the database server and result in many candidate answer objects returned from the database server to the location anonymizer.

In theory, let R_i be the cloaked spatial region for a querying user U at time t_i and $S(R_i)$ be a set of users located in R_i. Suppose U's query is first successfully cloaked at time t_1, it expires at time t_n, $U \in S(R_1)$, and $|S(R_1)| \geq k$. Without any additional information, the value of R_1's entropy, $H(R_1)$, is at least $\log |S(R_1)|$ which means that every user in R_1 has an equal chance of $1/|S(R_1)|$ to be U [60], i.e., R_1 is a k-anonymous cloaked spatial region for U. For U's cloaked spatial regions R_{i-1} and R_i generated at two consecutive times t_{i-1} and t_i ($1 < i \leq n$), respectively, if R_{i-1} is a k-anonymous cloaked spatial region and $S(R_{i-1}) \subseteq S(R_i)$, R_i is also a k-anonymous cloaked spatial region [60]. Thus, the group-based approach can ensure k-anonymity for the entire life span of a continuous location-based query.

4.3.1.2 Distortion-based Approach for Real-time Trajectory Data

The distortion-based approach aims at overcoming the drawbacks of the group-based approach. It only requires querying users to report their locations to the location anonymizer, and it also considers their movement directions and velocities to minimize cloaked spatial regions [46]. A distortion function is defined to measure

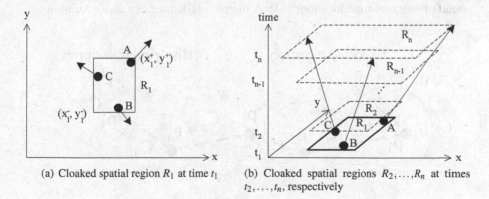

(a) Cloaked spatial region R_1 at time t_1 (b) Cloaked spatial regions $R_2, \ldots, R_n$ at times $t_2, \ldots, t_n$, respectively

Fig. 4.6 Query distortion for continuous spatial cloaking.

the temporal query distortion of a cluster of continuous queries. Figure 4.6 gives an example of how to determine query distortion. In this example, three users A, B and C that issue their continuous location-based queries at time t_1 constitute an anonymity set and their queries expire at time t_n. Their cloaked spatial region R_1 at time t_1 is a minimum bounding rectangle of the anonymity set, as represented by a rectangle (Figure 4.6a). Let (x_i^-, y_i^-) and (x_i^+, y_i^+) be the left-bottom and right-top vertices of a cloaked spatial region R_i at time t_i, respectively. The distortion for their queries with a cloaked spatial region R_i at time t_i is defined as:

$$\Delta(R_i) = \frac{(x_i^+ - x_i^-) + (y_i^+ - y_i^-)}{A_{height} + A_{width}}, \tag{4.1}$$

where A_{height} and A_{width} are the height and width of the minimum bounding rectangle of the entire system space, respectively. Assume their movement directions and velocities (represented by arrows in Figure 4.6b) remain the same from the time period t_1 to t_n, their subsequent cloaked spatial regions $R_2, R_3, \ldots, R_n$ at times $t_2, t_3, \ldots, t_n$ can be predicted, respectively. The distortion for their queries with respect to the time period from t_1 to t_n is defined as:

$$\int_{t_1}^{t_n} \Delta(R_i) = \frac{1}{A_{height} + A_{width}} \left\{ \int_{t_1}^{t_2} \Delta(R_1) dt + \int_{t_2}^{t_3} \Delta(R_2) dt + \ldots + \int_{t_{n-1}}^{t_n} \Delta(R_n) dt \right\},$$

Given a new continuous location-based query Q, greedy cloaking and bottom-up cloaking algorithms are designed to cluster Q with other $k-1$ outstanding queries into a group such that the group satisfies k-anonymity and their query distortion is minimized [46].

4.3.1.3 Predication-based Approach for Historical Trajectory Data

Another way to ensure k-anonymity is to use individuals' historical footprints, instead of their real-time locations [61]. A footprint is defined as a user's location col-

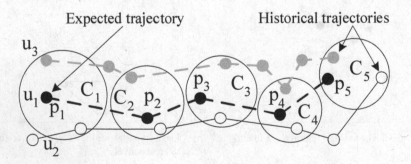

Fig. 4.7 Continuous spatial cloaking over historical trajectories.

lected at some point of time. Similar to the previous two approaches, a fully-trusted location anonymizer is placed between users and LBS service providers to collect users' footprints. Given a user's predicted trajectory (i.e., a sequence of expected footprints), the location anonymizer cloaks it with $k-1$ historical trajectories collected from other users. Figure 4.7 gives an example for continuous spatial cloaking over historical trajectories, where a user u_1 wants to subscribe continuous LBS from a service provider. u_1's predicted time-ordered footprints, $p_1 \rightarrow p_2 \rightarrow \ldots \rightarrow p_5$, are represented by black circles. If u_1's desired anonymity level is $k = 3$, the location anonymizer finds historical trajectories from other two users, u_2 and u_3. Then, each u_1's expected footprint p_i ($1 \leq i \leq 5$) is cloaked with at least one unique footprint of each of u_2's and u_3's trajectories to form a cloaked spatial region C_i. The sequence of such cloaked spatial regions constitutes the k-anonymized trajectory for u_1.

Given a k-anonymized trajectory $T = \{C_1, C_2, \ldots, C_n\}$, its resolution is defined as:

$$|T| = \frac{\sum_{i=1}^{n} Area(C_i)}{n}, \tag{4.2}$$

where $Area(C_i)$ is the area of cloaked spatial region C_i. For quality of services, $|T|$ should be minimized. Since the computation of an optimal T would be expensive, heuristic approaches are designed to find T [61]. Although using historical trajectory data gives better resolutions for k-anonymized trajectories, it would suffer from an observation attack. This is because an attacker may only see the querying user or less than k users located in a cloaked spatial region at its associated timestamp.

4.3.2 Mix-Zones

The concept of "mix" has been applied to anonymous communication in a network. A mix-network consists of normal message routers and mix-routers. The basic idea is that a mix-router collects k equal-length packets as input and reorders them randomly before forwarding them, thus ensuring unlinkability between incoming and outgoing messages. This concept has been extended to LBS, namely, *mix-zones* [4]. When users enter a mix-zone, they change to a new, unused pseudonym. In addition, they do not send their location information to any location-based application when they are in the mix-zone. When an adversary that sees a user U exits from the mix-zone cannot distinguish U from any other user who was in the mix-zone with U at the same time. The adversary is also unable to link people entering the mix-zone with those coming out of it. A set of users S is said to be k-anonymized in a mix-zone Z if all following conditions are met [45]:

1. The user set S contains at least k users, i.e., $|S| \geq k$.
2. All users in S are in Z at a point in time, i.e., all users in S must enter Z before any user in S exits.
3. Each user in S spends a completely random duration of time inside Z.

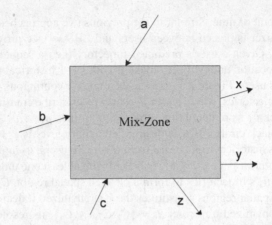

Fig. 4.8 A mix-zone with three users.

Table 4.3 An example of three-anonymized mix-zone.

User ID	Old Pseudonyms	New Pseudonyms	$time_{enter}$	$time_{exit}$	$time_{inside}$
α	a	y	2	9	7
β	c	x	5	8	3
γ	b	z	1	11	10

4. The probability of every user in S entering through an entry point is equally likely to exit in any of the exit points.

Table 4.3 gives an example of three-anonymity for the mix-zone depicted in Figure 4.8, where three users with real identities, α, β, and γ enter the mix-zone with old pseudonyms a, c, and b at times ($time_{enter}$) 2, 5, and 1, respectively. Users α, β, and γ exit the mix-zone with new pseudonyms y, x, and z at times ($time_{exit}$) 9, 8, and 11, respectively. Thus, they all are in the mix-zone during the time period from 5 to 8. Since they stay inside the mix-zone with random time periods (i.e., $time_{inside}$), there is a strong unlinkability between their entering order ($\gamma \rightarrow \alpha \rightarrow \beta$) and exiting order ($\beta \rightarrow \alpha \rightarrow \gamma$).

We can see that mix-zones require pseudonym change to protect user location privacy, so this technique can only support Category-II LBS. Mix-zones also impose limits on the services available to mobile users inside a mix-zone because they cannot update their locations until exiting the mix-zone. To minimize disruptions caused to users, the placement of mix-zones in the system should be optimized to limit the total number of mix-zones required to achieve a certain degree of anonymity [17].

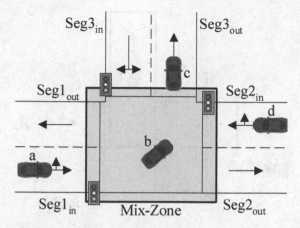

Fig. 4.9 A vehicular mix-zone.

4.3.3 Vehicular Mix-Zones

In a road network, vehicle movements are constrained by many spatial and temporal factors, such as physical roads, directions, speed limits, traffic conditions, and road conditions. Mix-zones designed for the Euclidean space are not secure enough to protect trajectory privacy in road networks [16, 45]. This is because an adversary can gain more background information from physical road constraints and delay characteristics to link entering events and exiting events of a mix-zone with a high degree of certainty. For example, a mix-zone (represented by a shaded area) is placed on an intersection of three road segments $Seg1$, $Seg2$, and $Seg3$, as depicted in Figure 4.9. If u-turn is not allowed in the intersection, an adversary knows that a vehicle with pseudonym c enters the mix-zone from either $Seg1_{in}$ or $Seg2_{in}$. Since a vehicle turning from $Seg1_{in}$ to $Seg3_{out}$ normally takes a longer time than turning from $Seg2_{in}$ to $Seg3_{out}$, the adversary would use this delay characteristic to link an exiting event at $Seg3_{out}$ to an entering event at $Seg1_{in}$ or $Seg2_{in}$. In addition, every vehicle may spend almost the same time during a short time period for a specific direction, e.g., u-turn, left, straight, or right. This temporal characteristic may violate the third necessary condition for mix-zones listed in Section 4.3.2.

An effective solution for vehicular mix-zones is to construct non-rectangular, adaptive mix-zones that start from the center of an road segment intersection on its outgoing road segments [45], as depicted in Figure 4.10. The length of each mix-zone on an outgoing segment is determined based on the average speed of the road segment, the time window, and the minimum pairwise entropy threshold. The dark shaded area should also be included in the mix-zone to ensure that an adversary cannot infer the vehicle movement direction (e.g., turn left or go straight in this example). The pairwise entropy is computed for every pair of users a and b in an anonymity set S by considering a and b to be the only members in S and determining the linkability between their old and new pseudonyms. Similar to mix-zones, vehic-

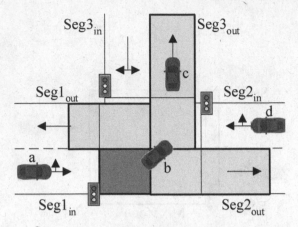

Fig. 4.10 Non-rectangular, adaptive vehicular mix-zones.

ular mix-zones require a pseudonym change, so they can only support Category-II LBS.

4.3.4 Path Confusion

Since consecutive location samples from a vehicle are temporally and spatially correlated, trajectories of individual vehicles can be constructed from a set of location samples with anonymized pseudonyms reported from several vehicles through target tracking algorithms [26]. The general idea of these algorithms is to predict the position of a target vehicle based on the last known speed and direction information and then decide which next location sample (or the one with the highest probability if there are multiple candidate location samples) to link to the same vehicle through Maximum Likelihood Detection [26].

The main goal of the path confusion technique is to avoid linking consecutive location samples to individual vehicles through target tracking algorithms with high confidence [27]. The degree of privacy of the path confusion technique is defined as the "time-to-confusion", i.e., the tracking time between two location samples where an adversary could not determine the next sample with sufficient tracking certainty. Tracking uncertainty is computed by $H = -\sum p_i \log p_i$, where p_i is the probability that location sample i belongs to a target vehicle. Smaller values of H means higher certainty or lower privacy. Given a maximum allowable time to confusion, *ConfusionTime*, and an associated uncertainty threshold, *ConfusionLevel*, a vehicle's location sample can be safely revealed if the time between the current time t and the last point of its confusion is less than *ConfusionTime* and the tracking uncertainty of its sample with all other location samples revealed at time t is higher than *ConfusionLevel*. To reduce computational overhead, the computation of tracking uncertainty can only consider the k-nearest location samples to a predicted location point (cal-

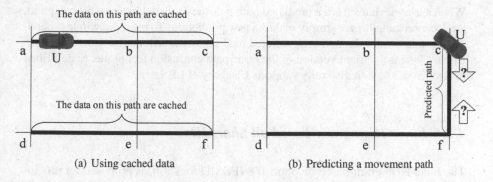

Fig. 4.11 Path confusion with mobility prediction and data caching.

culated by the target tracking algorithm), rather than all location samples reported at time t. Since a consistent user identity can link consecutive location samples, it cannot be revealed to any service provider; and hence, the path confusion technique only supports Category-II LBS.

4.3.5 Path Confusion with Mobility Prediction and Data Caching

The main idea of the path confusion technique with mobility prediction and data caching, called *CacheCloak*, is that the location anonymizer predicts vehicular movement paths, pre-fetches the spatial data on predicted paths, stores the data in a cache [40]. Figure 4.11a illustrates an example for *CacheCloak* where there are seven road segments, ab, bc, de, ef, ad, be, and cf, with six intersections, a to f. A bold road segment indicates that the data located on it are currently cached by the location anonymizer (e.g., data on road segments ab, bc, de, and ef are cached). Given a continuous location-based query from a user U (represented by a car in Figure 4.11), if U is moving on a path whose data are currently cached by the location anonymizer, the location anonymizer is able to return the query answer to U without contacting any LBS service provider, as depicted in Figure 4.11a. On the other hand, Figure 4.11b shows that U enters a road segment (i.e., cf) from an intersection c with no data in the cache. In this case, the location anonymizer predicts a path (which contains one road segment cf) for U that connects to an intersection of existing paths with cached data (i.e., f). Then, the location anonymizer issues a query Q to the service provider to get the data on the newly predicted path cf and caches the data. It finally returns the answer of Q to U.

CacheCloak prevents the LBS service provider from tracking any one user because the service provider can only see queries for a series of interweaving paths [40]. As shown in the example, the service provider cannot track the user who issues Q because the user could be turning in from road segment bc or from the other one ef (Figure 4.11b). *CacheCloak* can tolerate mispredictions or dynamic data.

When a user deviates from a predicted path p or the cached data for p have expired, the location anonymizer simply makes a new prediction. To preserve wireless bandwidth, stale data or data on mispredicted road segments will not be sent to the user. *CacheCloak* is a variant version of the basic path confusion technique, as described in Section 4.3.4, so it also only supports Category-II LBS.

4.3.6 Euler Histogram-based on Short IDs

The Euler Histogram-based on Short IDs (EHSID for short) is proposed for providing privacy-aware traffic monitoring services through answering aggregate queries [59]. EHSID supports two types of aggregate queries in road networks, namely, *ID-* and *entry-based* query types.

ID-based query type. ID-based queries ask for the count of unique vehicles inside a query region in a road network. Given a query region R, a *cross-border* query collects the count of unique vehicles that cross the boundary of R and a *distinct-object* query collects the count of unique vehicles that have been detected in R, including R's boundary. Figure 4.12 gives an example with two vehicles v_1 and v_2 whose trajectories intersect a rectangular query region R. Since there is only one vehicle, i.e., v_1, crossing the boundary of R, the *cross-border* query answer is one. On the other hand, both v_1 and v_2 are in R, so the *distinct-object* query answer is two.

Entry-based query type. Entry-based queries ask for the number of entries to a query region. If a vehicle has multiple entries to a query region (i.e., it enters the query region several times), its trajectory is divided into multiple sub-trajectories. Entry-based queries count the number of entries to the query region, even if the entries are from the same vehicle. Given a query region R, a *cross-border* query counts the number of entries that cross the boundary of R and a *distinct-object* query counts the number of distinct entries to R, including its boundary. Figure 4.12 depicts that v_1's trajectory can be divided into two sub-trajectories and each one crosses the boundary of the query region R, so the *cross-border* query answer is two. With an additional entry from v_2, the *distinct-object* query answer is three.

The basic idea of EHSID is based on Euler histograms, which were designed to count the number of rectangular objects in multidimensional space [50]. An Euler histogram is constructed by partitioning the space into a grid and maintaining a count for the number of intersecting objects of each face, edge, and vertex in the grid. Figure 4.13a gives an example of a 3×3 gird with three rectangular objects A, B, and C. Figure 4.13b shows the count of each face and vertex in the grid, and Figure 4.13c shows the count of each edge in the grid. Given a query region R, which is represented by a bold rectangle, the total number of distinct objects N in R is estimated by the equation $N = F + V - E$, where F is the sum of face counts inside R, V is the sum of vertex counts inside R (excluding its boundary),

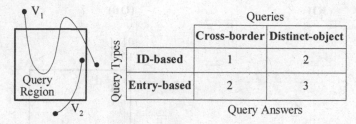

Fig. 4.12 ID- and entry-based aggregate queries in road networks.

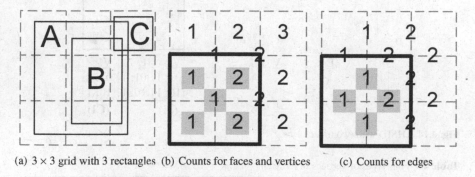

(a) 3 × 3 grid with 3 rectangles (b) Counts for faces and vertices   (c) Counts for edges

Fig. 4.13 Using an Euler histogram to count distinct rectangles in a query region.

and E is the sum of edge counts inside R (excluding its boundary). In this example, $F = 1 + 2 + 1 + 2 = 6$, $V = 1$, and $E = 1 + 1 + 1 + 2 = 5$; hence, $N = 6 + 1 - 5 = 2$, which is the exact number of distinct rectangles intersecting R.

To protect user trajectory privacy, each vehicle is identified by a short ID that is extracted from random bit positions from its full ID and EHSID periodically changes the bit positions. For example, if a vehicle's full ID is "110111011" and a bit pattern is 1, 3, 4, and 7, its short ID generated by the system is "1010". The random bit pattern is periodically updated. Let l_F and l_S be the length of the full and short IDs, respectively. The degree of k-anonymity of a short ID is measured as $k = 2^{l_F - l_S}$ [59].

Figure 4.14 gives a simple example with two vehicles for EHSID, where the road network consists of six vertices a to f and five road segments, ab, bc, cd, de, and ef. The short IDs of the two vehicles are "01" and "10". EHSID maintains a data list for each vertex and edge. An item in a data list is defined as (*short ID, the counter for the short ID*). The query region of a query Q is shown as a bold rectangle. In general, the query processing algorithm of EHSID has two main steps.

1. **Aggregating data at vertices and edges.** The algorithm identifies four types of relevant vertices to a query, as depicted in Table 4.4. An relevant edge is denoted as $E_{x,y}$, where x is the vertex type at one end and y is the vertex type at the other end. For example, $E_{jo,ob}$ denotes an edge with a *just-outside* vertex (V_{jo}) at one end and an *on-border* vertex (V_{ob}) at the other end. Then, the algorithm

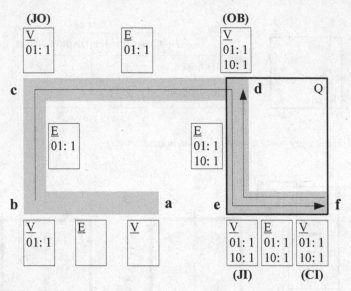

Fig. 4.14 EHSID with two trajectories.

Table 4.4 Four types of vertices in EHSID.

Vertex type	Relative position to query region R
Just Outside (V_{jo})	Located outside R and has a one-hop neighbor inside R
On Border (V_{ob})	Located inside R and has a one-hop neighbor outside R
Just Inside (V_{ji})	Located inside R and no one-hop neighbor is V_{jo} but has V_{ob} as an one-hop neighbor
Completely Inside (V_{ci})	Located inside R but is not V_{ob} or V_{ji}

Table 4.5 Relevant short IDs and edge datasets for a query.

Query type	Relevant short IDs	Relevant edge datasets
ID-based cross-border	$V_{jo} \cap V_{ob} \cap V_{ji}$	N/A
ID-based distinct-objects	$V_{ob} \cup V_{ji} \cup V_{ci}$	N/A
Entry-based cross-border	$V_{jo} \cap V_{ob} \cap V_{ji}$	$E_{jo,jo}, E_{ob,ob}, E_{ji,ji}, E_{jo,ob}, E_{ob,ji}$
Entry-based distinct-objects	$V_{ob} \cup V_{ji} \cup V_{ci}$	$E_{ob,ob}, E_{ji,ji}, E_{ci,ci}, E_{ob,ji}, E_{ji,ci}$

aggregates the data for each type of vertices and edges. If a short ID appears n times ($n > 0$) for the same vertex or edge type, its corresponding counter is simply set to n.

2. **Computing query answers.** Since a short ID of a vehicle may not be relevant to a query, the algorithm needs to find a set of relevant short IDs based on the definition of the query. The relevant set of short IDs and aggregate edge datasets to a query are defined in Table 4.5. EHSID only needs the relevant short IDs to compute answers for ID-based aggregate queries. If the query Q

Table 4.6 Privacy measures of the example in Figure 4.15.

Time (i)	t_1	t_2	t_3	t_4	t_5
Real trajectory (T_r)	(1,2)	(2,3)	(3,3)	(4,3)	(5,3)
Dummy (T_{d_1})	(1,4)	(2,3)	(2,2)	(2,1)	(3,1)
Dummy (T_{d_2})	(4,4)	(3,4)	(3,3)	(3,2)	(4,2)
$\|S_i\|$	3	2	2	3	3
$\frac{1}{n}\sum_{j=1}^{n} dist(T_r^i, T_{d_j}^i)$	2.80	0.71	0.71	2.12	2.12

depicted in Figure 4.14 is an ID-based cross-border query, the relevant short ID to Q is $V_{jo} \cap V_{ob} \cap V_{ji} = \{01\} \cap \{01, 10\} \cap \{01, 10\} = \{01\}$. Since there is only one relevant ID, the query answer is one. On the other hand, if Q is an ID-based distinct-objects query, the relevant short IDs to Q are $V_{ob} \cup V_{ji} \cup V_{ci} = \{01, 10\} \cup \{01, 10\} \cup \{01, 10\} = \{01, 10\}$. Since there are two relevant IDs, the query answer is two. In this example, EHSID finds the exact query answers for both the ID-based cross-border and distinct-objects queries.

EHSID uses a different method to compute answers for entry-based queries. After finding the relevant short IDs, for each relevant ID, we find the total count C_v for the ID in all the aggregate vertices (which are computed by the first step) and the total count C_e for the ID in all the relevant aggregate edge datasets. The query answer is computed as $C_v - C_e$. If Q is an entry-based cross-border query, there is only relevant ID, $\{01\}$. $\{01\}$ appears once in V_{jo}, V_{ob}, and V_{ji}, so $C_v = 3$. As $\{01\}$ appears once in $E_{jo,ob}$ and $E_{ob,ji}$, $C_e = 2$. Hence, the query answer is $3 - 2 = 1$. In case that Q is an entry-based distinct-objects query, we need to consider two relevant short IDs, $\{01, 10\}$. Each of these IDs appears once in V_{ob}, V_{ji} and V_{ci} and once in $E_{ob,ji}$ and $E_{ji,ci}$, so the number of entries for each of these IDs is $C_v - C_e = 3 - 2 = 1$. Hence, the query answer is two. EHSID also finds the exact query answers for both the entry-based cross-border and distinct-objects queries in this example.

4.3.7 Dummy Trajectories

Without relying on a trusted third party to perform anonymization, a mobile user can generate fake spatial trajectories, called *dummies*, to protect its privacy [33, 63]. Given a real user spatial trajectory T_r and a set of user-generated dummies T_d, the degree of privacy protection for the real trajectory is measured by the following metrics [63]:

1. **Snapshot disclosure (SD).** Let m be the number of location samples in T_r, S_i be the set of location samples in T_r and any T_d at time t_i, and $|S_i|$ be the size of S_i. SD is defined as the average probability of successfully inferring each true location sample in T_r, i.e., $SD = \frac{1}{m}\sum_{i=1}^{m}\frac{1}{|S_i|}$. Figure 4.15 gives a running

example of $n = 3$ trajectories and $m = 5$ where T_r is from source location s_1 to destination location d_1 (i.e., $s_1 \rightarrow d_1$), T_{d_1} is $s_2 \rightarrow d_2$, and T_{d_2} is $s_3 \rightarrow d_3$. There are two intersections I_1 and I_2. At time t_1, since there are three different locations, i.e., $(1,2)$, $(1,4)$ and $(4,4)$, $|S_1| = 3$. At time t_2, T_r and T_{d_1} share one location, i.e., $(2,3)$, so $|S_2| = 2$. Thus, $SD = \frac{1}{5} \times (\frac{1}{3} + \frac{1}{2} + \frac{1}{2} + \frac{1}{3} + \frac{1}{3}) = \frac{2}{5}$.

2. **Trajectory disclosure (TD).** Given n trajectories, where k trajectories have intersection with at least one other trajectory and $n - k$ trajectories do not intersect any other trajectory, let N_k be the number of possible trajectories among the k trajectories. TD is defined as the probability of successfully identifying the true trajectory among all possible trajectories is $\frac{1}{N_k + (n-k)}$. In the running example (Figure 4.15), there are two intersection points I_1 and I_2, $k = 3$ and $N_k = 8$, i.e., there are eight possible trajectories: $s_1 \rightarrow I_1 \rightarrow d_2$, $s_1 \rightarrow I_1 \rightarrow I_2 \rightarrow d_1$, $s_1 \rightarrow I_1 \rightarrow I_2 \rightarrow d_3$, $s_2 \rightarrow I_1 \rightarrow d_2$, $s_2 \rightarrow I_1 \rightarrow I_2 \rightarrow d_1$, $s_2 \rightarrow I_1 \rightarrow I_2 \rightarrow d_3$, $s_3 \rightarrow I_2 \rightarrow d_1$, and $s_3 \rightarrow I_2 \rightarrow d_3$. Hence, $TD = \frac{1}{8+(3-3)} = \frac{1}{8}$.

3. **Distance deviation (DD).** DD is defined as the average distance between the i-th location samples of T_r and each T_{d_j}, i.e., $DD = \frac{1}{m} \sum_{i=1}^{m} (\frac{1}{n} \sum_{j=1}^{n} dist(T_r^i, T_{d_j}^i))$, where $dist(p,q)$ denotes the Euclidean distance between two point locations p and q. In the running example, $DD = \frac{1}{5} \times (2.80 + 0.71 + 0.71 + 2.12 + 2.12) = 1.69$.

Given a real trajectory T_r and the three user-specified parameters SD, TD, and DD in a privacy profile, the dummy-based anonymization algorithm incrementally uses DD to find a set of candidate dummies and select one with the best matching to SD and TD until it finds a set of trajectories (including T_r and selected dummies) that satisfies all the parameters [63]. Since a user can use consistent identities for its

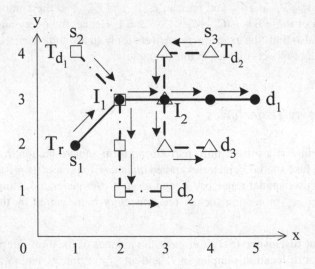

Fig. 4.15 One real trajectory T_r and two dummies T_{d_1} and T_{d_2}.

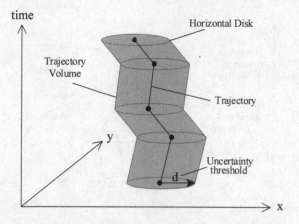

Fig. 4.16 The trajectory uncertainty model.

actual trajectory and other dummies, the dummy-based approach can support both Category-I and -II LBS, as depicted in Table 4.2.

4.4 Protecting Privacy in Trajectory Publication

In this section, we discuss privacy-preserving techniques for trajectory data publication. The anonymized trajectory data can be released to the public or third parties for answering spatio-temporal range queries and data mining. In the following sections, we present four trajectory anonymization techniques, namely, clustering-based, generalization-based, suppression-based and grid-based anonymization approaches, in Sections 4.4.1 to 4.4.4, respectively.

4.4.1 Clustering-based Anonymization Approach

The clustering-based approach [1] utilizes the uncertainty of trajectory data to group k co-localized trajectories within the same time period to form a k-anonymized aggregate trajectory. Given a trajectory T between times t_1 and t_n, i.e., $[t_1, t_n]$, and an uncertainty threshold d, each location sample in T, $p_i = (x_i, y_i, t_i)$, is modeled by a horizontal disk with radius d centered at (x_i, y_i). The union of all such disks constitutes the trajectory volume of T, as shown in Figure 4.16. Two trajectories T_p and T_q defined in $[t_1, t_n]$ are said to be co-localized with respect to d, if the Euclidean distance between each pair of points in T_p and T_q at time $t \in [t_1, t_n]$ is less than or equal to d. An anonymity set of k trajectories is defined as a set of at least k co-localized trajectories. The cluster of k co-localized trajectories is then transformed into an aggregate trajectory where each of its location points is computed by the

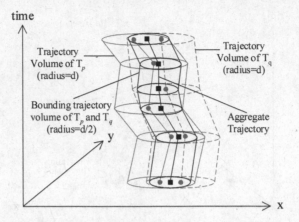

Fig. 4.17 Two-anonymized co-localized trajectories.

arithmetic mean of the location samples at the same time. Figure 4.17 gives the trajectory volumes of T_p and T_q that are represented by grey dotted lines, respectively. The trajectory volume with black lines is a bounding trajectory volume for T_p and T_q. The bounding trajectory volume is then transformed into an aggregate trajectory which is represented by the sequence of square markers.

The clustering-based anonymization algorithm consists of three main steps [1]:

1. **Pre-processing step.** The main task of this phase is to group all trajectories that have the same starting and ending times, i.e., they are in the same equivalence class with respect to time span. To increase the number of trajectories in an equivalence class, given an integer parameter π, all trajectories are trimmed if necessary such that only one timestamp every π can be the starting or ending point of a trajectory.
2. **Clustering step.** This phase clusters trajectories based on a greedy clustering scheme. For each equivalence class, a set of appropriate pivot trajectories are selected as cluster centers. For each cluster center, its nearest $k-1$ trajectories are assigned to the cluster, such that the radius of the bounding trajectory volume of the cluster is not larger than a certain threshold (e.g., $d/2$).
3. **Space transformation step.** Each cluster is transformed into a k-anonymized aggregate trajectory by moving all points at the same time to the corresponding arithmetic mean of the cluster.

4.4.2 Generalization-based Anonymization Approach

Since most data mining and statistical applications work on atomic trajectories, they are needed to be modified to work on aggregate trajectories generated by an anonymization algorithm (e.g., the clustering approach). To address this limitation,

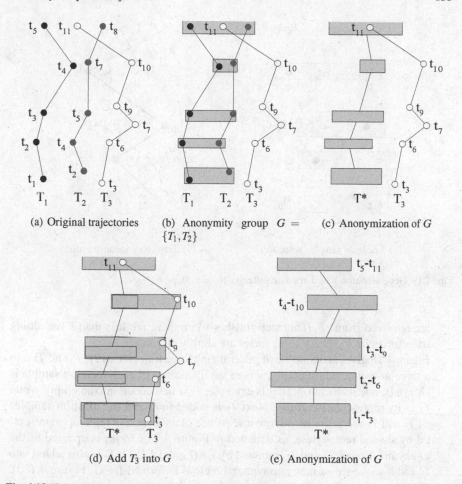

(a) Original trajectories

(b) Anonymity group $G = \{T_1, T_2\}$

(c) Anonymization of G

(d) Add T_3 into G

(e) Anonymization of G

Fig. 4.18 Generalization-based approach: Anonymization step.

the generalization-based algorithm first generalizes a trajectory data set into a set of k-anonymized trajectories, i.e., each one is a sequence of k-anonymized regions. Then, for each k-anonymized trajectory, the algorithm uniformly selects k atomic points from each anonymized region and links a unique atomic point from each anonymized region to reconstruct k trajectories [44]. More details about these two main steps are given below:

1. **Anonymization step.** Given a trajectory data set $\mathcal{T}$, each iteration of this step creates an empty anonymity group G and randomly samples one trajectory $T \in \mathcal{T}$. T is put into G as the group representative $Rep_G = T$. Then, the closest trajectory $T' \in \mathcal{T} - G$ to Rep_G is inserted into G and Rep_G is updated as the anonymization of Rep_G and T'. This anonymization process continues until G contains k trajectories. At the end of the iteration, the trajectories in G

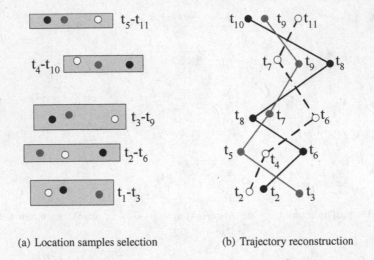

(a) Location samples selection   (b) Trajectory reconstruction

Fig. 4.19 Generalization-based approach: Reconstruction step.

are removed from $\mathscr{T}$. This step finishes when there are less than k remaining trajectories in $\mathscr{T}$. These trajectories are simply discarded.

Figure 4.18 gives an example of generalizing three trajectories T_1, T_2 and T_3 into a three-anonymized trajectory, where the timestamp of each location sample is shown beside its location. In this example, T_2 is first added into an empty group G as its representative Rep_G. Next T_1 is added to G and the location samples of T_1 and T_2 are generalized into a sequence of anonymized regions (represented by shaded rectangles), as depicted in Figure 4.18b. Rep_G is updated as the anonymization of T_1 and T_2, denoted by $T*$ (Figure 4.18c). T_3 is also added into G and a sequence of new anonymized regions is formed for G (Figure 4.18d). The time span of an anonymized region is the range from the smallest and largest timestamps of the location samples included in the region. Note that unmatched points (e.g., the location sample of T_3 at times t_7) are suppressed in this step. Since G already contains $k = 3$ trajectories, the anonymization process is done (Figure 4.18e).

2. **Reconstruction step.** Given a k-anonymized trajectory, k locations are uniformly selected in each of its anonymized regions, as illustrated in Figure 4.19a. Next, for each selected location, a timestamp is also uniformly selected from its associated time span. k trajectories are reconstructed by linking a unique location sample in each monitored region (Figure 4.19b).

The reconstructed trajectory data set can be released to the public or third parties for answering spatio-temporal queries and data analysis (e.g., data mining).

4.4.3 Suppression-based Anonymization Approach

Digital cash and electronic money have become very popular. People often use them to pay for their transportation and for their purchases at a wide variety of stores, e.g., convenient stores, grocery stores, restaurants and parking lots. Since a transaction is associated with the location of a particular store at a particular time, the sequence of a user's transactions can be considered as its trajectory information. Consider a digital cash or electronic money company publishes its original trajectory database T. A company A (e.g., 7-Eleven) that accepts its electronic payment service has part of trajectory information in T, i.e., the transactions are done by A. A can be considered as an adversary because A could use its knowledge T_A to identify its customers' private information [53].

Figure 4.20a shows an original trajectory database T that contains the location information of A's stores and other stores. A has part of the knowledge of T, T_A, i.e., the location information of its stores, as given in Figure 4.20b. By joining T_A to T, A knows that t_5^A actually corresponds to t_5 because t_5 is the only trajectory with $a_1 \rightarrow a_3$. Thus, A is 100% sure that the user of t_5^A visited b_1. Similarly, A knows that t_6^A corresponds to t_6, t_7 or t_8, so A can infer that the user of t_6^A visited b_2 with probability 66%, i.e., b_2 appears in t_7 and t_8.

The privacy protection for the above-mentioned linking attack is defined as: given an original trajectory database T and an adversary A's knowledge T_A, where T's locations take values from a data set P, T_A's locations take values from P_A and $P_A \subset P$, the probability that A can correctly identify the actual user of any location $p_i | p_i \in t_j \wedge p_i \notin t_k^A$ (where $t_j \in T$, $t_k^A \in T_A$, $p_i \in P$, and $p_i \notin P_A$) can be determined by the equation:

$$Pr(p_i, t_k^A, T) = \frac{|\{t' | t' \in S(t_k^A, T) \wedge p_i \in t' \wedge p_i \in P \wedge p_i \notin P_A\}|}{|S(t_k^A, T)|}, \quad (4.3)$$

where $S(t_k^A, T)$ denotes a set X of trajectories in T such that t_k^A is a subsequence of each trajectory in X. To protect user privacy, $Pr(p_i, t_k^A, T)$ should not be larger than a certain threshold δ. If $\delta = 50\%$, the publication of T makes privacy breaches, as given in the example (Figures 4.20a and 4.20b).

A greedy anonymization algorithm [53] is designed to iteratively suppress locations until the privacy constraint is met, i.e., $Pr(p_i, t_k^A, T) \leq \delta$ (computed by Equation 4.3) for each pair of $p_i \in P$ and $t_k^A \in T_A$. At the meantime, the utility of the published trajectory database is maximized. In the example, after the algorithm suppresses b_3 from t_2, a_1 from t_5, and b_3 from t_8, the transformed trajectory database T' ($\delta = 50\%$), as given in Figure 4.20c, can be published without compromising the user privacy.

One way to measure the utility is to compute the average difference between the original trajectories in T and the transformed ones in T' [53]. Let t be a trajectory in T and t' be its transformed form in T'. Let t_s and t_e (t_s' and t_e') be the starting and ending locations of t (t'), respectively. The difference between t and t', denoted by $diff(t, t')$, is computed as follows. For each location $p \in t$, it forms one component to the distance based on the following four cases:

ID	Trajectory
t_1	$a_1 \to b_1 \to a_2$
t_2	$a_1 \to b_1 \to a_2 \to b_3$
t_3	$a_1 \to b_2 \to a_2$
t_4	$a_1 \to a_2 \to b_2$
t_5	$a_1 \to a_3 \to b_1$
t_6	$a_3 \to b_1$
t_7	$a_3 \to b_2$
t_8	$a_3 \to b_2 \to b_3$

(a) Original database T

ID	Trajectory
t_1^A	$a_1 \to a_2$
t_2^A	$a_1 \to a_2$
t_3^A	$a_1 \to a_2$
t_4^A	$a_1 \to a_2$
t_5^A	$a_1 \to a_3$
t_6^A	a_3
t_7^A	a_3
t_8^A	a_3

(b) Adversary's knowledge T_A

ID	Trajectory
t_1'	$a_1 \to b_1 \to a_2$
t_2'	$a_1 \to b_1 \to a_2$
t_3'	$a_1 \to b_2 \to a_2$
t_4'	$a_1 \to a_2 \to b_2$
t_5'	$a_3 \to b_1$
t_6'	$a_3 \to b_1$
t_7'	$a_3 \to b_2$
t_8'	$a_3 \to b_2$

(c) Transformed database T'

Fig. 4.20 Trajectory databases (Figure 1 in [53]).

1. If p is before t_s', the corresponding component is $dist(p,t_s')$.
2. If p is after t_e', the corresponding component is $dist(p,t_e')$.
3. If p is in-between t_s' and t_e', the corresponding component is $dist(p,\overline{(p,t')})$, where $\overline{(p,t')}$ is the perpendicular projection of p onto the transformed trajectory t'.
4. If $t' = \emptyset$, the corresponding component is set to the maximum distance between two locations on the map.

The final step of computing $diff(t,t')$ is to sum up the square of each component and take the root of the sum. Figure 4.21 gives a trajectory $t : a_1 \to a_2 \to a_3 \to c_1$ that is transformed to $t' : a_2 \to c_1$ by suppressing a_1 and a_3 from t. For the first location a_1 in t, Figure 4.21a shows that a_1 is before t_s' (i.e., Case 1), so its corresponding component is $d(a_1,t_s') = d(a_1,a_2)$. Since a_2 is in-between t_s' and t_e' (i.e., Case 3), we need to find $\overline{(a_2,t')}$. As $\overline{(a_2,t')} = a_2$, the corresponding component of a_2 is $dist(a_2,a_2) = 0$. a_3 is also in-between t_s' and t_e' (i.e., Case 3), so its corresponding component is $dist(a_3,\overline{(a_3,t')})$. Similar to a_2, the corresponding component of c_1 is $dist(c_1,c_1) = 0$. Therefore, $diff(t,t') = \sqrt{[d(a_1,a_2)]^2 + [dist(a_3,\overline{(a_3,t')})]^2}$.

4.4.4 Grid-based Anonymization Approach

An grid-based anonymization approach is designed to anonymize user trajectories for privacy-preserving data mining [23]. This approach provides three anonymization features: spatial cloaking, temporal cloaking and trajectory splitting. Its basic idea is to construct a grid on the system space and partition the grid based on the required privacy constraint. Figure 4.22a shows part of the grid with 24 cells, i.e., $c_1, c_2, \ldots$. In this example, every four neighbor grid cells constitute a non-overlapping partition P. Generally speaking, we have to define a larger partition area for users who need a higher level of privacy protection, but a smaller partition

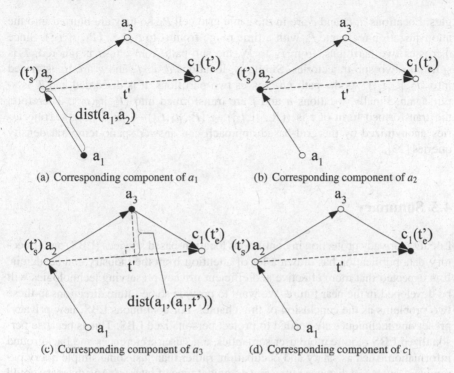

(a) Corresponding component of a_1 (b) Corresponding component of a_2

(c) Corresponding component of a_3 (d) Corresponding component of c_1

Fig. 4.21 The difference between an original trajectory t and its transformed trajectory t'.

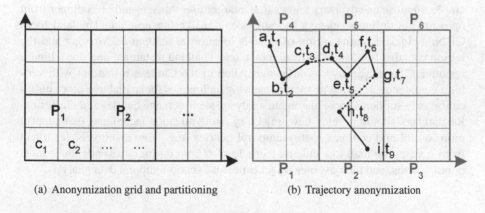

(a) Anonymization grid and partitioning   (b) Trajectory anonymization

Fig. 4.22 Grid-based anonymization for trajectories.

area for a lower level of privacy protection. However, larger partitions will lead to lower accuracy in the data mining results.

Figure 4.22b gives an example to illustrate how to anonymize a trajectory $T : \langle (a,t_1) \to (b,t_2) \to \ldots \to (i,t_9) \rangle$, where x and y in each pair (x,y) denote a location and its update timestamp, respectively, to a list of anonymization rectangles. Locations a, b, and c are in the same grid cell P_4, so they are blurred into the anonymization rectangle P_4 with a time range from t_1 to t_3, i.e., $(P_4, [t_1, t_3])$. Since t crosses two partitions, from P_4 to P_5, the sub-path from c to d is ignored. T is split into two sub-trajectories. Similarly, locations d, e, f, and g are transformed into $(P_5, [t_4, t_7])$. As the path gh crosses two partitions, it is discarded and T is split again. Finally, locations h and i are transformed into $(P_2, [t_8, t_9])$. Therefore, the transformed form of t is $\langle (P_4, [t_1, t_3]) \rightarrow (P_5, [t_4, t_7]) \rightarrow (P_2, [t_8, t_9]) \rangle$. Trajectories anonymized by the grid-based approach can answer spatio-temporal density queries [23].

4.5 Summary

Location privacy protection in continuous location-based services (LBS) and trajectory data publication has drawn a lot of attention from the industry and academia. It is expected that more effective and efficient privacy preserving technologies will be developed in the near future. We want to provide some future directions in these two problems as the conclusion of this chapter. For continuous LBS, new privacy-preserving techniques are needed to protect personalized LBS. This is because personalized LBS require more user semantics, e.g., user preferences and background information, such as salary and occupation, rather than just some simple query parameters, such as a distance range and an object type of interest. An adversary could use such user semantics to infer the user location with higher confidence. For example, suppose an adversary knows that a target user Alice usually has dinner from 6pm to 7pm during weekdays and she does not like Japanese and Thailand food. Given a cloaked spatial region of Alice's location at 6:30pm on Monday and the region contains two Japanese restaurants, one Thailand restaurant and one Chinese restaurant, the adversary can infer that Alice in the Chinese restaurant with very high confidence. Existing privacy-preserving techniques for spatial trajectory publication only support simple aggregate analysis, such as range queries and clustering. Researchers should develop new trajectory anonymization techniques that support more useful and complex spatio-temporal queries (e.g., how many vehicles travel from a shopping mall to a cinema from 1pm to 2pm during weekends, their most popular paths, and their average travel time) and spatio-temporal data analysis.

References

1. Abul, O., Bonchi, F., Nanni, M.: Never walk alone: Uncertainty for anonymity in moving objects databases. In: Proceedings of the IEEE International Conference on Data Engineering (2008)
2. Bamba, B., Liu, L., Pesti, P., Wang, T.: Supporting anonymous location queries in mobile environments with PrivacyGrid. In: Proceedings of the International Conference on World Wide Web (2008)

3. Bao, J., Chow, C.Y., Mokbel, M.F., Ku, W.S.: Efficient evaluation of k-range nearest neighbor queries in road networks. In: Proceedings of the International Conference on Mobile Data Management (2010)
4. Beresford, A.R., Stajano, F.: Location privacy in pervasive computing. IEEE Pervasive Computing **2**(1), 46–55 (2003)
5. Cheng, R., Zhang, Y., Bertino, E., Prabhakar, S.: Preserving user location privacy in mobile data management infrastructures. In: Proceedings of International Privacy Enhancing Technologies Symposium (2006)
6. Chow, C.Y., Bao, J., Mokbel, M.F.: Towards location-based social networking services. In: Proceedings of the ACM SIGSPATIAL International Workshop on Location Based Social Networks (2010)
7. Chow, C.Y., Mokbel, M., He, T.: A privacy-preserving location monitoring system for wireless sensor networks. IEEE Transactions on Mobile Computing **10**(1), 94–107 (2011)
8. Chow, C.Y., Mokbel, M.F.: Enabling private continuous queries for revealed user locations. In: Proceedings of the International Symposium on Spatial and Temporal Databases (2007)
9. Chow, C.Y., Mokbel, M.F., Aref, W.G.: Casper*: Query processing for location services without compromising privacy. ACM Transactions on Database Systems **34**(4), 24:1–24:48 (2009)
10. Chow, C.Y., Mokbel, M.F., Bao, J., Liu, X.: Query-aware location anonymization in road networks. GeoInformatica **15**(3), 571–607 (2011)
11. Chow, C.Y., Mokbel, M.F., Liu, X.: A peer-to-peer spatial cloaking algorithm for anonymous location-based services. In: Proceedings of the ACM Symposium on Advances in Geographic Information Systems (2006)
12. Chow, C.Y., Mokbel, M.F., Liu, X.: Spatial cloaking for anonymous location-based services in mobile peer-to-peer environments. GeoInformatica **15**(2), 351–380 (2011)
13. Dateline NBC: Tracing a stalker. http://www.msnbc.msn.com/id/19253352 (2007)
14. Duckham, M., Kulik, L.: A formal model of obfuscation and negotiation for location privacy. In: Proceedings of International Conference on Pervasive Computing (2005)
15. FoxNews: Man accused of stalking ex-girlfriend with GPS. http://www.foxnews.com/story/0,2933,131487,00.html (2004)
16. Freudiger, J., Raya, M., Felegyhazi, M., Papadimitratos, P., Hubaux, J.P.: Mix-zones for location privacy in vehicular networks. In: Proceedings of the International Workshop on Wireless Networking for Intelligent Transportation Systems (2007)
17. Freudiger, J., Shokri, R., Hubaux, J.P.: On the optimal placement of mix zones. In: Proceedings of International Privacy Enhancing Technologies Symposium (2009)
18. Fung, B.C.M., Wang, K., Chen, R., Yu, P.S.: Privacy-preserving data publishing: A survey of recent developments. ACM Computing Surveys **42**(4), 14:1–14:53 (2010)
19. Gedik, B., Liu, L.: Protecting location privacy with personalized k-anonymity: Architecture and algorithms. IEEE Transactions on Mobile Computing **7**(1), 1–18 (2008)
20. Ghinita, G., Kalnis, P., Khoshgozaran, A., Shahabi, C., Tan, K.L.: Private queries in location based services: Anonymizers are not necessary. In: Proceedings of the ACM Conference on Management of Data (2008)
21. Ghinita, G., Kalnis, P., Skiadopoulos, S.: PRIVÉ: Anonymous location-based queries in distributed mobile systems. In: Proceedings of the International Conference on World Wide Web (2007)
22. Ghinita1, G., Kalnis, P., Skiadopoulos, S.: MobiHide: A mobile peer-to-peer system for anonymous location-based queries. In: Proceedings of the International Symposium on Spatial and Temporal Databases (2007)
23. Gidófalvi, G., Huang, X., Pedersen, T.B.: Privacy-preserving data mining on moving object trajectories. In: Proceedings of the International Conference on Mobile Data Management (2007)
24. Google Geocoding API: http://code.google.com/apis/maps/documentation/geocoding/

25. Gruteser, M., Grunwald, D.: Anonymous usage of location-based services through spatial and temporal cloaking. In: Proceedings of the International Conference on Mobile Systems, Applications, and Services (2003)
26. Gruteser, M., Hoh, B.: On the anonymity of periodic location samples. In: Proceedings of the International Conference on Security in Pervasive Computing (2005)
27. Hoh, B., Gruteser, M., Xiong, H., Alrabady, A.: Achieving guaranteed anonymity in GPS traces via uncertainty-aware path cloaking. IEEE Transactions on Mobile Computing 9(8), 1089–1107 (2010)
28. Hong, J.I., Landay, J.A.: An architecture for privacy-sensitive ubiquitous computing. In: Proceedings of the International Conference on Mobile Systems, Applications, and Services (2004)
29. Hu, H., Lee, D.L.: Range nearest-neighbor query. IEEE Transactions on Knowledge and Data Engineering 18(1), 78–91 (2006)
30. Ilarri, S., Mena, E., Illarramendi, A.: Location-dependent query processing: Where we are and where we are heading. ACM Computing Surveys 42(3), 12:1–12:73 (2010)
31. Kalnis, P., Ghinita, G., Mouratidis, K., Papadias, D.: Preventing location-based identity inference in anonymous spatial queries. IEEE Transactions on Knowledge and Data Engineering 19(12), 1719–1733 (2007)
32. Khoshgozaran, A., Shahabi, C.: Blind evaluation of nearest neighbor queries using space transformation to preserve location privacy. In: Proceedings of the International Symposium on Spatial and Temporal Databases (2007)
33. Kido, H., Yanagisawa, Y., Satoh, T.: An anonymous communication technique using dummies for location-based services. In: Proceedings of IEEE International Conference on Pervasive Services (2005)
34. Ku, W.S., Zimmermann, R., Peng, W.C., Shroff, S.: Privacy protected query processing on spatial networks. In: Proceedings of the International Workshop on Privacy Data Management (2007)
35. LeFevre, K., DeWitt, D., Ramakrishnan, R.: Mondrian multidimensional k-anonymity. In: Proceedings of the IEEE International Conference on Data Engineering (2006)
36. Li, N., Li, T., Venkatasubramanian, S.: Closeness: A new privacy measure for data publishing. IEEE Transactions on Knowledge and Data Engineering 22(7), 943–956 (2010)
37. Ma, C.Y., Yau, D.K.Y., Yip, N.K., Rao, N.S.V.: Privacy vulnerability of published anonymous mobility traces. In: Proceedings of the ACM International Conference on Mobile Computing and Networking (2010)
38. Machanavajjhala, A., Kifer, D., Gehrke, J., Venkitasubramaniam, M.: l-diversity: Privacy beyond k-anonymity. ACM Transactions on Knowledge Discovery from Data 1(1), 3:1–3:52 (2007)
39. Marist Institute for Public Opinion (MIPO): Half of Social Networkers Online Concerned about Privacy. http://maristpoll.marist.edu/714-half-of-social-networkers-online-%concerned-about-privacy/. July 14, 2010
40. Meyerowitz, J., Choudhury, R.R.: Hiding stars with fireworks: Location privacy through camouflage. In: Proceedings of the ACM International Conference on Mobile Computing and Networking (2009)
41. Mokbel, M.F., Chow, C.Y., Aref, W.G.: The new casper: Query procesing for location services without compromising privacy. In: Proceedings of the International Conference on Very Large Data Bases (2006)
42. Mokbel, M.F., Levandoski, J.: Towards context and preference-aware location-based database systems. In: Proceedings of the ACM International Workshop on Data Engineering for Wireless and Mobile Access (2009)
43. Mouratidis, K., Yiu, M.L.: Anonymous query processing in road networks. IEEE Transactions on Knowledge and Data Engineering 22(1), 2–15 (2010)
44. Nergiz, M.E., Atzori, M., Saygin, Y., Güç, B.: Towards trajectory anonymization: A generalization-based approach. Transactions on Data Privacy 2(1), 47–75 (2009)

45. Palanisamy, B., Liu, L.: Mobimix: Protecting location privacy with mix zones over road networks. In: Proceedings of the IEEE International Conference on Data Engineering (2011)
46. Pan, X., Meng, X., Xu, J.: Distortion-based anonymity for continuous queries in location-based mobile services. In: Proceedings of the ACM SIGSPATIAL International Conference on Advances in Geographic Information Systems (2009)
47. Pfitzmann, A., Kohntopp, M.: Anonymity, unobservability, and pseudonymity - a proposal for terminology. In: Proceedings of the Workshop on Design Issues in Anonymity and Unobservability (2000)
48. Reid, D.: An algorithm for tracking multiple targets. IEEE Transactions on Automatic Control **24**(6), 843–854 (1979)
49. Samarati, P.: Protecting respondents identities in microdata release. IEEE Transactions on Knowledge and Data Engineering **13**(6), 1010–1027 (2001)
50. Sun, C., Agrawal, D., Abbadi, A.E.: Exploring spatial datasets with histograms. In: Proceedings of the IEEE International Conference on Data Engineering (2002)
51. Sweeney, L.: Achieving k-anonymity privacy protection using generalization and suppression. International Journal on Uncertainty, Fuzziness and Knowledge-based Systems **10**(5), 571–588 (2002)
52. Sweeney, L.: k-anonymity: A model for protecting privacy. International Journal on Uncertainty, Fuzziness and Knowledge-based Systems **10**(5), 557–570 (2002)
53. Terrovitis, M., Mamoulis, N.: Privacy preservation in the publication of trajectories. In: Proceedings of the International Conference on Mobile Data Management (2008)
54. USAToday: Authorities: GPS system used to stalk woman. http://www.usatoday.com/tech/news/2002-12-30-gps-stalker_x.htm (2002)
55. Voelcker, J.: Stalked by satellite: An alarming rise in gps-enabled harassment. IEEE Spectrum **47**(7), 15–16 (2006)
56. Wang, T., Liu, L.: Privacy-aware mobile services over road networks. In: Proceedings of the International Conference on Very Large Data Bases (2009)
57. Webroot Software, Inc.: Webroot survey finds geolocation apps prevalent amongst mobile device users, but 55% concerned about loss of privacy. http://pr.webroot.com/threat-research/cons/social-networks-mobile-security-071310.html. July 13, 2010
58. Xiao, X., Yi, K., Tao, Y.: The hardness and approximation algorithms for l-diversity. In: Proceedings of the International Conference on Extending Database Technology (2010)
59. Xie, H., Kulik, L., Tanin, E.: Privacy-aware traffic monitoring. IEEE Transactions on Intelligent Transportation Systems **11**(1), 61–70 (2010)
60. Xu, T., Cai, Y.: Location anonymity in continuous location-based services. In: Proceedings of the ACM Symposium on Advances in Geographic Information Systems (2007)
61. Xu, T., Cai, Y.: Exploring historical location data for anonymity preservation in location-based services. In: Proceedings of IEEE INFOCOM (2008)
62. Yiu, M.L., Jensen, C., Huang, X., Lu, H.: Spacetwist: Managing the trade-offs among location privacy, query performance, and query accuracy in mobile services. In: Proceedings of the IEEE International Conference on Data Engineering (2008)
63. You, T.H., Peng, W.C., Lee, W.C.: Protecting moving trajectories with dummies. In: Proceedings of the International Workshop on Privacy-Aware Location-Based Mobile Services (2007)
64. Zhang, C., Huang, Y.: Cloaking locations for anonymous location based services: A hybrid approach. GeoInformatica **13**(2), 159–182 (2009)

Chapter 5
Trajectory Pattern Mining

Hoyoung Jeung, Man Lung Yiu, and Christian S. Jensen

Abstract :
In step with the rapidly growing volumes of available moving-object trajectory data, there is also an increasing need for techniques that enable the analysis of trajectories. Such functionality may benefit a range of application area and services, including transportation, the sciences, sports, and prediction-based and social services, to name but a few. The chapter first provides an overview trajectory patterns and a categorization of trajectory patterns from the literature. Next, it examines relative motion patterns, which serve as fundamental background for the chapter's subsequent discussions. Relative patterns enable the specification of patterns to be identified in the data that refer to the relationships of motion attributes among moving objects. The chapter then studies disc-based and density-based patterns, which address some of the limitations of relative motion patterns. The chapter also reviews indexing structures and algorithms for trajectory pattern mining.

5.1 Introduction

We are witnessing a rapid and continued diffusion of mobile devices such as smartphones, personal navigation devices and tablet computers. Further, these devices are increasingly being geo-positioned using satellite navigation systems, e.g., GPS, systems that exploit the wireless communication infrastructure, and proximity-based systems, e.g., RFID-based systems.

Hoyoung Jeung
École Polytechnique Fédérale de Lausanne, Switzerland, e-mail: hoyoung.jeung@epfl.ch

Man Lung Yiu
Department of Computing, Hong Kong Polytechnic University, Hong Kong, e-mail: csmlyiu@comp.polyu.edu.hk

Christian S. Jensen
Department of Computer Science, Aarhus University, Denmark, e-mail: csj@cs.au.dk

The resulting *location-aware* devices find widespread use in various business and personal settings in society. As a consequence of this development, increasing volumes of position data are being accumulated, and the capability of analyzing large volumes of trajectory data is in increasing demand in a wide spectrum of applications.

Significant applications in various domains need to identify and utilize groups of trajectories that exhibit similar patterns from a collection of trajectories. Example applications include transportation optimization, prediction-enabled services, scientific and social analysis applications, sports analyses, as well as crowd and outlier analyses.

- Transportation optimization applications [25] need to find groups of similar trajectories that indicates that the corresponding objects traveled together. For instance, a car pooling application may connect drivers in the same trajectory group in order to reduce their travel expenses. A logistics application may examine the delivery trucks in the same group in order to achieve better planning.
- Prediction methods [50] may exploit knowledge of trajectory groups for the understanding of object behavior. Such knowledge can be used for offering effective notifications, for the delivery of advertisements to targeted audience, and for providing customized location-based services.
- Scientific studies may call for the identification of groups of animals that moved together. They are useful in discovering animal movement patterns (e.g., bumble bees, a variety of birds, sea turtles, whales, and fish) [2], in finding herds of animals, and in studying animal behavior patterns in habitats.
  Similarly, social analysis studies may aim to identify socio-economic patterns [23] from typical movement patterns of individuals.
- Team sports events [30, 1] (soccer, baseball, hockey, rugby, digital battle fields) also provide valuable trajectory data that capture the players' movements. By studying a game as groups of trajectories, it may be possible to better understand the game [30], to analyze the tactics used in the game, and to even extract the location and time of using a certain strategy.
- Traffic analysis applications may utilize trajectory groups for the study of crowds and outliers. In this scenario, a moving object can be either a vehicles on the road or a pedestrian on the street. A large trajectory group is likely to indicate a crowd behavior. By identifying crowds from the trajectories, a better understanding of crowds is possible, e.g., the times and places when and where crowds form and dissolve. Such information may be exploited for managing transportation infrastructures effectively.
  It is also of interest to mine outliers, which do not belong to any trajectory group. This may be used for detecting and removing errors in the trajectory data (e.g., finding a device with a malfunctioning GPS receiver). It may also be applied for identifying dangerous driving behaviors.

Trajectory pattern mining is an emerging and rapidly developing topic in the areas of data mining and query processing that aims at discovering groups of trajectories based on their proximity in either a spatial or a spatiotemporal sense. The literature contains a variety of recent proposals in this area.

Existing proposals represent different trade-offs in the expressiveness of the trajectory patterns studied. Considering only restricted patterns may result in not being able to identify interesting phenomena from the data, whereas considering quite relaxed patterns may lead to the reporting of insignificant patterns.

Existing proposals also come with their own index structures and mining algorithms that aim to enable efficient and scalable discovery of patterns in large trajectory datasets. This chapter presents an overview of the key concepts and discovery techniques in state-of-the-art studies in the mining of trajectory patterns.

The rest of the chapter is organized as follows. Section 5.2 introduces the concept of trajectory pattern and provides a categorization of patterns. We then study relative motion patterns in Section 5.3, presenting disc-based patterns in Section 5.4, and examining density-based patterns in Section 5.5. Section 5.6 covers distance measures and methods for mining trajectory patterns. We summarize the chapter in Section 5.7.

5.2 Overview of Trajectory Patterns

5.2.1 Pattern Discovery Process

A trajectory of a moving object is a continuous function from the time domain to the domain in which the movement occurs. In animal tracking, the objects typically move freely in the water, on the surface of the Earth, or in the air. In such cases, the movement domain is often modeled as two- or three-dimensional Euclidean space. In settings where the objects are vehicle, the movement domain is often modeled as a spatial graph that models a road network. Further, the spatial extent of a moving object is typically ignored so that the position of an object at a given time is modeled as a point.

Figure 5.1 illustrates the architectural context for managing trajectories. Multiple objects, each with a mobile device with a unique identifier id, contribute data. Each device samples the location of its object according to some policy. For example, devices may take a position sample at a fixed frequency such as every second. A location record has the format (t, x_t, y_t, id), where the sampled location at time t is (x_t, y_t).

The true trajectory of an object is unknown. The location records available for an object are used to create an approximation of the object's true trajectory. Specifically, existing techniques for trajectory pattern mining typically assume that the trajectory of an object is given by a polyline, i.e., a sequence of connected line segments.

A server is employed for storing and managing the collected trajectories from all devices. Important steps in the handling of trajectories are described next.

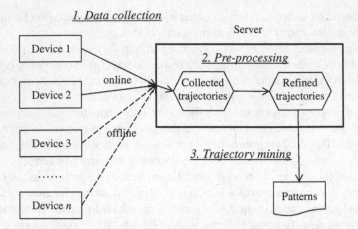

Fig. 5.1 Architecture.

1. *Data collection.*
   In this step, devices submit their location records to the server. Online devices may submit their data in real time or delayed, in batches. For example, a device may submit records as soon they become available on the device, or it may collect a small batch of records before submitting in order to save bandwidth.
   Offline devices accumulate their trajectory data, which is transferred to the server in batches via other means. For example, an offline navigation device in a vehicle may occasionally be connected to an online personal computer in order to transfer location records to the server.

2. *Pre-processing.*
   Due in part to imperfections in the positioning technologies employed and the use of different sampling policies, collected trajectories may be inhomogeneous and may not be of the desired quality. For instance, some of trajectories may lack location records for times where other trajectories posses such records.
   In pre-processing, the server "cleans" the data and converts it into a standard format in preparation for the data mining. As part of the pre-processing, the granularity and the representation of the refined trajectories need to be decided by the server. For example, a finest sampling rate is often chosen, and it is decided whether a trajectory be stored as a sequence of location records or should be approximated by a piecewise linear function from time to space.

3. *Mining trajectories.*
   The refined trajectories are made available for the mining algorithms to discover patterns (e.g., clusters of trajectories).

4. *Post-processing.*
   This step enables applications to analyze or tune the discovery process. For example, mined patterns can be displayed interactively using visualization tools, which can suggest more appropriate parameter values for the mining algorithms or finding further research issues.

5.2.2 Classification of Trajectory Pattern Concepts and Techniques

The term *trajectory pattern* covers many different types of patterns that can be mined from trajectory data [29]. As a result, the concepts and techniques underlying trajectory pattern discovery are classified by according to a variety of aspects [19]. In the following, we describe key classifications of trajectory pattern concepts and techniques while emphasizing the intuitions behind the classifications.

5.2.2.1 Mining Tasks on Trajectories

Two typical mining tasks on trajectories are *clustering* and *join*. Clustering aims at discovering groups of similar objects from a single trajectory collection. On the other hand, a join is a specific operation that computes pairs of similar objects from two trajectory collections.

Clustering of Trajectories

Clustering is the process of organizing objects into groups so that the members of a group are similar and so that members of distinct groups are dissimilar, according to some specific definition of similarity. Clustering is a fundamental concept in data mining, due to its wide spectrum of applications.

A clustering technique is classified according to the type of data that it is applicable to. Clustering techniques for different types of data are likely to vary significantly with respect to the underlying concepts and the specific techniques employed.

In the context of trajectory data, clustering attempts to group trajectories based on their geometric proximity in either spatial or spatiotemporal space. Clustering of trajectories is perhaps one of the most fundamental operations used in various types of trajectory pattern mining, since the discovery of trajectory patterns typically involves the process of grouping similar positions, trajectories, and objects.

Assuming the polyline representation of a trajectory, one fundamental approach to trajectory clustering treats a segment of a trajectory as a minimum unit when computing the distance between two trajectories. To this end, it is essential to design an effective distance function for measuring the distance between two trajectories (segment). We offer details on trajectory distance measures in Section 5.6.1. Another fundamental approach views a cluster of trajectories as a sequence of spatial clusters. Assuming that all objects have position samples at the same points in time, such spatial clusters can be obtained by applying some spatial clustering technique to the positions for each point in time.

This chapter describes in detail the core ideas and concepts of state-of-the-art algorithms for clustering of trajectories: *relative motion patterns, flock, TRACLAUS, moving cluster, convoy,* and *swarm.*

Trajectory Join

Some trajectory patterns are defined and computed by means of join queries. Given two data sets P_1 and P_2, spatio-temporal joins find pairs of elements from the two sets that satisfy a given predicate with both spatial and temporal attributes [31, 17]. The study of airplane or vessel trajectories with the objective of finding incidents may be accomplished using joins. Since joins may involve the comparison of all trajectories in data set P_1 with all trajectories in data set P_2, which is computationally expensive, a common approach for join processing involves the use if indexing techniques to avoid unnecessary distance computations. For example, Tao et al. [53] show how join queries are processed by using the time-parameterized methods [56, 52, 51].

The *close-pair join* [13] reports all object pairs (o_1, o_2) from $P_1 \times P_2$ with distance $D_\tau(o_1, o_2) \leq e$ within a time interval τ, where e is a user-specified distance. Plane-sweep techniques [6, 61] have been proposed for evaluating spatio-temporal joins. Zhou et al. [61] use join predicates that define a rectangular region in time and space. An index structure (MTSB-tree) is introduced to enable efficient retrieval of the pairs of trajectories that satisfy the join predicates. Instead of using an index, Arumugam and Jermaine [6] utilize MBR approximations of trajectory segments to reduce the computation of query processing.

As does the close-pair join, the *trajectory join* [10] aims at retrieving all pairs of similar trajectories in two data sets. Bakalov et al. [9, 10] represent trajectories as sequences of symbols, and apply sliding window techniques to measure the symbolic distance between possible pairs.

5.2.2.2 Spatial and Spatiotemporal Patterns

Groups of similar trajectories carry different semantics depending on whether they are grouped according to spatial or spatiotemporal geometric proximity. Figure 5.2 demonstrates the difference between *spatial* and *spatiotemporal* trajectory patterns. At time $t = 1$, two objects o_1 and o_2 start moving closely together in the direction of the upper-right corner. At time $t = 2$, a new object o_3 appears and starts to move from a close location to where o_1 and o_2 were at time $t = 1$. Meanwhile, o_1 and o_2 have reached the center of the movement space and continue to move. At time $t = 3$, o_3 keeps following a path similar to those that o_1 and o_2 have been following, while o_1 and o_2 complete their journeys. Finally, o_3 finishes its journey at time $t = 4$ close to where o_1 and o_2 stopped.

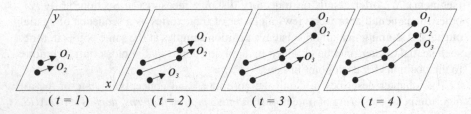

Fig. 5.2 Construction of three objects' trajectories during $t = [1,4]$.

Given the trajectories of o_1, o_2, and o_3 collected until $t = 4$, the embeddings of the trajectories into the movement space look similar. When we cluster the trajectories according to the similarity of the embeddings, the three trajectories may form one cluster, although o_3 has never traveled together with o_1 and o_2, but has merely followed the same route as these.

The above is an example of a *spatial trajectory pattern* (Figure 5.3(a)). Despite the fact that the concept of trajectory encompasses the time domain, a variety of applications do not need to consider the temporal aspects of trajectories.

For example, analyzing a set of hurricane trajectories collected over several years for forecasting the trajectories of future hurricanes may be done best by considering simply the embeddings into the movement space of the past hurricanes so that is becomes the similarity of the routes of the hurricanes that is studied. Put differently, the routes of the past hurricanes may be the most important aspect when aiming to offer pre-warnings to regions of potential damage.

In contrast to spatial trajectory patterns, *spatiotemporal trajectory patterns* take the time information in a trajectory into account. For example, o_3 in Figure 5.2 may not belong to the same cluster as the one o_1 and o_2 belong to (see Figure 5.3(b)), since o_3 always trailed behind o_1 and o_2. Having a time constraint, a spatiotemporal

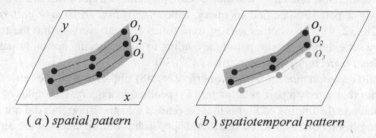

$(\,a\,)$ *spatial pattern* $(\,b\,)$ *spatiotemporal pattern*

Fig. 5.3 Spatial versus spatiotemporal trajectory patterns.

trajectory pattern is therefore stricter than a spatial trajectory pattern. This time constraint can help discover more specific patterns in many cases of mining trajectory data. For instance, identifying objects that have traveled together for the purpose of given car pooling recommendations should consider the temporal information of trajectories. Discovering places in a road network where congestion may occur based on vehicle trajectories should take into account the time information of the trajectories.

5.2.2.3 Granularity of Trajectory Patterns

The granularity of a trajectory pattern can be characterized by the time interval during which the pattern holds and the number of objects that are involved in the pattern.

Global Vs. Partial Patterns

Trajectory patterns can be classified based on the temporal extent of the trajectories that participate in the patterns. In *global trajectory patterns*, trajectories are viewed as non-decomposable, i.e., the basic unit of pattern discovery is a whole trajectory. For example, Gaffney et al. [24] propose a model-based clustering algorithm that groups trajectories with overall similar routes using a regression mixture model and the EM algorithm. This algorithm generate clusters of trajectories with respect to the overall distances among whole trajectories.

In contrast, *partial trajectory patterns* consider partial trajectories in the pattern discovery process. The idea behind this approach is that the clustering of whole trajectories may not detect trajectories with similar sub-trajectories. In general, a trajectory may be long and complex. Hence, even though some parts of trajectories are similar, the full trajectories might not be similar. Based in this view, Lee et al. [42] propose to partition trajectories into line segments and to build groups of close trajectory segments. Recent studies on trajectory clustering algorithms, such as moving clusters [36], flocks [11], convoys [35], and swarms [44], also consider patterns that appear in sub-trajectories.

Individual Vs. Group Patterns

Another distinction relates to whether a set of individual trajectories are retrieved that satisfy a pattern specified in query, called *individual trajectory pattern* retrieval [28, 32, 48], or whether sets of trajectories are retrieved so that the trajectories in a set exhibit a similar pattern according to some specific notion of pattern, called *group trajectory pattern* discovery [27, 33]

So-called *spatiotemporal pattern queries* [28, 48] illustrate well the retrieval of trajectories that satisfy a particular pattern as specified in a query. Examples of such queries include the finding of flights that descended into an airport, but did not land; identifying flights that had to make several approaches before entering the airport; and discovering vessels that changed course to avoid a collision.

The discovery of groups of trajectories that share patterns [57] is fundamentally different in that the objects returned are to be similar to each other according to a given notion of similarity rather than similar to a query pattern. The discovery of group patterns may enable different forms of sharing among the objects who have similar trajectories, e.g., in car pooling or delivery truck logistics. This class of patterns include concurrence [39], trend-setter [39], flock [11, 27, 26], leadership [5, 4], convergence [11], encounter [11], convoy [33], and swarm [44]. We describe each of these pattern concepts in detail in Sections 5.4 and 5.5.

5.2.2.4 Constrained Trajectory Patterns

Some trajectory patterns can be associated with constraints along the spatial and temporal dimensions.

Spatial Constraints: Movement on Spatial Networks

Many types of objects move in a spatial network, such as a road network, a rail network, or the kind of network made up by the corridors used by commercial aircraft. Since those objects are always located somewhere in the networks, raw trajectory data is typically modeled as or transformed to network-based trajectories, e.g., edge sequences in a road network graph [34, 60, 18]. As a result, trajectory patterns of network-constrained objects also have different forms and sometimes carry different semantics from the pattern types generally considered in unconstrained object trajectories.

As an example, Figure 5.4(a) illustrates a road network, with its graph model shown in Figure 5.4(b). All junctions form vertices, and each edge contains the corresponding road segment's information such as a weight w (distance). In this

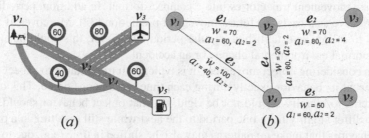

(a) (b)

Fig. 5.4 An example of road network-constrained trajectory model.

network model, an object's location is represented as a tuple $l = (e, d, t)$, meaning that the object is located on the edge $e = (v_i, v_j)$, at distance d from v_i at time t. Table 5.1 shows an example of network-constrained objects' trajectories over 2 days.

object	trajectory ID	sequence of network positions
o_1	1	$(e_1, 1.2, t_1), (e_1, 2.9, t_2), (e_2, 0.2, t_3)$
o_1	2	$(e_1, 1.3, t_2), (e_1, 3.4, t_3)$
o_2	3	$(e_3, 7.6, t_1), (e_5, 1.5, t_2), (e_5, 5.3, t_3)$
o_2	4	$(e_1, 1.2, t_1), (e_1, 2.9, t_2), (e_2, 0.2, t_3)$

Table 5.1 An example of network-constrained trajectories.

Given such network constrained trajectories, trajectory patterns reveal hidden, but useful, information. Finding popular routes, for example, can indicate reliable paths when drivers travel to unfamiliar destinations, as well as suggest higher-quality roads for truck delivery services [18]. They also enable location prediction that in turn can enable services that report relevant traffic conditions and upcoming points of interest (POIs) such as gas stations to users [12, 34]. In addition, Internet map services (e.g., Google Maps, Yahoo Maps) can be enriched based on trajectory patterns.

Temporal Constraints: Periodicity

In the real world, various types of objects exhibit periodic movement patterns. For example, many individuals go to work every weekday following the same or similar routes each day, public transportation is governed by time schedules, and animals annually migrate to reproduce or seek warmer climates. Findings in the literature suggest that car drivers tend to follow regular trajectories more than 70% of the time [37].

Periodic pattern mining of trajectory data concerns the discovery of periodic object behavior [47, 45], i.e., objects that follow the same routes (approximately) over regular time intervals. These periodic patterns provide an insight into, and concise explanation of, periodic behaviors (e.g., daily, weekly, monthly, and yearly) across long movement histories.

Periodic patterns are also useful for compressing movement data [14], since they summarize movement trajectories into a compact format. In addition, periodic patterns can serve as a basis for future-movement prediction [32]. Moreover, if an object fails to follow an established, regular periodic behavior, this could be a signal of an abnormal environmental change or an accident.

When considering object movement, it is typically unreasonable to expect an object to repeat its behavior exactly during each time period considered. This implies that patterns to identify should not be rigid, but that object behavior should be allowed to differ slightly from one period to the next while still resulting in a pattern. Next, behaviors that make up patterns may also be shifted in time (e.g., due to traffic delays).

The approximate nature of patterns in the spatiotemporal domain increases the complexity of mining tasks. In addition, the periods that yield patterns may be unknown. Further, multiple periods (e.g., *day* and *week*) may exist that yield different patterns in the same data. As a result, periodic pattern mining of trajectory data takes into account a wide variety of modeling approaches as well as efficient discovery algorithms.

5.3 Relative Motion Patterns

Given a collection of trajectories, it is challenging to capture and compare motion events of individuals and groups of individuals. To facilitate trajectory data analysis, the analysis concept RElative MOtion (REMO) by Laube *et al.* [39, 40, 38, 41] enables the identification of similar movements in a collection of moving object trajectories. The phrase "relative motion" refers to the relationships among motion attributes of different moving objects over space and over time.

REMO aims to enable the formulation and discovery of meaningful trajectory patterns based on *motion attributes* (i.e., speed, change of speed or motion azimuth), as extracted from raw trajectory data. Figure 5.5 demonstrates the process of building motion azimuth-based trajectories from raw trajectories. The left part of the figure shows trajectories of three objects moving in two-dimensional space. We then consider eight possible movement directions (motion azimuth): $\uparrow, \nearrow, \rightarrow, \searrow, \downarrow, \swarrow, \leftarrow, \nwarrow$. The right part shows the motion azimuth-based representation of the trajectories, where a trajectory is represented by a direction for each of four time points. This latter space of trajectories is called the analysis space in REMO.

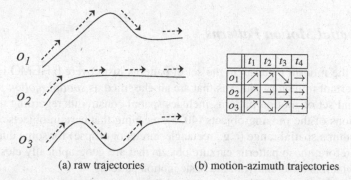

(a) raw trajectories   (b) motion-azimuth trajectories

Fig. 5.5 Motion azimuth-based trajectory representation.

REMO defines three sets of relative motion patterns: *basic motion patterns, spatial motion patterns*, and *aggregate/segregate motion patterns*. These pattern concepts serve as fundamental background for subsequent studies of moving objects patterns. This section summarizes the key ideas in REMO.

5.3.1 Basic Motion Patterns

The first set of REMO patterns describe motion events and patterns that explicitly disregard the absolute position of the moving objects. This means that a set of geographically distant objects can be identified as belonging to the same group, as long as they exhibit the same predefined motion properties. The basic motion patterns include (i) patterns over time, (ii) patterns across objects, and (iii) patterns over time and across objects. Such patters are exemplified next.

Constance
This pattern is defined as a sequence of equal motion attribute values for some consecutive time points. As an example in Figure 5.5, object o_2 moves with a constant motion azimuth $\rightarrow$ during the interval of $[t_2, t_4]$.

Concurrence
A concurrence motion pattern captures the incidence of having multiple objects with similar motion attributes. For example, in Figure 5.5, objects o_1, o_2 and o_3 move with the same motion azimuth $\nearrow$ at time t_1.

Trendsetter
This motion pattern attempts to find objects that anticipate a certain motion pattern that is shared by a set of other objects in future. Thus, this complex pattern combines

the simple patterns concurrence and constance. For example, o_2 anticipates at t_2 the motion azimuth $\rightarrow$ that is shared by all other objects at time t_4.

5.3.2 Spatial Motion Patterns

Based on the motion patterns in the set of patterns just covered, REMO identifies three important trajectory patterns that are all classified as *spatial motion patterns*. This second set of motion patterns includes spatial constraints regarding the absolute positions of the moving objects [40, 5], meaning that certain objects' motions appear within a spatial range (e.g., rectangle, circle, or ellipse) for some duration of time. Therefore, these patterns capture objects that are geographically close to one another, which is different from the basic motion patterns.

Figure 5.6 illustrates examples of the following three spatial motion patterns. Each spatial constraint is a circle.

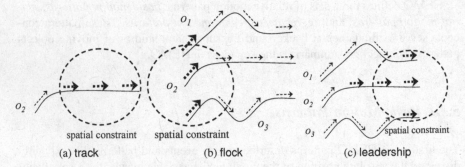

Fig. 5.6 Examples of REMO spatial motion patterns.

Track

Given a spatial range, e.g., a disc, a track motion pattern finds objects who travel within the range while keeping the same motion. For instance, an airplane that flies at constant speed and direction can be captured by the track motion pattern. Essentially, track combines constance pattern and a spatial constraint.

Flock

The concept of flock pattern in REMO covers not only a group of animals that live, travel, or feed together (e.g., a pride of lions, a school of fish, a gam of whales, a gaggle of geese, a murder of crows, or a swarm of insects), but also a group or crowd of people that move together.

In order to capture this concept using the basic motion patterns, the definition of flock consists of the concurrence pattern combined with a spatial constraint.

Intuitively, this patter may help identify groups of objects that travel together, i.e., within spatial proximity, for some duration of time.

Leadership

Leadership is a widespread phenomenon in social settings. In particular, the animal behavior research community studies the general topic of group decision-making in animals, searching for evidence for groups led in their activities by some dominant individuals, e.g., dominant breeding wolves frequently show significant frontal leadership, leading the pack during travel.

The leadership pattern corresponds to the trendsetter pattern, but includes a spatial constraint. For example, followers must lie within a given geographical region when they join the motion of the trendsetter. As a result, this pattern captures well the case where a group of people are under the leadership of one person.

5.3.3 Aggregate/Segregate Motion Patterns

The third set of relative motion patterns describe aggregation and segregation of objects' movements.

Convergence

This pattern describes a set of m objects during a time interval k that share motion azimuth vectors intersecting within a given spatial range, e.g., a disc with radius r. This pattern captures the behavior of a group of objects that converge in a certain region. An example of this pattern is wild animals that are heading in a synchronized fashion for a mating place.

Encounter

Suppose that some antelopes distributed over a field are heading for a location. At some later time, they will thus meet at the location, according to their current motions. To capture such an extrapolated (future) meeting within a spatial range, the encounter pattern is defined as a set of m objects that will arrive in a given spatial range r concurrently k time points later (i.e., the extrapolations of the objects' current motions intersect with r).

Divergence

This pattern is an opposite concept of convergence, integrating a spatial divergence pattern with the temporal constraint of a preceding meeting in a region. The graphical representation of the divergence pattern is that of "heading backwards" instead of forwards, relative to the direction of motion.

Breakup

Like the convergence pattern is an opposite concept of convergence, the concept of breakup is an opposite of the concept of encounter. This pattern captures objects' behaviors, such as departing from a meeting point.

5.3.4 Discussion

Although the REMO patterns constitute important mining concepts for trajectory data, they pose several open problems that should be addressed. We cover such open problems briefly.

- In the REMO framework, it is difficult to determine an absolute distance between two objects because the pattern discovery process is performed over the motion attributes (i.e. speed, change of speed, or motion azimuth) that are derived from raw trajectories. As a result, some pattern analysis tasks, such as finding the k nearest neighbor objects of a given object (trajectory), are difficult to support in REMO.
- Although the motion attributes in REMO encompass the objects' speeds and changes of speeds, the objects' motions are mainly analyzed considering the motion azimuths that capture the direction of a trajectory at a point in time. Full motion analysis considering all the motion attributes simultaneously is indeed not a simple task, but requires complex mechanisms and heavy computation.
- The default motion azimuths in REMO consists of eight different angles. Using a finer or coarser angle granularity would substantially impact the effectiveness and efficiency of trajectory pattern discovery. For example, the classification of motion azimuths into only the two classes East and West would reveal a lot of presumably meaningless constancy patterns. In contrast, every constancy pattern found with 360 azimuth classes may be unnecessary for typical applications. It is non-trivial to determine an appropriate angle granularity for a given trajectory data set.
- During the detection of relative motion patterns, time intervals with uncertain or missing data points may reduce the accuracy and effectiveness of pattern discovery significantly. In addition, the REMO framework assumes all trajectories have the same sampling rate as the granularity used in the analysis task, which may not hold for real-world trajectory data. Before employing the concept of relative motion patterns, pre-processing steps are required (e.g., interpolating missing samples and smoothing trajectories) in order to ensure a uniform time granularity.

5.4 Disc-Based Trajectory Patterns

Subsequent to the introduction of relative motion patterns by Laube *et al.* , a substantial body of studies have continued the study of trajectory data pattern mining [11, 35, 27, 54, 44, 8, 5, 4]. These subsequent studies have redefined, extended, and further developed the concepts of trajectory patterns substantially. The advances over the original relative motion pattern concepts relate mainly to three aspects.

1. *Distance-based trajectory analysis*: the core idea of REMO—i.e., extracting and utilizing motion features of trajectories for pattern analysis—is no longer

considered in further studies. Instead of using motion attributes, the subsequent studies consider Euclidean distance for measuring the proximity of positions of trajectories at time points. As a result, the importance of the basic relative motion patterns constance, concurrence, trend-setter decreases in trajectory analysis.

2. *Disc-based spatial range*: while REMO allows regions with arbitrary geometric shapes (e.g., ellipse, rectangle, disc) to be used as spatial ranges in motion patterns (Section 5.3.2), the subsequent studies of REMO take into account only circular ranges, i.e., discs. This approach is easy to apply, and discs capture intuitively the ranges of moving-object groups.

3. The REMO framework lacks time constraints, meaning that the relative motion patterns do not include an explicit parameter for specifying the time lengths of motion patterns. A relative motion pattern that, e.g., lasts for a few seconds may be too short to identify a common behavior among objects. Yet such a pattern can be identified with REMO. Subsequent studies include an explicit time duration constraint in order to alleviate this problem.

Built on the above principles, a rich body of studies have refined or redefined the original concepts and definitions of relative motion patterns. They have also introduced a set of new pattern types, mainly based on the REMO concept.

We categorize these newly introduced patterns into two classes:

- *Prospective patterns* that are likely to occur some time in the near future. This class covers encounter and convergence.
- *Flock-based* patterns that correspond to the original flock pattern of REMO. This class includes redefinitions of flock, meet, and leadership.

This section discusses in detail concepts that relate to these patterns.

5.4.1 Prospective Patterns

Gudmundsson *et al.* [27] revisit the encounter and convergence patterns in REMO, providing generic definitions based on the geometric arrangements of the moving objects. The resulting new patterns thus do not require the motion attributes in the REMO framework. The new pattern definitions consider the future trajectories of moving objects, meaning that the new encounter and convergence patterns capture future events that are likely to occur, based on the current motions of moving objects. Specifically, they are defined as follows.

Encounter (m, r)
Satisfied by a group of at least m objects that will arrive simultaneously in a disc with radius r, assuming that they keep their current speeds and directions.

Convergence (m, r)
Satisfied by a group of at least m objects that will pass through a disc with radius r

(not necessarily at the same time), assuming that they keep their current movement directions.

Figure 5.7 demonstrates the current and future trajectories obtained from five objects o_1, o_2, o_3, o_4, and o_5. The arrows show the current positions and their current motions, i.e., speed and direction; the dotted lines indicate the objects' future trajectories based on the current motions. According to the future trajectories, o_2, o_3, o_4, and o_5 will pass the disc r, forming a convergence pattern if given an m that does not exceed 45. If these objects arrive at r at the same time, they also satisfy the encounter pattern with the same parameters.

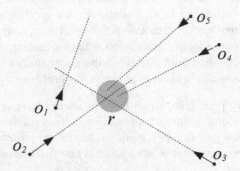

Fig. 5.7 Examples of the encounter and convergence patterns.

5.4.2 Flock-Driven Patterns

Gudmundsson *et al.* [11, 27, 26, 3, 54] have studied extensions of the flock pattern introduced by Laube *et al.* [39, 40] intensively. The studies have resulted in a wide range of interesting trajectory pattern concepts and discovery techniques. This section summarizes key ideas in relation to the extensions and variants of flock.

Flock (m, r, k)
In one study [11], two types of flock patterns are defined: one concerns continuous object movement, while the other concerns discrete object movement. The latter is widely used in database and data mining research. A discrete flock occurs if at least m objects move together for at least k consecutive time points while staying within a disc with radius r.

Based on this flock concept, two interesting variants, meet [11] and leadership [5], are defined:

Meet (m, r, k)
A meet pattern occurs if at least m objects stay together in a *stationary* disc with

radius r for at least k consecutive time points. Unlike the definition of flock, the disc specified in meet has a fixed location. Thus, the concept of meet resembles a past variant of encounter.

Figure 5.8 illustrates the concepts of flock and meet. In Figure 5.8(a), a flock of objects o_1, o_2, and o_3 is found during the time points $[t_7, t_8$, and t_9, while the meet pattern of objects o_1 and o_2 is identified during the time points $t_3, \ldots, t_8$ (Figure 5.8(b)).

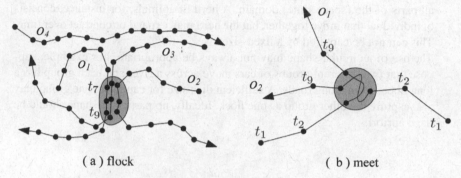

(a) flock   (b) meet

Fig. 5.8 Examples of the flock and meet patterns.

Leadership (m, r, k)

Andersson *et al.* [4] present a clear concept and definition of a leadership pattern. Their leadership pattern is essentially an extension of flock. Yet it captures the spatial constraint of the leader being ahead of the followers. Formally, the pattern occurs when a set of at least m objects move together for at least k consecutive time points while staying within a disc with radius r, and when at least one of the objects is/was heading in the leader's direction.

Andersson *et al.* [5] also introduce a variant of leadership where two new parameters are added to the definition of leadership. In Leadership (m, r, k, α, β), parameter α influences the spatial extent of a pattern, and parameter β determines the spatial characteristics of a pattern.

5.4.3 Discussion

The disc-based trajectory patterns are satisfied by groups of objects that move together within a disc with some user-specified extent. As a result, the chosen disc extent has a substantial effect on the result of the discovery process. Although the disc concept is intuitive, it raises potential problems when specifying trajectory patterns, as discussed next.

- The selection of a proper disc size turns out to be difficult, as situations can occur where objects that intuitively belong together or do not belong together are not quite within any disk of the given size or are within such a disk. In Figure 5.9, for example, all objects travel together in a natural group. However, object o_4 does not enter the disc and is not discovered as a member of the flock. This problem occurs because what constitutes a flock is very sensitive to the user-specified disc size, which does not take the data distribution into account. Indeed, if a flock must contain at least 4 objects, no flock would be found in this example.
- For some data sets, no single appropriate disc size may exist that works well for all parts of the (space, time) domain. A herd of animals, for instance, consists of individual that move together, but the herd may expand or contract over time. This can not be captured by a fixed-size disc.
- The use of a circular shape may not always be appropriate. For example, suppose that two different groups of cars move across a river and each group has a long linear form along roads. A sufficient disc size for capturing one group may also capture the other group as one flock. Ideally, no particular shape should be fixed apriori.

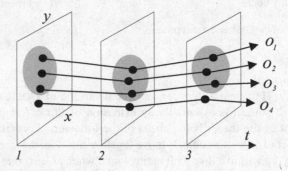

Fig. 5.9 Lossy-flock problem.

5.5 Density-Based Trajectory Patterns

In order to avoid the size and shape restrictions inherent to disc-based trajectory patterns, *density-based trajectory patterns* have been introduced. These employ density concepts [22] that allow the capture of generic trajectory pattern of any shape and any extent. In this section, we first review the concept of density-based clustering and then describe in detail the key concepts and definitions relating to density-based trajectory patterns.

5.5.1 Density Notions

As a precursor to defining density-based trajectory patterns, we need to understand the notion of density connection [22]. The main advantage of this concept is that it is able to capture clusters of any size and shape, as long as the cluster members meet certain distance-related conditions.

- Given a distance threshold e and a set of points S, the *e-neighborhood* of a point p is given as $NH_e(p) = \{q \in S \mid D(p,q) \leq e\}$.
- Given a distance threshold e and an integer m, a point p is *directly density-reachable* from a point q with respect to e and m if $p \in NH_e(q)$ and $|NH_e(q)| \geq m$.
- A point p is said to be *density-reachable* from a point q with respect to e and m if there exists a chain of points $p_1, p_2, ..., p_n$ in set S such that $p_1 = q$, $p_n = p$, and p_{i+1} is directly density-reachable from p_i with respect to e and m, $i \in \{1, ..., n-1\}$.
- Given a set of points S, a point $p \in S$ is *density-connected* to a point $q \in S$ with respect to e and m if there exists a point $x \in S$ such that both p and q are density-reachable from x with respect to e and m.

Figure 5.10(a) exemplifies a density-connected cluster when assuming that $m = 4$. Each dashed circle indicates an e-neighborhood of some object. For instance, o_1 is directly density reachable from o_3 with respect to threshold e and $m = 4$ because it belongs to the e-neighborhood of o_3 that contains a total of 4 objects. Also, o_1 and o_9 are density-connected through the following chain of points: $\langle o_3, o_4, o_6, o_7 \rangle$. The collective extent of the cluster can exceed e, and the cluster can have arbitrary shape. Since the parameter m is used to determine whether a point is directly density reachable from another, the size of a density-connected cluster is at least m. In addition, the cluster is preserved at time $t = 2$ in Figure 5.10(b), although the size of the cluster and the topology of the cluster members are changed.

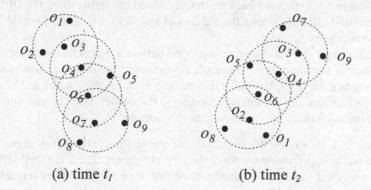

(a) time t_1 (b) time t_2

Fig. 5.10 An example of a density-connected cluster.

As shown in the above example, the definition of density-connection permits us to capture a group of "connected" points with arbitrary shape and extent. It thus allows us to overcome the main drawbacks of the concept of disc-based trajectory patterns. We proceed to introduce trajectory patterns built upon the concepts of density-connected objects.

5.5.2 Moving Objects Clustering

Recently, a wide variety of clustering concepts and algorithms for trajectory data have appeared in the literature. The different proposals generally assume different data and application requirements, and they have different advantages and disadvantages. In this section, we present a clear, systematic, and comprehensive overview of recent works on the clustering of trajectories in order to offer a good foundation for subsequent studies of trajectory pattern mining.

5.5.2.1 TRACLUS

Lee *et al.* introduce *trajectory clustering* [42] for the purpose of grouping trajectory segments. Specifically, they proposed a partition-and-group framework TRACLUS that clusters trajectories in the following three steps:

- *Partitioning*: each trajectory is partitioned into a set of line segments. This process finds the points where the behavior of a trajectory changes, called its characteristic points. The process adopts the minimum description length (MDL) principle, which is used widely in information theory.
- *Grouping*: trajectory segments that are close to each other according to a certain distance measure are grouped into a cluster. For this process, Lee *et al.* define a set of distance measures for trajectory segments (details are presented in Section 5.6.1) that are used for density-based clustering using the DBSCAN algorithm [22]. This allows the clusters of trajectories obtained by TRACLUS to form any shape and size.
- *Representing*: given a cluster , this process derives a representative trajectory for the cluster that describes the overall movement of the trajectory partitions that belong to the cluster. Basically, this process averages the lengths and angles of the entire trajectory segments belonging to the cluster, thus constructing a new trajectory segment, i.e., a representative trajectory.

Figure 5.11 illustrates the result of the above three processing steps of the partition-and-group framework for trajectory clustering. The trajectory segments shown in grey in Figure 5.11(b) denote those that are not included in the result clusters c_1 and c_2. The dotted line segments "in" each cluster in Figure 5.11(b) show the representative trajectory segments of c_1 and c_2.

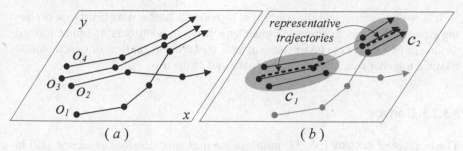

Fig. 5.11 An example of trajectory clustering: (a) raw trajectories; (b) clustered trajectory segments.

Notice that TRACLUS does not consider the temporal aspects of the trajectories while partitioning and grouping trajectories. As a result, some objects can belong to the same group even though they have never traveled close together (i.e., traveled at the same time). That is, TRACLUS performs spatial trajectory pattern clustering as opposed to spatio-temporal trajectory pattern clustering, as described in Section 5.2.2.2.

5.5.2.2 Moving Cluster

Kalnis *et al.* introduce the concept of moving cluster [36] that is defined as a set of objects that move close to each other for a time duration. It is a sequence of spatial clusters appearing during consecutive time points, such that the portion of common objects in any two consecutive clusters is not below a given threshold θ, i.e., $\frac{|c_t \cap c_{t+1}|}{|c_t \cup c_{t+1}|} \geq \theta$, where c_t denotes a cluster at time t. As an example, o_1, o_2, and o_3 in Figure 5.12 form a moving cluster if $\theta \geq \frac{3}{4}$ (i.e., requiring at least 75% overlapping objects between two consecutive clusters in time).

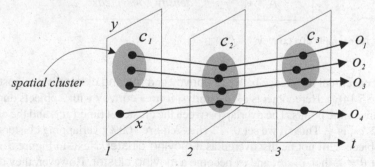

Fig. 5.12 An example of moving cluster.

It is worth noticing that the concept of moving cluster is not effective at capturing groups of objects that traveled together, since a moving cluster can be formed as long as two snapshot clusters have at least θ overlap, regardless of which objects traveled together from the beginning to the end of the trip.

5.5.2.3 Convoy

The concept of convoy [35, 33] employs the notion of density connection [22] in order to enable the formulation of arbitrary shapes of groups. More specifically, given a set of trajectories O, an integer m, a distance value e, and a lifetime k, a convoy is defined as a group that has at least m objects who are density-connected with respect to distance e and cardinality m during k consecutive time points. Each convoy is associated with a time interval during which the objects in the group traveled together.

Figure 5.13 shows polylines that represent the trajectories of three objects o_1, o_2, and o_3, during the time interval from t_1 to t_4. Consider the convoy specified by the parameters $m = 2$ and $k = 3$ issued over the trajectories in the figure. The result is $\langle o_2, o_3, [t_1, t_3] \rangle$, meaning that o_2 and o_3 form a convoy during the consecutive time points from t_1 to t_3.

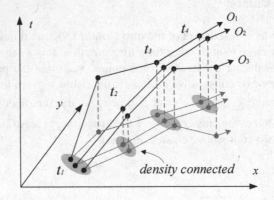

Fig. 5.13 An example of a convoy.

The conceptual difference between convoy and moving cluster is demonstrated in Figure 5.14(a). Here, objects o_2, o_3, and o_4 form a convoy with 3 objects during 3 consecutive time points. The overlap between the cluster at time 1, c_1, and the cluster at time 2, c_2, is $\frac{3}{4}$. Thus, if we set $\theta = 1$ (i.e., require 100% overlapping clusters), the above objects will not be discovered as a moving cluster. Next, in Figure 5.14(b), if we set $\theta = \frac{1}{2}$ then c_1, c_2, and c_3 become a moving cluster. However, they do not form a convoy.

convoy Variants
Since the concept of convoy was originally introduced [35], subsequent studies

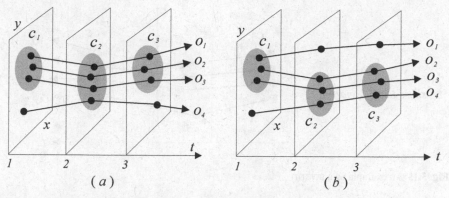

Fig. 5.14 Comparison between moving cluster and convoy.

have supplemented the original concept. Aung and Tan [8] introduce two variants of convoys: **dynamic convoys** allow their members to be absent briefly during the convoy lifetime, while **evolving convoys** are allowed to grow and shrink in cardinality during their lifetimes, which reduces the number of convoys that have large sets of overlapping objects and similar lifetimes.

In addition, Yoon and Shahabi [59] report that the original concept of **convoy** may miss some convoys that still satisfy with the definition, when two separate convoys are merged into one larger, union convoy. To capture this case, they introduce a variant of the convoy concept, termed a **valid convoy**.

5.5.2.4 Swarm

The **convoy** concept has a strict restriction regarding time: all convoy objects must be together for all of the consecutive time points in a lifetime of their convoy. For example, consider the group of moving objects o_1, o_2, o_3, and o_4 shown in Figure 5.15, where the first three objects belong to the same cluster for three time points considered and where object o_4 was apart from the others only during the second time point. Due to the strict time constraint, o_4 cannot belong to the same convoy as the first three objects although all the objects may be viewed as traveling together.

In order to overcome this drawback of **convoy**, Li *et al.* introduce a new trajectory pattern type, **swarm** [44], that extends the concept of **convoy** by relaxing the consecutive-time constraint. Specifically, the definition of **swarm** replaces the parameter k of convoy with k_{min}, such that k_{min} denotes a minimum of time duration to form a moving object cluster, regardless of the consecutiveness in time. This allows an individual moving object to temporarily leave its group as long as it is close to other group members for most of the time.

The concept of **swarm** is similar to that of **dynamic convoy** [8]; however, it differs in its discovery technique. **swarm** also has a definition similar to that of **group pattern** [57] in the sense of identifying a group of objects that travel together while

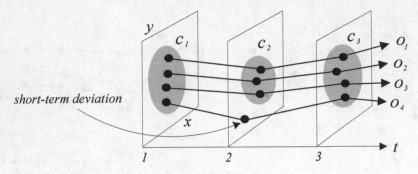

Fig. 5.15 An example of a swarm.

allowing relaxation of the time constraint. Nevertheless, group pattern belongs to the class of disc-based trajectory patterns.

Variants of Swarm

The discovery of swarms in a large collection of trajectories may result in a number of similar or redundant clusters in time or objects. To avoid finding such redundant swarms, the concept of closed swarm is introduced [44].

A follower trajectory pattern [46] identifies objects that follow a swarm with a certain time lag, which may be useful to capture a case such as when a animal follows a herd at a certain distance. This is similar to the concept of leadership in the sense that some objects spatially intersect a certain cluster with some time delay.

5.5.3 Discussion

Table 5.2 summarizes the key features of each concept of moving object clustering.

	Pattern space	Proximity-based	Online adaptation	Variants
TRACLUS	spatial	trajectory segments	difficult	TraClass [43]
Moving cluster	spatiotemporal	snapshot points	easy	
Convoy (CMC)	spatiotemporal	snapshot points	easy	dynamic/evolving
(CuTS)	spatiotemporal	trajectory segments	difficult	/valid convoy
Swarm	spatiotemporal	snapshot points	difficult	closed swarm, follower

Table 5.2 Summary of moving object clustering algorithms.

5.6 Methods for Mining Trajectory Patterns

We review distance functions for measuring the distance between two trajectories and then examine techniques for the mining of patterns from trajectories.

5.6.1 Trajectory Distance Measures

We use the notation $P_{1..n}$ for a trajectory that consists of a sequence of time-referenced point locations $P_1, P_2, \cdots, P_n$, where P_t is the point at time t (in the interval $[1,n]$). A trajectory pattern mining task (e.g., trajectory clustering) often needs to compute the "closeness" of two trajectories, which is captured by a distance function. We classify existing distance functions as global distance measures or local distance measures. In the sequel, we consider measures in each class in turn and discuss how the individual measures capture the characteristics of trajectories.

5.6.1.1 Global Distance Measures

A global distance function defines the overall distance between two trajectories with respect to all points in those trajectories.

Euclidean Distance

A simple approach to measuring the distance between two trajectories is to compute the sum of the Euclidean distances between all corresponding pairs of point locations in the two. This definition assumes that a pair of argument trajectories P and Q have the same length—i.e., they are sampled at the same times.

Specifically, the Euclidean distance between such trajectories $P_{1..n}$ and $Q_{1..n}$, $D_{Euclid}(P_{1..n}, Q_{1..n})$, is defined as the sum of their point distances at each sampling time t, i.e.,

$$D_{Euclid}(P_{1..n}, Q_{1..n}) = \sum_{t=1}^{n} ||P_t - Q_t|| \, ;$$

where the Euclidean distance between two points is defined as:

$$||P_t - Q_t|| = \sqrt{(P_t.x - Q_t.x)^2 + (P_t.y - Q_t.y)^2}$$

In cases where only relative distances, not absolute distance, are important, the squared Euclidean distance may be used as an alternative to the Euclidean distance. The squared Euclidean distance preserves the ordering of the Euclidean distance and is attractive because it is easier to compute.

Alignment-based Distance

The Euclidean distance measure is affected by noise and distortion that may obscure a trajectory. Thus, it may be argued that the Euclidean distance is unable to capture the inherent distance between trajectories. As a result, various alignment-based distance measures have been proposed for computing the distance between a pair of trajectories. These measures also lift the rigid length assumption inherent to the Euclidean distance and permit the comparison of trajectories with different lengths.

Dynamic Time Warping (DTW) [58] is defined by the following recurrence equation. It attempts to align two trajectories in such a way that their overall distance is minimized. The computation of DTW requires dynamic programming and it takes $O(n \cdot m)$ time.

$$DTW(P_{1..n}, Q_{1..m}) = ||P_n - Q_m|| + \min \begin{cases} DTW(P_{1..n-1}, Q_{1..m-1}) \\ DTW(P_{1..n-1}, Q_{1..m}) \\ DTW(P_{1..n}, Q_{1..m-1}) \end{cases}$$

In the equation, $P_{1..n-1}$ denotes the sub-trajectory of $P_{1..n}$ that covers the time points from 1 to $n-1$ only.

The dynamic time warping distance is not a metric, as it does not satisfy the triangle inequality. In contrast, a recent, related measure called the Edit Distance on Real sequence (EDR) [16] is a metric. It satisfies the triangle inequality, so it can be exploited for pruning unnecessary trajectories effectively during query processing.

When compared to the above distance measures, the Longest Common SubSequence measure (LCSS) [55] is more robust to noise. However, it requires the user to specify two parameters δ and ε as tolerances for two points to match with respect to time and space. This measure is defined by the recurrence equation next.

$$LCSS(P_{1..n}, Q_{1..m}) = \begin{cases} 0 & \text{if } n = 0 \lor m = 0 \\ 1 + LCSS(P_{1..n-1}, Q_{1..m-1}) & \text{if } |n - m| \leq \delta \\ & \land ||P_n - Q_m|| \leq \varepsilon \\ \max\{LCSS(P_{1..n-1}, Q_{1..m}), \} \\ \quad LCSS(P_{1..n}, Q_{1..m-1}) & \text{otherwise} \end{cases}$$

A recent distance measure called the Edit distance with Real Penalty (ERP) [15] is also designed to handle noise and it is even better than LCSS.

5.6.1.2 Local Distance Measures

Global distance measures capture the overall similarity between a pair of trajectories, not their local similarity during some short time interval. We proceed to consider local distance functions that capture the similarity between sub-trajectories.

MBR-Based Distance

A commonly used trajectory distance measure is derived based on the use of Minimum Bounding Rectangles (MBRs), which are often used to approximate trajectory segments and provide fast computation of trajectory distances. Let B_1 and B_2 be the MBR of the line segment L_1 and L_2, respectively. The distance $D_{min}(B_1, B_2)$ represents the minimum distance between any pair of points in B_1 and B_2. It is defined as:

$$D_{min}(B_1,B_2) = \sqrt{(\Delta(B_1.[x_l,x_u],B_2.[x_l,x_u]))^2 + (\Delta(B_1.[y_l,y_u],B_2.[y_l,y_u]))^2}\,,$$

where the distance between two intervals is defined as:

$$\Delta([l_1,u_1],[l_2,u_2]) = \begin{cases} 0 & [l_1,u_1]\cap[l_2,u_2] \neq \emptyset \\ l_2-u_1 & \text{if } u_1 < l_2 \\ l_1-u_2 & \text{if } u_2 < l_1 \end{cases}$$

Trajectory-Hausdorff Distance

Lee *et al.* [42] propose a distance function that is a weighted sum of three terms:

$$D_{Hausdorff} = w_\perp \cdot d_\perp + w_\parallel \cdot d_\parallel + w_\theta \cdot d_\theta\,,$$

where $w_\perp$, $w_\parallel$, and w_θ are the weights of the following components. They suggest that different applications may require different weights.

- The aggregate perpendicular distance ($d_\perp$) that measures the separation between two trajectories. where $d_{\perp,a}$ and $d_{\perp,b}$ are two perpendicular distances between
- The aggregate parallel distance ($d_\parallel$) that captures the difference in length between two trajectories.
- The angular distance (d_θ) that reflects the orientation difference between two trajectories.

Figure 5.16 illustrates these distance components between two line segments L_1 and L_2. Specifically, we have:

$$d_\perp = \frac{d_{\perp,a}^2 + d_{\perp,b}^2}{d_{\perp,a} + d_{\perp,b}}$$

$$d_\parallel = \min\{d_{\parallel,a}, d_{\parallel,b}\}$$

$$d_\theta = \|L_2\| \times \sin\theta$$

where $d_{\perp,a}, d_{\perp,b}$ are two perpendicular distances between L_1 and L_2, $d_{\parallel,a}, d_{\parallel,b}$ are two parallel distances between L_1 and L_2, and θ is the angle between L_1 and L_2.

Trajectory-Segment Distance

Jeung *et al.* [35] use two simple measures as the distance between two trajectory segments l'_1 and l'_2. These measures are designed to support efficient trajectory clustering.

The line distance $D_{LL}(\cdot,\cdot)$ is simply the shortest distance Euclidean distance between a pair of points located on the two argument line segments, and it is thus defined as follows.

$$D_{LL}(l'_1,l'_2) = \min_{p_i\in l'_1, b_j\in l'_2} \|p_i - b_j\|$$

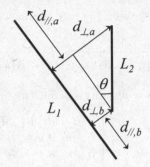

Fig. 5.16 Trajectory-Hausdorff distance.

Figure 5.17(a) shows two line segments l'_1 and l'_2. Here, l'_1 has the endpoints p'_1 and p'_4, corresponding to its locations at times t_1 and t_4. Similarly, l'_2 has endpoints b'_3 and b'_5, for its locations at times t_3 and t_5. The shortest distance between l'_1 and l'_2 is given by $D_{LL}(l'_1, l'_2)$. Note that this distance is purely spatial and ignores the temporal dimension.

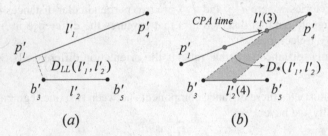

$$(a) \qquad\qquad\qquad (b)$$

Fig. 5.17 Trajectory-segment distance examples.

In contrast, the tightened distance $D_*(\cdot, \cdot)$ is spatio-temporal and thus aligns two line segments based on time. It then measures the shortest distance between two trajectories as of the same time point and is defined as follows.

$$D_*(l'_1, l'_2) = \min_{p_i \in l'_1, b_j \in l'_2, i=j} ||p_i - b_j||$$

In the example in Figure 5.17(b), the tightened distance $D_*(l'_1, l'_2)$ is the distance between the locations $l'_1(3)$ and $l'_2(3)$, which are the derived locations of the two trajectories at time 3.

It is possible to compute the tightened distance $D_*(l'_1, l'_2)$ efficiently. Let $l'_p = \{p_u, p_v\}$ be a line segment having a time interval $l'_p.\tau = [u, v]$. The location of a point l'_p in the segment as of time $t \in [u, v]$ is defined as:

$$l'_p(t) = p_u + \frac{t - u}{v - u}(p_v - p_u)$$

Note that the terms $l'_p(t)$, p_u, and $(p_v - p_u)$ are 2D vectors representing locations.

We then introduce the *Closest Point of Approach* time, called the CPA time (t_{CPA}) [6]. This is the time when the distance between two dynamic objects is the shortest, considering their velocities. Let $l'_q = \{q_w, q_z\}$ be another line segment during $l'_q.\tau = [w,z]$. The CPA time for l'_p and l'_q is computed by:

$$t_{CPA} = \frac{-(p_u - q_w) \cdot (l'_p(t) - l'_q(t))}{|l'_p(t) - l'_q(t)|^2},$$

where, $l'_q(t)$, q_w, and $(q_w - q_z)$ are also location vectors.

Considering again the example in Figure 5.17(b), the gray region indicates that the common time interval of l'_1 and l'_2 is $[t_3, t_4]$. The *tightened* shortest distance $D_*(l'_1, l'_2)$ between the two segments is then taken as:

$$D_*(l'_1, l'_2)) = \begin{cases} D(l'_1(t_{CPA}), l'_2(t_{CPA})) & \text{if } t_{CPA} \in (l'_1.\tau \cap l'_2.\tau) \\ \infty & \text{if } l'_1.\tau \cap l'_2.\tau = \emptyset \end{cases}$$

hen the time intervals of the two argument segments do not intersect, i.e., $l'_1.\tau \cap l'_2.\tau = \emptyset$, their distance is ∞.

The above equation can be expressed as a quadratic equation of t, and its minimum value can be found in constant time, regardless of the length of the common time interval $l'_1.\tau \cap l'_2.\tau$.

Observe that $D_*(l'_1, l'_2)$ is longer than $D_{LL}(l'_1, l'_2)$; hence, the line segments in Figure 5.17(b) have a lower probability of forming a cluster than do those in Figure 5.17(a). The tightened distance bounds thus improve the effectiveness of a filtering step that is applied as part of trajectory clustering.

5.6.2 Techniques for Efficient Pattern Discovery

It is difficult to discover trajectory patterns from a large trajectory database in an efficient manner. This process generally calls for the computation of sets of objects, which is more expensive than, e.g., spatio-temporal joins [10] that compute pairs of objects. In this section, we present various techniques from the literature that enable efficient discovery of trajectory patterns.

5.6.2.1 Raw Data Transformation

The REMO Matrix
The REMO framework enables the comparisons of the motion attributes of point objects over space and time, and it thus makes it possible to relate one object's motion to the motions of all other objects. More specifically, the given raw trajectory data is first transformed into a REMO matrix featuring motion attributes (i.e. speed,

change of speed, or motion azimuth). The REMO framework then uses the REMO matrix to identify notable individual motion behaviors as well as events of distinct group motion behavior, e.g., patterns like constancy, concurrence, and trend-setter as discussed in Section 5.3.

Table 5.3 illustrates an example of a REMO matrix that supports efficient discovery of patterns as illustrated by the following examples.

- The constancy pattern appears as a consecutive row of identical values, e.g., object o_1 from time t_1 to t_4.
- The concurrence pattern appears as a column (not necessarily consecutive) of identical values, e.g., objects o_2, o_3, o_4 at time t_2.
- The trend-setter pattern is a constancy pattern followed by a concurrence pattern, e.g., object o_4 from time t_4 to t_6, then objects o_4, o_5, o_6 at time t_6.

	t_1	t_2	t_3	t_4	t_5	t_6
o_1	↑	↑	↑	↑	→	↗
o_2	↗	↓	↑	↖	←	↘
o_3	→	↓	←	↘	↗	
o_4	↘	↓	→	↙	↙	↙
o_5	←	↗	↖	↗	↑	↙
o_6	↖	↘	↓	↗	↖	↙

Table 5.3 An example REMO matrix.

In this approach, raw data is approximated and transformed into an analysis format (i.e., the motion matrix) optimized for pattern discovery. This accelerates the pattern discovery process substantially when compared with directly accessing raw trajectory data to retrieve relative motion patterns.

Trajectory Simplification

The work on convoy discovery [35, 33] applies the filter-and-refinement paradigm in order to reduce the overall computational cost. In the filtering step, the original trajectories are simplified and a clustering algorithm is applied on the simplified trajectories in order to obtain convoy candidates. The goal is to retrieve efficiently a superset of the actual convoys. In the refinement step, each candidate convoy is considered in turn, and clustering is performed on the original trajectories of the objects involved so as to determine whether they indeed form an actual convoy.

Specifically, given a trajectory represented as a polyline $o = \langle p_1, p_2, \cdots, p_T \rangle$, and a simplification tolerance δ, the Douglas-Peucker algorithm (DP) [20] is applied to derive a *simplified trajectory* o' such that o' has fewer points and o' deviates from o by at most δ at any point. Initially, DP composes the line $\overline{p_1 p_T}$ and finds the point $p_i \in o$ farthest from the line. If the distance $D_{PL}(p_i, \overline{p_1 p_T}) \leq \delta$, then the segment $\overline{p_1 p_T}$ is reported as the simplified trajectory o'. Otherwise, DP recursively examines the sub-trajectories $\langle p_1, \cdots, p_i \rangle$ and $\langle p_i, \cdots, p_T \rangle$, reporting the concatenation of their simplified trajectories as the simplified trajectory o'. Figure 5.18(a) illustrates

three original trajectories. Their simplified trajectories (by using DP) are shown in
Figure 5.18(b).

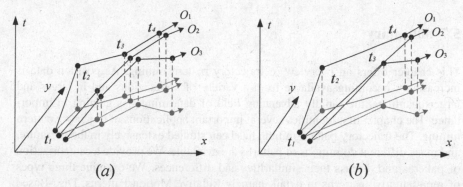

(a)   (b)

Fig. 5.18 An example of trajectory simplification.

5.6.2.2 Indexing

Indexing is a popular approach to accelerate the discovery process in data mining.
Here we present a couple of examples used in the discovery of the trajectory patterns
presented earlier in this chapter.

Kalnis *et al.* [36] employ a grid index G_t at each time point t for storing the data
points at that time. The density-based clustering algorithm DBSCAN [22] is then
applied on the grid index G_t in order to identify the clusters at time t.

Gudmundsson *et al.* [27] utilize the quadtree [49], the compressed quadtree [7],
and the skip-quadtree [21] for the fast discovery of the flock, leadership, convergence, and encounter patterns. Specifically, they first index the spatial data points
at each time point using the compressed quadtree. They then position a disc with
radius r into the non-empty entries of the index and enlarge the disc to have radius
$(1 + \delta) \cdot r$ when necessary. The relative motion patterns are then retrieved by repeating the above process at each time point, and comparing the objects in the disc with
the others in the neighbor time points.

5.6.2.3 The Apriori Approach

The Apriori approach has been applied to discover trajectory patterns efficiently [57, 44]. First, some (short) candidate trajectory patterns are generated, and their
frequencies are counted by scanning the trajectory database. Then unpromising patterns (with low frequencies) are removed from consideration. The remaining candidates are combined together to form larger candidate patterns. This process is
repeated until no further candidates are generated. The Apriori approach is able to

reduce the search space significantly in the first few iterations, allowing trajectory patterns to be mined efficiently.

5.7 Summary

This chapter offers an overview of trajectory pattern mining. As position data is increasingly becoming available from a variety of applications and in larger and larger quantities, the rapidly advancing field of data mining is gaining in importance. The chapter first identifies several important applications of trajectory pattern mining. The trajectory pattern mining has been studied extensively in the literature, and many different definitions of patterns are available. We provide a categorization of patterns and discuss their similarities and differences. We examine three types of representative patterns in detail, namely Relative Motion Patterns, Disc-Based Trajectory Patterns, and Density-Based Trajectory Patterns. In addition, we cover distance measures and data transformation methods, indexing methods, and mining techniques for the discovery of trajectory patterns.

References

1. Digital battle field. http://www.defenselink.mil/news/newsarticle.aspx?id=45084
2. Porcupine caribou herd satellite collar project. http://www.taiga.net/satellite/.
3. Al-Naymat, G., Chawla, S., Gudmundsson, J.: Dimensionality reduction for long duration and complex spatio-temporal queries. In: Proceedings of the ACM symposium on Applied computing, pp. 393–397 (2007)
4. Andersson, M., Gudmundsson, J., Laube, P., Wolle, T.: Reporting leadership patterns among trajectories. In: SAC, pp. 3–7 (2007)
5. Andersson, M., Gudmundsson, J., Laube, P., Wolle, T.: Reporting leaders and followers among trajectories of moving point objects. GeoInformatica 12(4), 497–528 (2008)
6. Arumugam, S., Jermaine, C.: Closest-point-of-approach join for moving object histories. In: Proceedings of the IEEE International Conference on Data Engineering, p. 86 (2006)
7. Arya, S., Mount, D.M., Netanyahu, N.S., Silverman, R., Wu, A.Y.: An optimal algorithm for approximate nearest neighbor searching. In: SODA, pp. 573–582 (1994)
8. Aung, H.H., Tan, K.L.: Discovery of evolving convoys. In: Proceedings of the 22nd international conference on Scientific and statistical database management, pp. 196–213 (2010)
9. Bakalov, P., Hadjieleftheriou, M., Keogh, E., Tsotras, V.J.: Efficient trajectory joins using symbolic representations. In: Proceedings of the international conference on Mobile data management, pp. 86–93 (2005)
10. Bakalov, P., Hadjieleftheriou, M., Tsotras, V.J.: Time relaxed spatiotemporal trajectory joins. In: Proceedings of the ACM international symposium on Advances in geographic information systems, pp. 182–191 (2005)
11. Benkert, M., Gudmundsson, J., Hbner, F., Wolle, T.: Reporting flock patterns. Computational Geometry 41(1), 111125 (2008)
12. Brilingaité, A.: Location-related context in mobile services. Ph.D. dissertation, Aalborg University (2006)

13. Brinkhoff, T., Kriegel, H.P., Seeger, B.: Efficient processing of spatial joins using r-trees. In: SIGMOD Conference, pp. 237–246 (1993)
14. Cao, H., Mamoulis, N., Cheung, D.W.: Discovery of periodic patterns in spatiotemporal sequences. TJDE **19**, 453–467 (2007)
15. Chen, L., Ng, R.: On the marriage of lp-norms and edit distance. In: Proceedings of the international conference on Very large data bases, pp. 792–803 (2004)
16. Chen, L., Özsu, M.T., Oria, V.: Robust and fast similarity search for moving object trajectories. In: Proceedings of the ACM SIGMOD international conference on Management of data, pp. 491–502 (2005)
17. Chen, Y., Patel, J.M.: Design and evaluation of trajectory join algorithms. In: GIS, pp. 266–275 (2009)
18. Chen, Z., Shen, H.T., Zhou, X.: Discovering popular routes from trajectories. In: ICDE, pp. 900–911 (2011)
19. Dodge, S., Weibel, R., Lautenschütz, A.K.: Towards a taxonomy of movement patterns. Information Visualization **7**, 240–252 (2008)
20. Douglas, D., Peucker, T.K.: Algorithms for the reduction of the number of points required to represent a line or its character. The American Cartographer **10**(42), 112–123 (1973)
21. Eppstein, D., Goodrich, M.T., Sun, J.Z.: The skip quadtree: a simple dynamic data structure for multidimensional data. In: SCG, pp. 296–305 (2005)
22. Ester, M., Kriegel, H.P., Sander, J., Xu, X.: A density-based algorithm for discovering clusters in large spatial databases with noise. In: Proceedings of the ACM SIGKDD International Conference on Knowledge Discovery and Data Mining, pp. 226–231 (1996)
23. Frank, A., Raper, J., Cheylan, J.P.: Life and motion of spatial socio-economic units. Taylor & Francis, London (2001)
24. Gaffney, S., Smyth, P.: Trajectory clustering with mixtures of regression models. In: KDD, pp. 63–72 (1999)
25. Gidófalvi, G., Pedersen, T.B.: Cab-sharing: An effective, door-to-door, on-demand transportation service. In: Proceedings of the 6th European Congress on Intelligent Transport Systems and Services (2007)
26. Gudmundsson, J., van Kreveld, M.: Computing longest duration flocks in trajectory data. In: Proceedings of the ACM international symposium on Advances in geographic information systems, pp. 35–42 (2006)
27. Gudmundsson, J., van Kreveld, M., Speckmann, B.: Efficient detection of motion patterns in spatio-temporal data sets. In: Proceedings of the ACM international symposium on Advances in geographic information systems, pp. 250–257 (2004)
28. Hadjieleftheriou, M., Kollios, G., Bakalov, P., Tsotras, V.J.: Complex spatio-temporal pattern queries. In: VLDB, pp. 877–888 (2005)
29. Han, J., Li, Z., Tang, L.A.: Mining moving object, trajectory and traffic data. In: DASFAA, pp. 485–486 (2010)
30. Iwase, S., Saito, H.: Tracking soccer player using multiple views. In: Proceedings of the IAPR Workshop on Machine Vision Applications (2002)
31. Jeong, S.H., Paton, N.W., Fernandes, A.A., Griffiths, T.: An experimental performance evaluation of spatiotemporal join strategies. Transactions in GIS **9**(2), 129–156 (2005)
32. Jeung, H., Liu, Q., Shen, H.T., Zhou, X.: A hybrid prediction model for moving objects. In: ICDE, pp. 70–79 (2008)
33. Jeung, H., Shen, H.T., Zhou, X.: Convoy queries in spatio-temporal databases. In: Proceedings of the IEEE International Conference on Data Engineering, pp. 1457–1459 (2008)
34. Jeung, H., Yiu, M.L., Zhou, X., Jensen, C.S.: Path prediction and predictive range querying in road network databases. The VLDB Journal **19**(4), 585–602 (2010)
35. Jeung, H., Yiu, M.L., Zhou, X., Jensen, C.S., Shen, H.T.: Discovery of convoys in trajectory databases. Proceedings of the VLDB Endowment **1**(1), 1068–1080 (2008)
36. Kalnis, P., Mamoulis, N., Bakiras, S.: On discovering moving clusters in spatio-temporal data. In: Proceedings of the International Symposium on Spatial and Temporal Databases, pp. 364–381 (2005)

37. Karimi, H.A., Liu, X.: A predictive location model for location-based services. In: Proceedings of the ACM International Symposium on Advances in Geographic Information Systems, pp. 126–133 (2003)
38. Laube, P., Imfeld, S.: Analyzing relative motion within groups of trackable moving point objects. In: GIScience, pp. 132–144 (2002)
39. Laube, P., Imfeld, S., Weibel, R.: Discovering relative motion patterns in groups of moving point objects. International Journal of Geographical Information Science 19(6), 639–668 (2005)
40. Laube, P., van Kreveld, M., Imfeld, S.: Finding remo - detecting relative motion patterns in geospatial lifelines. In: Proceedings of the International Symposium on Spatial Data Handling, pp. 201–214 (2004)
41. Laube, P., Purves, R.S.: An approach to evaluating motion pattern detection techniques in spatio-temporal data. Computers, Environment and Urban Systems 30(3), 347–374 (2006)
42. Lee, J., Han, J., Whang, K.: Trajectory clustering: a partition-and-group framework. In: Proceedings of the ACM SIGMOD international conference on Management of data, pp. 593–604 (2007)
43. Lee, J.G., Han, J., Li, X., Gonzalez, H.: *TraClass*: trajectory classification using hierarchical region-based and trajectory-based clustering. PVLDB 1(1), 1081–1094 (2008)
44. Li, Z., Ding, B., Han, J., Kays, R.: Swarm: mining relaxed temporal moving object clusters. PVLDB 3, 723–734 (2010)
45. Li, Z., Ding, B., Han, J., Kays, R., Nye, P.: Mining periodic behaviors for moving objects. In: SIGKDD, pp. 1099–1108 (2010)
46. Li, Z., Ji, M., Lee, J.G., Tang, L.A., Yu, Y., Han, J., Kays, R.: MoveMine: Mining moving object databases. In: Proceedings of the ACM SIGMOD international conference on Management of data, pp. 1203–1206 (2010)
47. Mamoulis, N., Cao, H., Kollios, G., Hadjieleftheriou, M., Tao, Y., Cheung, D.W.: Mining, indexing, and querying historical spatiotemporal data. In: Proceedings of the ACM SIGKDD International Conference on Knowledge Discovery and Data Mining, pp. 236–245 (2004)
48. Sakr, M.A., Shams, A.: Spatiotemporal pattern queries. Geoinformatica 14 (2010)
49. Samet, H.: Foundations of Multidimensional and Metric Data Structures (The Morgan Kaufmann Series in Computer Graphics and Geometric Modeling). Morgan Kaufmann Publishers Inc. (2005)
50. Sumpter, N., Bulpitt, A.: Learning spatio-temporal patterns for predicting object behaviour. Image Vision Computing 18, 697V704 (2000)
51. Tao, Y., Faloutsos, C., Papadias, D., Liu, B.: Prediction and indexing of moving objects with unknown motion patterns. In: Proceedings of the ACM SIGMOD International Conference on Management of Data, pp. 611–622 (2004)
52. Tao, Y., Papadias, D., Sun, J.: The tpr*-tree: An optimized spatio-temporal access method for predictive queries. In: Proceedings of the International Conference on Very Large Data Bases, pp. 790–801 (2003)
53. Tao, Y., Sun, J., Papadias, D.: Analysis of predictive spatio-temporal queries. ACM Transaction on Database Systems 28(4), 295–336 (2003)
54. Vieira, M.R., Bakalov, P., Tsotras, V.J.: On-line discovery of flock patterns in spatio-temporal data. In: Proceedings of the 17th ACM SIGSPATIAL International Conference on Advances in Geographic Information Systems, pp. 286–295 (2009)
55. Vlachos, M., Gunopoulos, D., Kollios, G.: Discovering similar multidimensional trajectories. In: Proceedings of the IEEE International Conference on Data Engineering, pp. 673–684 (2002)
56. Šaltenis, S., Jensen, C.S., Leutenegger, S.T., Lopez, M.A.: Indexing the positions of continuously moving objects. In: Proceedings of the ACM SIGMOD International Conference on Management of Data, pp. 331–342 (2000)
57. Wang, Y., Lim, E.P., Hwang, S.Y.: Efficient mining of group patterns from user movement data. DKE 57, 240–282 (2006)

58. Yi, B.K., Jagadish, H.V., Faloutsos, C.: Efficient retrieval of similar time sequences under time warping. In: Proceedings of the IEEE International Conference on Data Engineering, pp. 201–208 (1998)
59. Yoon, H., Shahabi, C.: Accurate discovery of valid convoys from moving object trajectories. In: IEEE International Conference on Data Mining Workshops, pp. 636–643 (2009)
60. Zheng, K., Trajcevski, G., Zhou, X., Scheuermann, P.: Probabilistic range queries for uncertain trajectories on road networks. In: EDBT, pp. 283–294 (2011)
61. Zhou, P., Zhang, D., Salzberg, B., Cooperman, G., Kollios, G.: Close pair queries in moving object databases. In: Proceedings of the ACM international symposium on Advances in geographic information systems, pp. 2–11 (2005)

Chapter 6
Activity Recognition from Trajectory Data

Yin Zhu, Vincent Wenchen Zheng and Qiang Yang

Abstract In today's world, we have increasingly sophisticated means to record the movement of humans and other moving objects in the form of trajectory data. These data are being accumulated at an extremely fast rate. As a result, knowledge discovery from these data for recognizing activities has become an important problem. The discovered activity patterns can help us understand people's lives, analyze traffic in a large city and study social networks among people. Trajectory-based activity recognition builds upon some fundamental functions of location estimation and machine learning, and can provide new insights on how to infer high-level goals and objectives from low-level sensor readings. In this chapter, we survey the area of trajectory-based activity recognition. We start from research in location estimation from sensors for obtaining the trajectories. We then review trajectory-based activity recognition research. We classify the research work on trajectory-based activity recognition into several broad categories, and systematically summarize existing work as well as future works in light of the categorization.

6.1 Introduction

Mining the activities and patterns from the moving objects' trajectory data is an important challenge in today's society. This problem is becoming increasingly necessary as we accumulate large amounts of trajectory data in our daily lives. Recent development of social networks and maturing sensing technologies makes it possible to record sequences of location related data. Activity information hidden in these data can power services such as location-based recommendation systems, e-health and intelligent transportation. Activity recognition is a process to extract high-level activity and goal related information from low-level sensor readings through ma-

Yin Zhu, Vincent Wenchen Zheng and Qiang Yang
Hong Kong University of Science and Technology
e-mail: {yinz,vincentz,qyang}@cse.ust.hk

chine learning and data mining techniques. In this survey, we review and summarize existing literature on trajectory-based activity recognition, and project into the future in this active research area.

A natural categorization of trajectory-based activity recognition can be done by considering the different levels in which we place the activities. First, at the lowest level, we consider location-estimation techniques that take into sensor readings and produces estimates on the object's locations. This layer provides the important information input for the further activity recognition. Second, we survey representative methods that can determine trajectory segments, transportation modes and activities. The target output includes actions and goals. At the third level, we consider various applications that can be built based on activity data. Figure 6.1 shows a process view of these three levels.

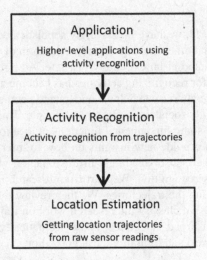

Fig. 6.1 A process view of trajectory-based activity recognition.

The activity recognition layer is essentially conducting information mining and compression from the huge spatial trajectories. To give a concrete impression the functionality of activity recognition, we use GeoLife, a location-based social-networking service developed by Microsoft Research Asia[1], as an example. GeoLife enables users to share their life experiences and build connections among each other using trajectory history data. One important component in Geolife is a transportation mode detection system [47], which classifies a segment of a GPS trajectory into one of { "walking", "driving", "biking" and "onBus"}. In this application, knowledge can be discovered from the trajectory data in the form of transportation mode. To the GeoLife system, users may ask queries like "how many times did I drive to work in this month?" Such questions can be answered by keeping a diary for the user. If an intelligent system can answer them directly from the user's GPS trajectories,

---

[1] http://research.microsoft.com/jump/79442

much manual effort is saved and mistakes avoided. Activity recognition also serves as an important component for higher-level applications in the GeoLife system. For example, a main goal of Geolife is to provide a trajectory sharing system that connects people using their trajectories as a medium. Transportation modes and other activities on the trajectories can be strong indicators for the similarity among people. This reveals another aspect of activity recognition: activity recognition serves as an important embedded component for other applications.

In this chapter, we consider **trajectory-based activity recognition** as a task that takes the sequences of sensor readings and context as input, and produces predictions of actions, goals, and plans. In particular, we first introduce location estimation techniques and focus on learning-based models. Then we consider various forms of activity recognition from trajectory data. We classify the current work into different categories, and survey existing research work under each category. We also present our views into the future.

6.2 Location Estimation for Obtaining the Trajectory Data

Before inferring activities for moving objects, we need to get their trajectory as a sequence of locations. In this section, we review location estimation techniques. In general, there are two categories of location estimation methods, including propagation models and learning based models.

6.2.1 Propagation Models for Outdoor Location Estimation

Global Positioning System (GPS) is the most widely used positing system in outdoor environments. It is based on a global navigation satellite system with at least 24 geo-synchronized satellites. A GPS receiver with four or more satellites in sight is able to know its location on the earth. In particular, the signal received from each satellite includes two pieces of information, including the coordinate of the satellite and the time stamp when the package is sent out. The distance between the GPS receiver and the satellite can be calculated by using the time difference between the sending time stamp and the receiving time stamp. After the GPS receiver knows the distances to four satellites, its location could be calculated by solving a distance equation system with three known parameters (latitude, longitude and altitude). This mathematical procedure is also called trilateration, where numerical root finding algorithms such as Newton-Raphson [4] are often used. The positioning accuracy for GPS can reach ten meters in an open area [14]. It is more accurate in an open area, and less accurate in cities with tall buildings. Because walls and other barriers block the satellite signals, GPS usually does not work in indoor environments. Besides G-

PS, there are also some other similar satellite positioning systems, such as Beidou[2] from China, and Galileo[3] from the European Union.

Though the GPS system is easy to use and relatively accurate in outdoor environments, it has obvious disadvantages. Most notably, it has difficulty working in indoor environments where signals from satellites are blocked or weakened. Other outdoor location estimation systems include GSM based localization, which uses cell-towers and their positions as basis for propagation model calculation.

6.2.2 Indoor Location Estimation using Learning-based Models

Since GPS hardly works in the indoor environment, researchers have considered alternative methods that work indoors and other complex environments. Among these methods, WiFi localization [1, 19] is one of the most mature and widely adopted solutions in research and practice. Even though, in theory, radio signal strength from a WiFi access point decays linearly with log distance, which allows a triangulation based method to identify the client mobile devices, in practice, it is difficult to obtain an accurate signal-propagation model in an indoor environment. This is because the physical characteristics of an environment, such as walls, furniture and even human activities, add significant noise to radio signal strength measurements. One particular problem in indoor environment is that the signals can reflect on the surfaces of certain materials, such as some walls. That means the signal strength the device receives is highly dependent on the local environment where the WiFi access points and the receiving device are. This scenario is referred to as the multiple path problem [1].

One intuitive idea to tackle this problem is to remember the signal strength behavior at different places in the environment, and train a learning-based method on these data. This method is commonly referred to as the fingerprinting technique. The following is an example illustrating this basic idea. When at place A, the device remembers the signal strength pattern. When this device arrives at A again, it knows that the scanned signal is most similar to one that is stored before. Therefore the device is able to infer its current location. However, there is much to be improved in this setting, e.g. how to collect fewer labeled points and achieve at the same accuracy, how to utilize the floor structure, and how to deal with the instability of WiFi signals.

A WiFi-enabled device, e.g. a laptop or a smartphone, can usually receive signals from multiple access points, which are installed at fixed positions in the environment, although their absolute positions may not be known. A WiFi access point is uniquely identified by its Media Access Control address (MAC address). The signal strength is called radio signal strength (RSS), which is a value usually spans from -120 dBm to -50 dBm. Once a mobile device scans the WiFi nearby, it gets a list of

---

[2] http://www.beidou.gov.cn/
[3] http://www.esa.int/esaNA/galileo.html

access points and the signal strength values to them. These received signal strength values can be represented as a vector $\mathbf{x} \in \mathbb{R}^d$, where d is the number of total access points and x_j is the RSS to the j-th access point. The fingerprints can be formally represented as a signal strength data set with locations as their labels $X = \{\mathbf{x}_i, y_i\}_{i=1}^N$, where N is the total number of data examples. y_i denotes a location, which can be a discrete label indicating a "block" in an environment or a continuous coordinate pair such as $(p, q) \in \mathbb{R}^2$ in the 2-D case. There are some public WiFi data sets that can be used for indoor localization study. For example, one collection of data is provided by the IEEE 2007 ICDM data mining contest [43]. It contains a set of $5,333$ fingerprint points collected from 247 locations in an office area of the HKUST academic building.

6.2.2.1 Nearest Neighbor Based Methods

RADAR [1] is one of the earliest system on WiFi localization. This project started as a localization system using only triangulation. But the multi-path problems led the researchers to use a fingerprint or learning-based method, which later became successful and highly influential. RADAR works in a test bed which is a 43.5m by 22.5m office area as shown in Figure 6.2. There are three base stations $BS1$, $BS2$ and $BS3$, which are similar to WiFi access points. A laptop is used as the mobile device to collect labeled signal strength at each location. Besides the signal strength SS, the device used in this project also uses one more strength called the signal-to-noise-ratio (SNR). For each location point (l_x, l_y) and each facing direction of the person holding the laptop $d \in \{N, E, S, W\}$, at least 20 pairs of SS and SNR to the three base stations are collected.

An important data preprocessing step in the RADAR system is that, it uses mean, standard deviation, and median of the signal strength values at each location for the location. After this, the data become less noisy. The data set contains many data tuples, each of which is represented as $(l_x, l_y, d, ss_i, snr_i)$ with $i = 1, 2, 3$. ss_i and snr_i are the mean values averaged over all the signal points collected under (l_x, l_y, d). The proposed location estimation method is based on nearest-neighbor search. A signal vector is matched to the nearest fingerprint vector sets in the feature space given some distance measure. Since these fingerprint points are already labeled, their location labels are used by average to predict the test signal vector's location. RADAR is an important early localization system with the meter-level accuracy, and it supports both learning-based method and radio propagation method for localization.

6.2.2.2 Bayesian Methods

Ladd et al. provide a major improvement to the RADAR system by further reducing the localization error to one meter for a similar test bed [19]. They propose a Bayesian model for indoor localization. Assume that the indoor space consists of n location states $S = \{s_1, \cdots, s_n\}$, where each state represents a location coordinate

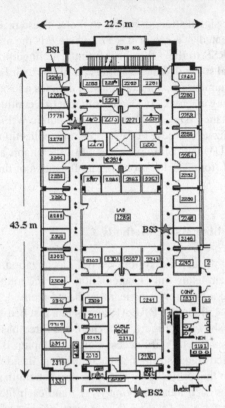

Fig. 6.2 The indoor map used in RADAR [1].

(l_x, l_y) with the facing direction d of the mobile user who holds a laptop in data collection. Given an observation o from a WiFi scan, the probability that it belongs to a state s_j is calculated as:

$$P(o|s_j) = \left(\prod_{j=1}^{N} P(f_j|s_i) \right) \cdot \left(\prod_{j=1}^{N} P(\lambda_j|b_j, s_i) \right),$$

where $P(f_j|s_i)$ is the probability that the frequency for j-th WiFi access point's frequency count f_j. $P(\lambda_j|b_j, s_i)$ measures the probability of received signal strength λ_j given the access point b_j and the state s_i. With this likelihood formula, one can calculate the posterior probability of the observation o on the location states:

$$P(s_i|o) = \frac{P(s_i) \cdot P(o|s_i)}{\sum_{j=1}^{n} P(s_i) \cdot P(o|s_i)}.$$

Ladd et al. further improve this Bayesian prediction by using the sequential constraints with a Hidden Markov Model (HMM) [34]. In practice, two consecutive locations in an object's spatial trajectory are close to each other. Such a constraint can be naturally encoded in the HMM model as state transitions. An HMM is a state transition model $P(s_i|s_j)$ with observations $P(o|s)$. Given a sequence of observations, the states are hidden variables. The inference process gives the most probable state sequences, which is known as decoding. At inference time, observations and hidden states are both location states. HMM decoding is used as a smoothing procedure. In most applications, observations and hidden states are different. For example, sensor readings are the observations and the unknown activities are the hidden states. Other techniques such as Kalman filters and particle filters [37] are can also be used to model such sequential information. Interested readers are referred to Chapter 9 of [18] and [9, 8] for more details. The experimental results from this improved system also validate that using an HMM for post processing the predicted location states improves the error by 40%.

6.2.2.3 Summary

There are several directions to improve the above WiFi localization systems. For example, Pan et al. [31] present a manifold based method to recover both the mobile-device locations and the access point locations from labeled and unlabeled trajectories. Because unlabeled signal points are much easier to collect, this semi-supervised approach also reduces the manual labeling effort. One observation is that mobile devices and access points are spatially close to each other if their signal vectors are similar on some manifold structure. An illustration of the experimental result is shown in Figure 6.3. With only a few labeled WiFi signal points, the structure of the office area can be discovered by using many other unlabeled WiFi trajectories.

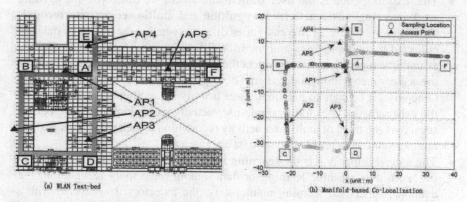

(a) WLAN Test-bed　　　　　　　　　　　(b) Manifold-based Co-Localization

Fig. 6.3 Indoor environment layout and the manifold learning result in [31].

There has also been some work that uses additional sources such as signal sniffers with known locations or GPS to automatically annotate the data. For example, LANDMARC [30] is an indoor localization system that uses RFID sensors as reference points, whose locations are known. They are used to collect signal calibration data. The VTrack system [39] can use GPS to calibrate a WiFi localization system for outdoor usage. GPS is believed to be accurate in an outdoor environment. But there are cases where GPS is temporarily unavailable or the mobile device is forced to turn off GPS to save energy. Commercial systems such as Google WiFi localization and Skyhook also use the similar techniques. Table 6.1 compares the different learning-based localization methods regarding their calibration algorithms and the methods to collect labeled data.

Table 6.1 Comparison of the learning-based localization methods.

Project	Calibration algorithm	Method to collect data
RADAR [1]	KNN	manual
Improved RADAR [19]	Naive Bayes + HMM	manual
LANDMARK [30]	KNN	reference points
Geometry-based [31]	Semi-supervised learning	reduced manual
VTrack [39]	KNN + HMM	GPS Vehicle

6.3 Trajectory-based Activity Recognition

After getting the spatial trajectories, we need to build models to extract useful activity information from them. In this chapter, we categorize the existing work for trajectory-based activity recognition along two possible dimensions.

- The first dimension is the **user dimension**, where we categorize the existing work into **single-user activity recognition** and **multi-user activity recognition**. In particular, these two categories differ on whether the trajectory data are collected by multiple users and the user difference is modeled. If a model uses multiple users' data and considers the difference among the users in training the activity recognition model, then it belongs to multi-user activity recognition. Otherwise, it belongs to the single-user activity recognition category. Most of the existing work goes to the single-user activity recognition category, but we see a growing trend of multi-user activity recognition in the recent years as there are more and more users' activity data accumulate.

- The second dimension is the **learning method dimension**. Most of the existing work uses machine learning or data mining for activity recognition. For example, supervised learning methods use the trajectory data labeled with a predefined set of activity labels for training an activity recognition model. Unsupervised learning and frequent pattern mining methods aim to extract useful activity patterns directly from the trajectory data.

There are also some other possible criteria to categorize the existing trajectory-based activity recognition work, including venue (e.g. indoor and outdoor), sensors (e.g.

GPS, WiFi, cameras.), etc. However, the categories generated by using these criteria may not be mutually exclusive to each other. For example, an activity model that works in indoor environments may also work in outdoor environments. Therefore, we do not take these criteria into account, and only focus the user and learning method dimensions in categorization. We list some existing work that falls into these two categories in Table 6.2. In the following sections, we shall introduce some representative work within each category, and hopefully shed some light on the emerging research directions.

Table 6.2 Trajectory Activity Recognition Categorization.

User	Supervised methods	Unsupervised methods	Frequent pattern mining
Single-user	[44, 32, 25, 24, 45, 35, 47, 41]	[12, 10, 16]	[21, 29]
Multiple-user	[27, 42, 46, 23, 17]	?	?

6.3.1 Single-user Activity Recognition

In this section, we focus on single-user activity recognition and further categorize the existing work into three learning paradigms: supervised learning, unsupervised learning and frequent pattern mining.

6.3.1.1 Supervised Learning

In activity recognition, standard supervised learning algorithms take trajectory data, as well as their corresponding activity labels, as inputs. By training some classification (or regression) model with the given inputs, the supervised learning algorithm can use it to predict the activity labels for some test trajectory observations as output. We shall introduce some typical supervised learning models used by the existing work in this section.

Decision Tree with Sequence Smoothing and Hidden Markov Model

Sequence information is important in modeling the trajectory data for activity recognition. For example, Zheng et al. propose to use decision tree and sequence smoothing to detect a mobile user's transportation mode, such as "Drive", "Bus", "Bike" and "Walk", based on her GPS trajectory [47]. In order to predict the transportation modes, a GPS trajectory is segmented into a sequence of fix-sized time slices. Within each time slice t, raw GPS points denoted as latitude and longitude coordinates are used to calculate some features for a feature vector $\mathbf{x}_t$, such as

- *Heading change rate* measures the percentage of GPS points in a window that change their heading directions. As shown in Figure 6.4, for example, being constrained by the road, people driving a car or taking a bus cannot change their heading directions as flexible as if they are walking or cycling. Such heading direction changes can be different for different transportation modes.
- *Stop rate* measures the percentage of GPS points whose velocity values to their previous points are less than a threshold. This feature is inspired by the heuristic that the stop rate is usually higher for walking and lower for driving or taking a bus. Besides, a bus could also take more stops than a car, and thus taking a bus still has a higher stop rate than driving.
- *Velocity change rate* measures the percentage of GPS points with a velocity change percentage above a certain threshold within a unit distance as shown in Figure 6.5. This feature, together with the stop rate above, can be used to differentiate the transportation modes.

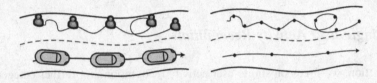

Fig. 6.4 Heading change rate of different modes.

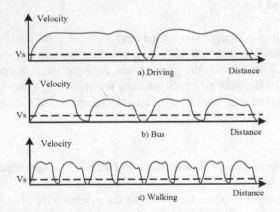

Fig. 6.5 Illustration for velocity change rate.

After the features $\mathbf{x}_t$ are extracted for each time slice, a decision tree is applied first to predict the transportation mode denoted as $y_t \in \{Drive, Bus, Bike, Walk\}$ in it. As such decision tree model does not take the sequence information into account, Zheng et al. further propose to model the location transition probability so that, if two locations have high transition probability value, then the transportation modes for a mobile user to travel from one location to the other can be more consistent. Such location transition probabilities can be learned by clustering on the given GPS trajectory data in training. Finally, sequence smoothing is used to adjust possible misclassification in the inference step. The experimental results show that the system's overall accuracy archives 76.2% over the four transportation modes.

The proposed decision tree with sequence smoothing model can be seen as a variant of the Hidden Markov Model introduced in Section 6.2.2.2, except HMM learns the state emission probability $p(\mathbf{x}_t|y_t)$ and the state transition probability $p(y_{t+1}|y_t)$ at the same time. HMM is also extensively used in trajectory-based activity recognition. For example, HMM can be used to predict the taxi's activity state such as occupied and non-occupied. In particular, a taxi's GPS trajectory is segmented into a sequence of time slices. At each time slice, similar to the above mentioned transportation mode detection algorithm, some features such as the point-of-interest information around the trajectory, taxi velocity and heading change rate are extracted. Then, an HMM is used to formulate the taxi state transition probabilities and the state emission probabilities. An overall recognition accuracy of 75% is observed by using HMM, and it is close to the recognition accuracy from some human experts.

Dynamic Bayesian Networks

Dynamic Bayesian network is a natural extension to the standard Hidden Markov Model, and is also extensively used in many trajectory-based activity recognition problems. For example, Yin et al. develop a location-based activity recognition algorithm, which models the sensor-location-activity dependencies with some dynamic Bayesian network model [44]. In particular, as shown in Figure 6.6, they design a two-level Bayesian network that uses a sensor model to estimate the locations at the lower level and a goal recognition model to predict the user goals at the higher level. The shaded nodes SS denote some user's received signal strengths from the WiFi access points installed in the environment as shown in Figure 6.7. All the other variables, including the user's physical location L, the user action A and the user goal G, are all hidden. Their values are to be inferred from the raw signal data.

At the lower level of the DBN model, given the WiFi received signal strength values $SS_t = <ss_{t,1},\ldots,ss_{t,N}> \in \mathbb{R}^N$ from N access points at time t, the proposed DBN model first predicts the mobile user's location L by using some Naive Bayes model $P(SS_t|L_{t,i}) = \prod_{j=1}^{N} P(ss_{t,j}|L_{t,i})$. Then, based on a series of location predictions, the model infers the user's action such as going to "Hall way 2" or "Elevator 3", etc. Such inference can be obtained by Expectation-Maximization as shown in [44]. At the higher level, the action sequences are taken as input to predict goals G. As the complexity of a DBN model is exponential in the number of hidden variables, the above estimations of these conditional probabilities can be computationally expensive. An alternative choice to infer these goals in the DBN is to use a separate N-gram model. Suppose that the actions $A_1,\cdots,A_T$ are inferred from the raw WiFi signal sequences. The most likely goal G^* can be obtained as follows:

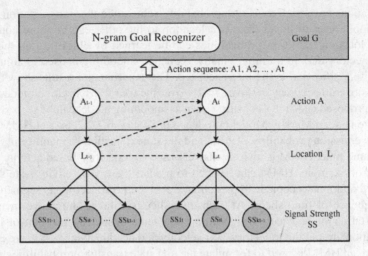

Fig. 6.6 The DBN + N-Gram model for activity recognition [44].

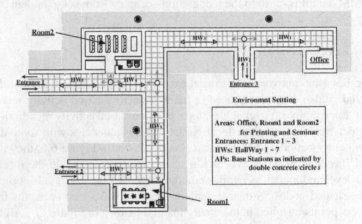

Fig. 6.7 The map for the office area of HKUST CSE department [45].

$$G^* = \underset{G_k}{\operatorname{argmax}} P(G_k|A_1, A_2, \cdots, A_t) = \underset{G_k}{\operatorname{argmax}} P(G_k|A_{1:t}).$$

By applying Bayes Rule, the above formula becomes:

$$G^* = \underset{G_k}{\operatorname{argmax}} \frac{P(A_{1:t}|G_k)P(G_k)}{P(A_{1:t})} = \underset{G_k}{\operatorname{argmax}} P(A_{1:t}|G_k)P(G_k),$$

where the conditional probability $P(A_{1:t}|G_k)$ can be simplified with a N-gram model:

$$P(A_{1:t}|G_k) = P(A_t|A_{t-1}, \ldots, A_1, G_k) \cdot P(A_{t-1}|A_{t-2}, \ldots, A_1, G_k) \cdots P(A_1|G_k).$$

Hence, an action A_t only depends on the goal G and its previous actions A_{t-1}, A_{t-2}, $\ldots, A_{t-n+1}$. When $n = 2$, the Bigram model is as follows:

$$G^* = \underset{G_k}{\operatorname{argmax}} \, P(G_k) P(A_1|G) \prod_{i=2}^{T} P(A_i|A_{i-1}, G_k).$$

The above inference's computational complexity is reduced to be linear in the number of goals and in the length of an action sequence, and thus can be efficiently performed.

Conditional Random Fields

Conditional Random Fields (CRF) is another important sequence learning model. Different from the generative models such as Hidden Markov Model and Dynamic Bayesian Networks, CRF is a discriminative model, and it is shown to perform well in various sequence labeling tasks including natural language processing and activity recognition. In this part, we introduce how CRF can be used in location-based activity recognition with GPS data [24].

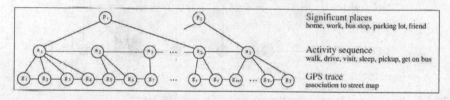

Fig. 6.8 Recognizing activities and significant places from the GPS trajectories.

As shown in Figure 6.8, Liao et al. show that they can employ a hierarchical model for location-based activity recognition. There are three levels in the model:

- **GPS trajectories** are sequences of some latitude and longitude coordinates in the map. As the raw GPS points can be noisy and less informative, a map-matching algorithm based on linear-chain CRF is employed to group the consecutive raw GPS points within every 10 meters into some segment, and further align it to some street patch in the map. Later, these street patches are used for feature extraction in recognizing the activities and significant places.
- **Activities** are estimated for the street patches obtained from the segmented G-PS trajectory. The activities such as "walk", "drive" and "sleep", after being recognized, are further used to infer the significant places.
- **Significant places** are those locations that play a significant role in the activities of a person. Such places include a person's home and work place, the bus stops and parking lots the person typically uses, the homes of friends, stores the person frequently shops in, and so on.

The whole activity and significant place recognition task is decomposed into two sub-tasks, including a map matching process which groups the raw GPS data into some informative street patches, and a joint recognition process which detects the activities and the significant places together from the trajectory data. In each subtask, a CRF model is employed to formulate the dependencies among the GPS trajectory features and the hidden variables of activities and significant places.

Step 1: Map matching.

As the raw GPS data can be noisy, the observed user location may not be always accurate. For example, a user's position can be shown to be deviated from a street even she is in the middle of it. Consequently, some data processing with map matching becomes necessary. Liao et al. propose to employ a CRF model to implement the map matching. As a discriminative model, one of the most important things in using the CRF model is to design reasonable feature functions, which can capture the domain knowledge. Each feature function is defined on a variable node clique of the graphical model as shown in Figure 6.9.

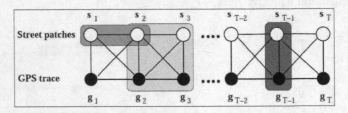

Fig. 6.9 Three types of features functions designed for map matching.

In particular, three types of feature functions are designed for this map matching subtask:

- **Measurement feature function** formulates the distance between a street patch s_{T-1} and a GPS point g_{T-1} for some time moment $t = T - 1$, as shown in the dark grey part in the figure:

$$f_{meas}(g_t, s_t) = \frac{|g_t - s_t|^2}{\sigma^2},$$

where *sigma* > 0 is used to control the scale of the distance. Generally, if the GPS point g_t is closer to the street patch center s_t, then the distance f_{meas} is smaller.

- **Consistency feature function** captures the temporal consistency of sequential GPS points, as shown in the light grey part in the figure:

$$f_{cons}(g_t, g_{t+1}, s_t, s_{t+1}) = \frac{|(g_{t+1} - g_t) - (s_{t+1} - s_t)|^2}{\sigma^2}.$$

This feature means that, if two GPS points (g_{t+1} and g_t) are close, their associated street patches (s_{t+1} and s_t) should be close too.

- **Smoothness feature function** constrain two consecutive street patches to be consistent, as shown in the medium grey part in the figure:

$$f_{smooth}(s_t, s_{t+1}) = \delta(s_t.street, s_{t+1}.street) \cdot \delta(s_t.direction, s_{t+1}.direction),$$

where $\delta(u, v)$ is an indicator function which equals 1 when $u = v$ and 0 otherwise.

With these feature functions, a CRF model formulates the likelihood function as:

$$p(s|g) = \frac{1}{Z} \exp \left\{ \sum_{t=1}^{T} w_m \cdot f_{meas}(g_t, s_t) \right.$$
$$\left. + \sum_{t=1}^{T-1} \left(w_c \cdot f_{cons}(g_t, g_{t+1}, s_t, s_{t+1}) + w_s \cdot f_{smooth}(s_t, s_{t+1}) \right) \right\},$$

where the model parameters w_m, w_c and w_s are to be learned by maximizing this likelihood function w.r.t. all the GPS sequences.

Step 2: Activity and significant place recognition.

After the map matching is ready, various useful features, such as average velocity, time of the day and so, can be extracted from each street patch. Such features are denoted as *evidence* in Figure 6.10, and further used to infer the activities and significant places with a new CRF model. This CRF, different from the one-layer CRF model used in map matching, contains two layers of labeling information about the activities and the significant places. Similarly, some feature functions are defined to capture the data dependencies. For example, one can design a measurement feature function for an activity a_i and the evidence's time feature t_i by $f(a_i, t_i) = \delta(a_i, Work) \cdot \delta(t_i, Morning)$. Similarly, a smoothness feature function can be defined on the transition of different activities: $f(a_i, a_{i+1}) = \delta(a_i, OnBus) \cdot \delta(a_{i+1}, Car)$. After having these features functions, one can train the CRF model by maximizing the data likelihood like map matching.

However, a major challenge in training this new CRF model is that in practice only the activity labeling is available while the significant place labeling is not, because the user usually does not label the significant places for his data. In other words, the training data for this CRF model only contain GPS street patches as the observations and the activities as the labels. In order to discover the significant places as well, Liao et al. propose to use an expectation-maximization (EM) algorithm to iteratively solve the problem. In each iteration, some initial guesses of the significant places are provided based on clustering with the observed activities. Then, these discovered significant places, together with the activities and the street patch evidences, are used to train the whole CRF model. The iteration continues until the inferred activities and the significant places do not change much.

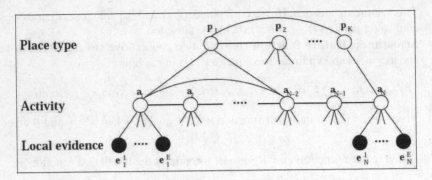

Fig. 6.10 The CRF for activity and significant place recognition [24].

Summary

Early work on supervised learning focuses on designing sophisticated learning models, e.g. the multi-layer DBNs in [25, 45]. A recent trend is paying more attention to incorporating domain knowledge as features, e.g. the heading change feature in [47] and the accelerometer data features [35]. It is believed that good features can lead to good activity recognition results, such as higher classification accuracy. Taking transportation mode as an example, Liao et al. designed a baroque DBN model to encode domain knowledge [25] while the recent research [47, 35] uses more domain-specific features. An additional advantage of the latter approach is that it is easier to implement and standardized.

6.3.1.2 Unsupervised Learning

Unsupervised learning is also used in trajectory-based activity recognition in order to discover some useful activity patterns. In general, this category of learning models take only the trajectory data as input and do not use any activity label information. Therefore, the outputs of these models are some data clusters [16] or low-dimensional feature representations [11]. Such outputs may not correspond to exact activity labels, but may imply some underlying activity patterns. In this section, we introduce three typical methods in this category: K-means clustering, Principal Component Analysis (PCA), and Latent Dirichlet Allocation (LDA). We use a representative project to introduce each kind of methods.

Clustering Methods

One intuitive unsupervised learning model is clustering, which can uncover structure in a set of data examples by grouping them according to some distance metric. Clustering methods are also used in activity recognition. For example, Huynh et al.

study how to use K-means clustering to rank individual features of accelerometer trajectory data according to their discriminative power in activity recognition [16]. In particular, the movement trajectory sensed by 3D accelerometers is usually segmented into a sequence of time slices, and the basic statistics values such as mean and variance are calculated as the features for each time slice. For different window sizes, many features could be defined. Huynh et al. use K-means clustering method to evaluate the effectiveness of each feature. Ideally, the data examples in each cluster should come from the same activity class. Such an intuition can be defined as:

$$P_{i,j} = \frac{|C_{i,j}|}{\sum_j |C_{i,j}|},$$

where $C_{i,j}$ is the set of examples in cluster i labeled with activity j. The cluster precision for activity j is defined as the weighted sum over different clusters:

$$p_j = \frac{\sum_i p_{i,j} |C_{i,j}|}{\sum_i |C_{i,j}|}.$$

If an activity has a cluster precision close to one, this indicates that there are many clusters mainly consisting of samples for this activity. Therefore, such a clustering process can help to decide which features to use for further activity recognition.

Principal Component Analysis

The MIT RealityMining project [11] aims to model conversation context, proximity sensing, and temporospatial location throughout large communities of individuals. Among others, it provides a very valuable dataset, which contains cellphone usage logs of one hundred MIT faculty members and students over nine months. The dataset is available online for other researchers to use[4]. For each user, the log records phone calls and short messages, phone status information, proximity to Bluetooth devices, and celltower IDs. The celltower ID indicates a coarse location of each user, based on which a user spatial trajectory for each user can be built. Figure 6.11 illustrates a single user's daily trajectories over 113 days. The celltower IDs are further associated with activity and status labels: Work, Home, No Signal, etc.

The Eigenbehavior project studies the patterns in the location trajectories for each user. For each user, a behavior dataset $\Gamma = \{\Gamma_1, \Gamma_2, \ldots, \Gamma_D\}$ is built where D is the number of days and Γ_i is a H-dimensional binary vector. Each bit in the behavior vector Γ_i indicates whether one of the above five status states is one in a specific hour. Similar to face recognition research such as eigenface learning [40], we can get the average behavior of a user via Principle Component Analysis (PCA). In this analysis, we have $\Phi = \frac{1}{D} \sum_{i=1}^{D} \Gamma_i$, and we build difference behavior vectors as $\Phi_i = \Gamma_i - \Phi$. An eigenbehavior is then the eigenvectors of the covariance matrix AA^T where the

---

[4] http://reality.media.mit.edu/download.php.

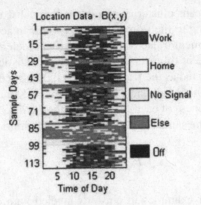

Fig. 6.11 A user's daily trajectories over 113 days [10].

matrix $A = [\Phi_1, \Phi_2, \ldots, \Phi_D]$. Figure 6.12 shows the top three eigenbehaviors of one subject in the dataset. The length of each vector is H, and is decomposed into five segments for visualization. We can clearly see that Eigenbehavior #1 represents a normal routine: at home from 20pm to 8am, at work from 8am to 20pm, and occasionally being at elsewhere. Thus, Eigenbehavior #1 represents a typical workday pattern. In contrast, Eigenbehavior #2 has large values at "elsewhere" part of the vector, which represents a weekend pattern.

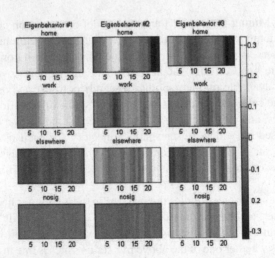

Fig. 6.12 The three eigenvectors generated by PCA [10].

Latent Dirichlet Allocation

Farrahi and Gatica-Perez [12] apply Latent Dirichlet Allocation (LDA), which is an unsupervised model, to extract the activity routines from the RealityMining data set. LDA is a probabilistic modeling method for unsupervised learning. It is originally developed to perform topic modeling on a text corpus [3]. Later, researchers found it useful for non-text data mining as well; e.g., it can be used for tagging images [2] and for accelerometer based routine recognition [15]. LDA is usually explained in the language of a text domain. Given a collection of text documents, where each document contains a set of words, the output of a LDA algorithm is to find K latent topics $Z = \{z_1, z_2, \ldots, z_K\}$, which each topic z_i is a distribution over words w_j via a conditional probability $P(w_j|z_i)$. Different topics favor different words. For example, the *politics* topic may have a larger value for the probability for words such as "president" and "war", while the *computer* topic may have a larger probability value for words such as "java" and "Microsoft".

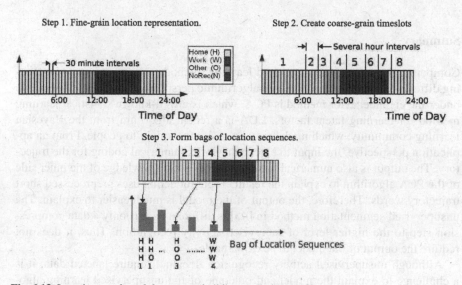

Fig. 6.13 Location words and documents building procedures [12].

In activity recognition, each day's sensor data is analogous to a document and each time slot represents a word. As shown in Figure 6.13, Farrahi and Gatica-Perez divide a day into 30-minute time slots and assigned each slot with a single location label that occurred for the longest duration. Here, the location labels include home, work, other and no-reception. Consequently, they treat each day's location data as a document, which consists of several location words. Each location word is a four-dimensional vector, containing three consecutive locations within a coarse time slot as well as the time slot ID. Then, the LDA algorithm was applied to find the activity routines, or so-called topics, from the data. Each day can be seen as a mixture of the topics. The transition pattern is encoded as a word. The more frequent a word occurs, the more likely the pattern it encodes. Table 6.3 shows an example result of topic modeling. For example, the pattern "WOO7" means that there is a high transition probability from Work to other activities during the time of 7pm to 9pm. The words in each topic are also similar; for example the pattern words in topic 2 mainly represent Work-to-Other activity during 7-9pm.

Table 6.3 Activity routines illustrated in four topics [12].

Topic 2 - LDA		Topic 3 - LDA		Topic 23 - LDA		Topic 183 - LDA		Topic 171 - LDA	
Word	$p(w\|z)$	Word	$p(w\|z)$	Word	$p(w\|z)$	Word	$p(w\|z)$	Word	$p(w\|z)$
W W W 6	0.548	W W W 5	0.462	H H H 2	0.528	W W W 1	0.920	W W O 4	0.300
W O O 7	0.212	W W O 6	0.255	H W W 3	0.212	W W W 2	0.020	O W W 4	0.290
W W O 7	0.196	W O O 6	0.231	H H W 3	0.201	W W O 2	0.013	W O W 4	0.273
O O W 3	0.003	O W W 4	0.003	H H H 3	0.022	W O O 2	0.008	O O W 3	0.046

Work-Out 7-9pm   Work-Out 5-7pm   Home-Work 9-11am   Work in morning   Work-Out 9am-2pm

Summary

Comparing the different unsupervised learning methods, we can see some interesting differences among them. From an algorithmic perspective, the working principle under the eigenbehavior method is PCA, which is a well-known statistical learning method for learning latent factors. LDA is a recent algorithm from the Bayesian learning community, which model may be easier to explain to people. From an application perspective, the input to eigenbehavior is numerical coding for the trajectory. The output is also numerical, which requires good knowledge of the inner side of the PCA algorithm to explain the result. Topic modeling uses preprocessed short trajectory words. Therefore, the output of the model is much easier to explain. The unsupervised segmentation method in [45] is different as it is only a data compression step to the higher level of supervised activity recognition. Thus, it does not require the output of the model to be explainable.

Although unsupervised activity recognition does not require labeled data, it is a challenge to explain the model and outcome of the unsupervised learning algorithm to people. For example, in [10], the outcome of the PCA model is a set of eigenvectors, which indicates that each day's user behavior can be seen as a linear combination of the some activity routines such as midnight to 9:00 at home, 10:00 to 20:00 at work, etc. This is a key insight why PCA can be used to extract and explain the behaviors. The transformation of daily activity representation in Figure 6.13 to the binary vectors is the key idea for PCA. In LDA , a challenge lies in how to define the pattern word.

6.3.1.3 Frequent Pattern Mining

Over the years, there has been much research work using association rule and frequent pattern mining for spatial and temporal data [7, 29]. Sequential activity pattern mining from trajectory data also belongs to this category. Here we shall take a recent study on mining the animal periodic patterns from trajectories [22, 21] as an example to start introducing this area. Figure 6.14 shows the location history of a bald eagle. Li et al. aim to analyze such a spatial trajectory and find some periodic activity patterns of the eagle [22]: for example, from December to March the eagle stays in the New York area, and some time later, it may leave to some other place etc.

Fig. 6.14 One trajectory from the bald eagle [22].

In order to achieve this goal, Li et al. propose to first discover meaningful reference spots where the eagle often stays, and then try to detect the periods from the binary movement sequence (indicating whether the eagle is in this reference spot at some time t) within each reference spot. Finally, all the movement sequences across the reference spots with the same period will be put into some clustering algorithm in order to find the frequent patterns. The details for each step are given as follows:

- *Finding reference spots.* In order to get the reference spots, they divide the map into $w \times h$ cells of the same size. For each cell c, the GPS data density can be calculated as:

$$f(c) = \frac{1}{n\gamma^2} \sum_{i=1}^{n} \frac{1}{2\pi} \exp(-\frac{|c - loc_i|^2}{2\gamma^2}),$$

  where $|c - loc_i|$ is the distance between cell center c and location loc_i. γ is a smoothing factor defined as $\gamma = \frac{1}{2}(\sigma_x^2 + \sigma_y^2)^{1/2} n^{-1/6}$. Therefore, if there are more GPS points from the eagle in this cell, the density is higher. Besides, if the GPS points are closer to the cell center, the density is also higher. After obtaining the density values, an reference spot can be defined by a contour line on the map, as shown in Figure 6.15(a), which joins the cells of the equal density value, with some density threshold. The threshold can be determined as the top-$p\%$ density value among all the density values of all cells. The larger the value p is, the bigger the size of reference spot is.

- *Detecting periods on binary movement sequence.* Given a single reference spot, the eagle's movement sequence can be transformed into a binary sequence $B = b_1 b_2 \ldots b_n$, where $b_i = 1$ when the eagle is within the reference spot at time i and 0 otherwise. Such a sequence can be used to find the sequence period patterns (e.g. in days) by combining Fourier transform and autocorrelation. We refer readers to the details of such period discovery in the paper [22].

- *Mining periodic patterns.* Let $O_T = \{o_1, \ldots, o_d\}$ denote reference spots with the same period T such as a day. Given $LOC = loc_1 \ldots loc_n$, one can generate the corresponding symbolized movement sequence $S = s_1 \ldots s_n$, where $s_i = j$ if loc_i is within o_j. S is further segmented into $m = \lfloor \frac{n}{T} \rfloor$ segments so that each segment can represent a period such as a "day" if $T = 24$. Finally, the periodic activities can be found by using hierarchical agglomerative clustering to group these segments.

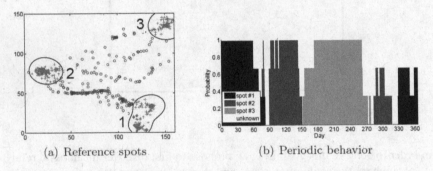

(a) Reference spots   (b) Periodic behavior

Fig. 6.15 Reference spots and pattern distribution [22].

Figure 6.15(b) shows the daily behavior pattern of the bald eagle over one year, indicating that this eagle stays in New York area (i.e., reference spot 1) from December to March. In March, it flies to Great Lakes area (i.e., reference spot 2) and stays there until the end of May. It flies to Quebec area (i.e., reference spot 3) in the summer and stays there until late September. Then it flies back to Great Lake again staying there from mid-October to mid-November and goes back to New York in December.

6.3.2 Multiple-user Activity Recognition

As mobile devices and sensors become extensively available, there are more and more data collected from different users. In the past, people have considered using all the users data together to train an activity recognition model. For example, Liao et al. pool all the users' GPS trajectory data together and train a hierarchical activity model to predict some transportation routine activities [25]. Such a method may work well if there is no big difference among different users' activities. However, this is not always true. For example, in WiFi-based activity recognition, a user visits the coffee shop for meal and the other just enjoys sitting in its outdoor couches to read research paper. These two users are very likely to observe similar WiFi signals, but their activities are quite personalized. This motivates us to consider the user-user relationship in utilizing such multi-user data for activity recognition. In the following, we introduce some existing work that falls into this multi-user activity recognition category.

Coupled Hidden Markov Model for Concurrent Activity Recognition

Wang et al. [42] provide a concurrent activity recognition system based on coupled Hidden Markov Model. In particular, each user u_i, $1 \leq i \leq n$, has a sensor observation sequence $\{o_1, o_2, \cdots, o_T\}$ of length T. Each observation o_i has a label from a set of m activities $L = \{y_1, y_2, \cdots, y_m\}$. One can use a standard HMM to model each user's activity with the hidden states as the activity labels and the corresponding observations as the observations. To model the interaction and influence between users along the time, Wang et al. use the Coupled Hidden Markov Models (CHMM) to formulate the state transitions among different users. Figure 6.16 shows a CHMM for two users A and B. The two hidden state sequences $\{a_i\}$ and $\{b_i\}$ represent the activity sequences of user A and user B. The states from A and B are cross linked to capture the interactions between the two users. For example, the activity of B at time t is influenced not only by B's state at time $t-1$, but also A's state at time $t-1$.

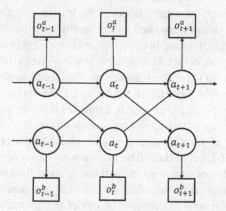

Fig. 6.16 A two-user Coupled Hidden Markov Model (CHMM) for activity recognition [42].

After such a CHMM model is built from the multi-user training data, the most probable label sequence for an observation sequence is $\arg\max_S P(S|O)$, which can be effectively solved by dynamic programming. Notice that in this work, although the training phase takes the concurrent activities of different users into consideration, the inference procedure only works for a single observation sequence.

Ensemble Learning

The DarwinPhone project takes a different approach towards multi-user activity recognition [27]. Different from [42], its inference is based on multi-user, i.e. when doing online activity recognition on one user's phone, it also uses the information from all the other nearby phones' data. In practice, different phones may have very different sensing information even at the same environment because of their positions and settings. For example, a phone inside a handbag may sense that its environment is quiet, but the other phone open in the air may sense a louder environment. In this case, using the activity recognition models on other nearby phones is very helpful to get more stable and more accurate recognition result. In the proposed method, each phone has a separate model for some activity. For some activity event, its nearby phones can pass the extracted features from their own phone sensors to it. Then, each received feature vector are used to generate an activity prediction, and finally a voting mechanism is used to decide which activity it is.

Transfer Learning

An important problem in activity recognition is that, every time when we want to build an activity recognition model for a new environment, we have to collect a labeled data set in it. This is because most activity recognition models are specific to each data set on which they are trained. If the training data set changes, for example given different sensors installed and different users having different activity patterns, the learned model in one house may not be applicable in another house. An interesting question to ask is that, is it possible to use some existing labeled data sets of various houses to help learn the parameters of a model applied in a new house? If possible, many data labeling efforts can be saved. Kasteren et al. give a positive answer to the above question by providing a transfer learning solution [17].

In the transfer learning based activity recognition task, there is a target house for which there is little or no training data available, and a number of source houses for which there is a lot of labeled data. The activities to be recognized in each house are the same, while the sensors used in each house are different. The task's goal is to use the source houses' data, together with the target house's possibly available training data, to train an activity recognition model for the target house. In order to achieve this goal, two challenges need to be addressed, including:

- The sensors used in different houses can be different;
- The activity patterns from multiple users in different houses can be different.

The first challenge leads to different feature spaces in different houses' data. In order to address this challenge, Kasteren et al. propose to introduce some meta-feature mapping function, which can map different sensor sets into a single common feature set that can be used for all houses. In particular, each sensor is described by one or more meta features, for example, a sensor on the microwave might have one meta feature describing that the sensor is located in the kitchen, and another that the sensor is attached to a heating device. The second challenge describes the activity differences given a series of similar sensor observations. For example, one person might often have cereal for breakfast, while another prefers toast, though they can trigger a series of similar sensor events w.r.t. their similar moving trajectories in the houses. Such activity differences require different sets of parameters to allow the model to recognize the corresponding activities. Therefore, Kasteren et al. propose to use a separate model for each house, and further assume that the target house's model and the source houses' model parameters share some common prior distribution. By learning the prior distribution from the source houses, one can use it to provide a reasonable initial value for the target house's model, and meanwhile make the new model be able to capture the unique activity patterns in the target house.

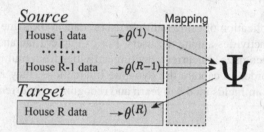

Fig. 6.17 The illustration for the transfer learning algorithm [17].

The detailed transfer learning model is illustrated in Figure 6.17. In each house, a Hidden Markov Model (HMM) is used to model the sensor sequence data together with the activities. Let us denote x_t as a sensor feature vector at time t, and y_t as its corresponding activity label. Here, each house's sensors are mapped to some common features; in other words, the feature spaces of x at each house, after the meta-feature mapping, are now the same. A HMM formulates the generative probability

$$p(\mathbf{y}_{1:T}, \mathbf{x}_{1:T}) = p(y_1) \prod_{t=1}^{T} p(\mathbf{x}_t|y_t) \prod_{t=2}^{T} p(y_t|y_{t-1}),$$

where $p(y_1)$ is the prior state distribution parameterized by π, $p(\mathbf{x}_t|y_t)$ measures the emission probability parameterized by B and $p(y_t|y_{t-1})$ measures the transition probability parameterized by A. Therefore, a HMM is denoted as $\theta = \{\pi, A, A\}$. Notice that each house has a HMM parameterized as θ_j, the proposed transfer learning model assigns some common prior distributions to these HMM parameters θ_j. For example, Kasteren et al. design each prior state probability as $p(y_1) = \prod_{i=1}^{K} \pi_i^{\delta(y_1-i)}$, where K is the number of states, $\delta(s) = 1$ if $s = 1$ and 0 otherwise. Then, they force such prior state probabilities from different HMM's to share a same Dirichlet prior distribution, which is given by $Dir(\pi|\eta) = \frac{\Gamma(\Sigma_{k=1}^{K}\eta_k)}{\Gamma(\eta_1)...\Gamma(\eta_K)}\prod_{k=1}^{K}\pi_k^{\eta_k-1}$ with parameters η. Here, $\Gamma(\cdot)$ is a Gamma distribution and we refer readers to the details in the paper. As the parameters η are shared across different houses, one can use different houses' data to learn them and thus be able to estimate the prior state distribution $\pi^{(j)}$ at each house j. Similarly, Kasteren et al. also assign a beta prior distribution to the emission probability and a Dirichlet prior distribution to the transition probability. These prior distributions' parameters are then learned using multiple houses' data and used to estimate the HMMs for each house [17].

Finally, three real world data sets are used to evaluate the proposed transfer learning solution. The experimental results show that, the proposed method can give good performance in activity recognition for a house with little or no labeled data, and generally outperforms some competing baselines. The datasets collected by the authors are also available on line[5].

Factorial Conditional Random Fields

One important application of using multiple users' data at the same time is to model the concurrent activities among the users. For example, Lian and Hsu develop a methodology to recognize concurrent chatting activities from multiple users' audio streams [23]. In order to capture the dynamic interactions, they adopt a factorial Conditional Random Fields model to learn and recognize concurrent chatting activities.

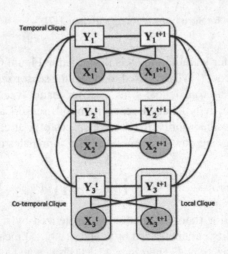

Fig. 6.18 A sample FCRF of 3 concurrent chatting activities [23].

Figure 6.18 shows a sample FCRFs model for the recognition of concurrent chatting activities among three users. Denote X as the acoustic feature variables and Y as the set of chatting activity variables. Let $x_i^t \in X$ denote an observed acoustic sensor feature value at time t for chatting activity i from a user. Let y_i^t be the corresponding state of chatting activity i. Let $G = (V, E)$ be an undirected graph structure shown in Figure 6.18, with V as the node set and E as the edge set. For any given time slice t and chatting activity i, Lian and Hsu build edges $(Y_i^t, Y_i^{t+1}) \in E$ to represent the possibility of activity state transition across time slices. They also build edges $(X_i^t, Y_i^t) \in E$ to represent the possible relationships between activity labels and acoustic observations. Besides, edges $(Y_i^t, Y_j^t) \in E$ are built to represent the possibility of co-temporal relationships between any two concurrent chatting activities i and j. That is, all the hidden nodes within the same time slice are fully connected.

An FCRF model allows us to design various feature functions in formulating the conditional probabilities $p(y|x)$ for a sensor feature x and its chatting activity state y. In particular, based on the propoesd FCRF model shown in Figure 6.18, Lian and Hsu propose to design three types of feature functions:

- Local potential function $\phi_i^A(x_i^t, y_i^t, t) = \exp\left(\sum_{p=1}^P w_i^{(p)} f_i^{(p)}(x_i^t, y_i^t, t)\right)$, where $f_i^{(p)}$ is a function indicating whether the state values are equal to the p-th state combination within the local clique (x_i^t, y_i^t). $w_i^{(p)}$ are some weights to learn later.
- Temporal potential function

$$\phi_i^B(x_i^t, y_i^t, x_i^{t+1}, y_i^{t+1}, t) = \exp\left(\sum_{q=1}^Q w_i^{(q)} f_i^{(q)}(x_i^t, y_i^t, x_i^{t+1}, y_i^{t+1}, t)\right),$$

where $f_i^{(q)}$ is a function indicating whether the state values are equal to the q-th state combination within the temporal clique $(x_i^t, y_i^t, x_i^{t+1}, y_i^{t+1})$. $w_i^{(q)}$ are some weights to learn later.
- Co-temporal potential function

$$\phi_i^A(x_i^t, y_i^t, x_j^t, y_j^t, t) = \exp\left(\sum_{r=1}^R w_{ij}^{(r)} f_{ij}^{(r)}(x_i^t, y_i^t, x_j^t, y_j^t, t)\right),$$

where $f_{ij}^{(r)}$ is a function indicating whether the state values are equal to the r-th state combination within the co-temporal clique $(x_i^t, y_i^t, x_j^t, y_j^t)$. $w_{ij}^{(r)}$ are some weights to learn later.

Finally, an FCRF model formulate the following conditional probability for N users' concurrent chatting activities:

$$p(y|x, w) = \frac{1}{Z(x)} \cdot \left(\prod_{t=1}^T \prod_{i=1}^N \phi_i^A(x_i^t, y_i^t, t)\right) \cdot \left(\prod_{t=1}^{T-1} \prod_{i=1}^N \phi_i^B(x_i^t, y_i^t, x_i^{t+1}, y_i^{t+1}, t)\right)$$

$$\cdot \left(\prod_{t=1}^T \prod_{i,j} \phi_i^A(x_i^t, y_i^t, x_j^t, y_j^t, t)\right),$$

where w denote the parameter set of $w_i^{(p)}$, $w_i^{(q)}$ and $w_{ij}^{(r)}$. $Z(x)$ is a normalization factor. By learning the parameters w based on the observed audio sequence data, one can predict the chatting activity state of a particular user at each time stamp. We refer interested readers to the technical details of learning and inference of such an FCRF model in [23]. Some experiments based on two concurrent chatting activity data sets show that, the proposed FCRF model is consistently better than some competing baselines such as Coupled Hidden Markov Model [5] in the comparison of F-score.

Latent Aspect Model

In [46], Zheng and Yang propose a user-dependent aspect model to help the users collaboratively build an activity recognition model that can give personalized predictions. Rather than simply pooling multiple users' data together, the proposed model introduces user aspect variables to capture the user grouping information from their data. As a result, for a targeted user, the data from her similar users in the same group can also help with her personalized activity recognition. In this way, one can greatly reduce the need of much valuable and expensive labeled data required in training the personalized recognition model.

In a WiFi environment, multiple users collect wireless signal data with activity annotations for around a month. The data format is in a set of quads: $\{\langle a_i, u_i, f_i, t_i \rangle | i = 1, ..., L\}$, where a is an activity, u is a user, and f is a feature observed at time t. In the WiFi case, a feature corresponds to a wireless access point (AP) that the mobile device can detect. A data record quad indicates that a user u is doing an activity a at time t, and meanwhile her wireless device detects some AP f. The goal is to build a personalized activity recognition model by using these data, so that with a user's WiFi observations at some time, we can predict what she is doing.

The proposed model extends the standard aspect model [13] by introducing user aspects, as well as time aspects and feature aspects to model personalized activity recognition from time-dependent sensor data. Figure 6.19 depicts the graphical model. The shadow nodes for user variables u, time variables f, feature variables f and activity variables a are observations. The blank nodes inside the rectangle are latent aspect variables. The user latent aspects $Z_u \in \{z_u^1, z_u^2, ..., z_u^{D_u}\}$ are discrete variables, indicating D_u user clusters. The model adopts such a *user-cluster-activity* hierarchy to help the users to collaboratively build an activity recognizer. In contrast to a two-tier *user-activity* hierarchy where each user can only rely on herself to do activity recognition, the proposed model can make the users from a same cluster to contribute together to train the recognizer from their feature and time observations. Therefore, even if some user has limited data to train an activity recognition model, she can still benefit from other similar users in the same group(s). As each user can belong to multiple user clusters at the same time with different probabilities, they actually contribute differently to each user cluster in training the recognition model and consequently get different predictions in real-time recognition. This helps to achieve the model personalization.

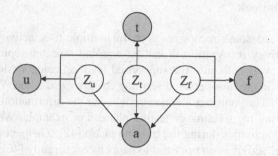

Fig. 6.19 User-dependent aspect model.

Some other latent aspects $Z_f \in \{z_f^1, z_f^2, ..., z_f^{D_f}\}$ and $Z_t \in \{z_t^1, z_t^2, ..., z_t^{D_t}\}$ are also introduced to encode the data observations on feature and time. They are used to capture the dependency between activities and observations, considering that similar feature observations at similar time periods are likely to imply some same activity. Note that these aspects do not necessarily rely on users. One can also take all the users' data as input, and only use them (i.e. Z_f and Z_t) to build a user-independent model for general activity recognition. However, such a user-independent model, as shown in their empirical experiment, does not perform as well as the user-dependent model. The user latent aspects help to achieve personalization, so it is called a user-dependent aspect model.

In general, the proposed aspect model is a generative model, which uses the latent aspect variables to explain the observations. It specifies a joint probability of the observed random variables:

$$P(a, u, f, t) = \sum_{Z_u, Z_f, Z_t} P(a, u, f, t, Z_u, Z_f, Z_t), \tag{6.1}$$

where $P(a, u, f, t, Z_u, Z_f, Z_t)$ is expanded, according to the graphical model, as follows:

$$\begin{aligned} P(a, f, u, t, Z_u, Z_f, Z_t) &= P(Z_u)P(Z_f)P(Z_t) \\ &\quad P(u|Z_u)P(f|Z_f)P(t|Z_t)P(a|Z_u, Z_f, Z_t). \end{aligned} \tag{6.2}$$

Here, the user variables u, feature variables f and activity variables a are all discrete in nature, so their conditional probabilities on latent aspects can be modeled easily by multi-nominal distributions. One exception is the time, which could be continuous. To formulate $P(t|Z_t)$, we discretize the time t with two possible strategies. One is "ByHour", which segments the time into hours. The other is "ByPeriod", which segments it into larger time periods; for example, one can define five periods, including morning (7am~11am), noon (11am~2pm), afternoon (2pm~6pm), evening (6pm~12am) and night (12am~7am).

Summary and Outlook

We have introduced some recent research in multiple-user activity recognition. Multiple-user activity recognition is still an ongoing research topic. Most of the existing research work falls into the supervised learning category, leaving a lot of future work to do in the unsupervised learning category and so on. We give a summary in Table 6.4. The summary focuses on whether the information from multiple users is used during the training or testing phase. For example, Wang et al. only model concurrent activities during the training phase [42]. Zheng et al. model multiple users, but the activities do not need to happen concurrently [46]. In Miluzzo et al.'s work [27], the multiple user's data are only used in the inference (or, testing) phase. Lian et al.'s work [23] [46] supports multi-user modeling for training and testing phases. However, the model in [46] is more general as it does not require the activity data to be observed at the same time. Kasteren et al.'s work [17] is different from the other four. It follows the transfer learning stetting, where training data and testing data are of different distributions.

Table 6.4 Comparison of different multi-user activity recognition projects.

Research work	Training	Testing
Wang et al. [42]	Multi-user	Single-user
Miluzzo et al. [27]	Single-user	Multi-user
Kasteren et al.[17]	Multi-user	Single-user
Lian et al.[23]	Multi-user	Multi-user
Zheng et al.[46]	Multi-user	Single-user

Much future work can be done in multi-user activity recognition. One particularly interesting direction is social activity recognition, which aims to use the social media information as input and predict the users' physical activities. In social activity recognition, the scales on both users and data can be much larger than those using limited sensor or spatial trajectory data. For example, Sakaki et al. treat the Twitter users as social sensors, and use these sensors for event detection [36]. In their work, the words and links in the Twitter messages are used as sensor readings, and a spatial-temporal model is built on the Twitter message streams. They build an earthquake notification system which uses these Twitter messages as the input, and outputs the earthquake notification when there is one.

6.4 Summary

This chapter surveys the recent research in trajectory-based activity recognition. We start with introducing the learning-based localization methods, which aim to generate the spatial trajectories. Many machine learning techniques used in localization also reappear in the trajectory activity recognition. For example, HMM can be either used to smooth the location estimations [19] or used in sequential classifications for transportation mode detection [35]. The main part of this chapter discusses trajectory-based activity recognition for a single user and for multiple users. We have also categorized the previous research work according to the learning paradigms: supervised learning, unsupervised learning and frequent pattern mining.

Table 6.5 gives a summary of the applications based on trajectory activity recognition. Some of the applications are direct activity recognition problems, while the others can be more high-level and are based on activity recognition such as GeoLife. The second category of applications are usually built on the first category.

Table 6.5 Applications of Trajectory-based Activity Recognition

Applications
transportation mode detection. [47, 35]
goal recognition. [45]
animal moving patterns. [26]
daily living patterns. [12]
earthquick detection. [36]
smart houseing. [17]

Collecting the experimental data in trajectory activity recognition is usually tedious and expensive. For example, the MIT RealityMining data set has 100 subjects to collect their daily behavior over a period of nine months, which costs a lot of money and time. Fortunately some of data sets have been made public, and Table 6.6 lists some of them related to this chapter.

Table 6.6 Open datasets

Dataset
Geolife [48]
RealityMining [11]
SmartHome [29]
HouseTransfer [17]
WiFiLocalization [43]

Trajectory-based activity recognition is a great application area for many machine learning algorithms. Table 6.7 lists the algorithms used in the previous research work covered by this chapter. Most of the algorithms listed in the table are for sequential data. Some non-sequential algorithms such as KNN and Decision trees can also be used to give predictions for each trajectory segment independently, and also shown to have reasonable performances.

One contribution of this article is the categorization of trajectory-based activity recognition. We believe that multiple-user activity recognition will be an important future direction in this area. Besides digital sensors, social media such as Twitter contains very rich information about the users' activities, and thus can be used for further activity recognition. Introducing the social media data can greatly relieve

Table 6.7 Machine learning and data mining algorithms

Algorithm	Used in	Classical reference
KNN	[1]	
Decision Tree	[35]	[33, 6]
HMM	[19, 47, 35]	[34]
CHMM	[42]	[5]
DBN	[45]	[28]
CRF	[24]	[20]
FCRF	[23]	[38]

the data sparsity in activity recognition, but it also brings the large-scale data computation problem. Some cloud computing solution could be an option. Besides, as the social media data are very noisy, more careful data cleaning and information retrieval are vital.

References

1. Bahl, P., Padmanabhan, V.N.: Radar: An in-building rf-based user location and tracking system. In: Proc. The Annual IEEE International Conference on Computer Communications (INFOCOM), pp. 775–784 (2000)
2. Barnard, K., Duygulu, P., Forsyth, D.A., de Freitas, N., Blei, D.M., Jordan, M.I.: Matching words and pictures. Journal of Machine Learning Research **3** (2003)
3. Blei, D.M., Ng, A.Y., Jordan, M.I.: Latent dirichlet allocation. Journal of Machine Learning Research **3**, 993–1022 (2003)
4. Boyd, S., Vandenberghe, L.: Convex optimization. Cambridge Univ Pr (2004)
5. Brand, M., Oliver, N., Pentland, A.: Coupled hidden markov models for complex action recognition. In: Proceedings of the 1997 Conference on Computer Vision and Pattern Recognition (CVPR '97), CVPR '97, p. 994. IEEE Computer Society, Washington, DC, USA (1997)
6. Breiman, L.: Classification and regression trees. Chapman & Hall/CRC (1984)
7. Cao, H., Mamoulis, N., Cheung, D.: Mining frequent spatio-temporal sequential patterns. In: Proc. of IEEE International Conference on Data Mining (ICDM), pp. 8–pp. IEEE (2005)
8. Doucet, A., De Freitas, N., Gordon, N.: Sequential Monte Carlo methods in practice. Springer Verlag (2001)
9. Doucet, A., Godsill, S., Andrieu, C.: On sequential monte carlo sampling methods for bayesian filtering. Statistics and computing **10**(3), 197–208 (2000)
10. Eagle, N., Pentland, A.: Eigenbehaviors: Identifying structure in routine. Behavioral Ecology and Sociobiology **63**(7), 1057–1066 (2009)
11. Eagle, N., (Sandy) Pentland, A.: Reality mining: sensing complex social systems. Personal Ubiquitous Comput. **10**, 255–268 (2006)
12. Farrahi, K., Gatica-Perez, D.: Discovering routines from large-scale human locations using probabilistic topic models. ACM Transactions on Intelligent Systems and Technology (TIST) **2**(1), 3 (2011)
13. Hofmann, T., Puzicha, J.: Latent class models for collaborative filtering. In: Proc. of the 16th International Joint Conference on Artificial Intelligence (IJCAI '99), pp. 688–693 (1999)
14. Hofmann-Wellenhof, B., Lichtenegger, H., Collins, J.: Global positioning System. Theory and Practice. (1993)
15. Huynh, T., Fritz, M., Schiele, B.: Discovery of activity patterns using topic models. In: Proc. of International Conference on Ubiquitous Computing (UbiComp), pp. 10–19 (2008)

16. Huynh, T., Schiele, B.: Analyzing features for activity recognition. p. 159C163. Smart objects and ambient in- telligence: innovative context-aware services (2005)
17. van Kasteren, T., Englebienne, G., Kröse, B.J.A.: Transferring knowledge of activity recognition across sensor networks. In: Proc. of the International Conference of Pervasive Computing, pp. 283–300 (2010)
18. Krumm, J. (ed.): Ubiquitous Computing Fundamentals. Chapman and Hall/CRC, Boca Raton, FL (2010)
19. Ladd, A.M., Bekris, K.E., Rudys, A., Kavraki, L.E., Wallach, D.S., Marceau, G.: Robotics-based location sensing using wireless ethernet. In: Proc. of The Annual International Conference on Mobile Computing and Networking (MobiCom), pp. 227–238 (2002)
20. Lafferty, J.D., McCallum, A., Pereira, F.C.N.: Conditional random fields: Probabilistic models for segmenting and labeling sequence data. In: Proc. The International Conference on Machine Learning (ICML), pp. 282–289 (2001)
21. Li, Z., Ding, B., Han, J., Kays, R., Nye, P.: Mining periodic behaviors for moving objects. In: Proc. of the ACM Conference on Knowledge Discovery and Data Mining (KDD), pp. 1099–1108 (2010)
22. Li, Z., Han, J., Ji, M., Tang, L.A., Yu, Y., Ding, B., Lee, J.G., Kays, R.: Movemine: Mining moving object data for discovery of animal movement patterns. ACM Transactions on Intelligent Systems and Technology (ACM TIST) (Special Issue on Computational Sustainability) (Aug. 2010)
23. Lian, C.C., Hsu, J.Y.j.: Probabilistic models for concurrent chatting activity recognition. In: Proc. of the 21st International Jont Conference on Artifical Jntelligence (IJCAI '09) (2009)
24. Liao, L., Fox, D., Kautz, H.A.: Extracting places and activities from gps traces using hierarchical conditional random fields. I. J. Robotic Res. 26(1), 119–134 (2007)
25. Liao, L., Patterson, D.J., Fox, D., Kautz, H.A.: Learning and inferring transportation routines. Artif. Intell. 171(5-6), 311–331 (2007)
26. Liu, S., Liu, Y., Ni, L.M., 0002, J.F., Li, M.: Towards mobility-based clustering. In: Proc. of the ACM Conference on Knowledge Discovery and Data Mining (KDD), pp. 919–928 (2010)
27. Miluzzo, E., Cornelius, C., Ramaswamy, A., Choudhury, T., Liu, Z., Campbell, A.T.: Darwin phones: the evolution of sensing and inference on mobile phones. In: Proc. The Annual International Conference on Mobile Systems (MobiSys), pp. 5–20 (2010)
28. Murphy, K.: Dynamic bayesian networks: representation, inference and learning. Ph.D. thesis, UC Berkeley, Computer Science Division (2002)
29. Nazerfard, E., Rashidi, P., Cook, D.J.: Using association rule mining to discover temporal relations of daily activities
30. Ni, L.M., Liu, Y., Lau, Y.C., Patil, A.P.: Landmarc: Indoor location sensing using active rfid. Wireless Networks 10(6), 701–710 (2004)
31. Pan, J.J., Yang, Q., Pan, S.J.: Online co-localization in indoor wireless networks by dimension reduction. In: Proc. of National Conference on Artificial Intelligence (AAAI), pp. 1102–1107 (2007)
32. Patterson, D.J., Liao, L., Fox, D., Kautz, H.A.: Inferring high-level behavior from low-level sensors. In: Proc. of International Conference on Ubiquitous Computing (UbiComp), pp. 73–89 (2003)
33. Quinlan, J.: C4. 5: programs for machine learning. Morgan Kaufmann (1993)
34. Rabiner, L.: A tutorial on hidden markov models and selected applications in speech recognition. Proceedings of the IEEE 77(2), 257–286 (1989)
35. Reddy, S., Mun, M., Burke, J., Estrin, D., Hansen, M.H., Srivastava, M.B.: Using mobile phones to determine transportation modes. ACM Transactions on Sensor Networks (TOSN) 6(2) (2010)
36. Sakaki, T., Okazaki, M., Matsuo, Y.: Earthquake shakes twitter users: real-time event detection by social sensors. In: Proc. of International World Wide Web Conference, pp. 851–860 (2010)
37. Schulz, D., Fox, D., Hightower, J.: People tracking with anonymous and id-sensors using rao-blackwellised particle filters. In: Proc. of International Joint Conferences on Artificial Intelligence (IJCAI), pp. 921–928 (2003)

38. Sutton, C.A., Rohanimanesh, K., McCallum, A.: Dynamic conditional random fields: factorized probabilistic models for labeling and segmenting sequence data. In: Proc. The International Conference on Machine Learning (ICML) (2004)
39. Thiagarajan, A., Ravindranath, L., LaCurts, K., Madden, S., Balakrishnan, H., Toledo, S., Eriksson, J.: Vtrack: accurate, energy-aware road traffic delay estimation using mobile phones
40. Turk, M., Pentland, A.: Eigenfaces for recognition. Journal of cognitive neuroscience 3(1), 71–86 (1991)
41. Vail, D.L., Veloso, M.M., Lafferty, J.D.: Conditional random fields for activity recognition. In: Proc. the International Conference on Autonomous Agents and Multiagent Systems (AAMAS), p. 235 (2007)
42. Wang, L., Gu, T., Tao, X., Lu, J.: Sensor-based human activity recognition in a multi-user scenario. In: Proc. of the International Joint Conference on Ambient Intelligence (AmI), pp. 78–87 (2009)
43. Yang, Q., Pan, S.J., Zheng, V.W.: Estimating location using wi-fi. IEEE Intelligent Systems 23(1), 8–13 (2008)
44. Yin, J., Chai, X., Yang, Q.: High-level goal recognition in a wireless lan. In: Proc. of National Conference on Artificial Intelligence (AAAI), pp. 578–584 (2004)
45. Yin, J., Shen, D., Yang, Q., Li, Z.N.: Activity recognition through goal-based segmentation. In: Proc. of National Conference on Artificial Intelligence (AAAI), pp. 28–34 (2005)
46. Zheng, V.W., Yang, Q.: User-dependent aspect model for collaborative activity recognition. In: In Proc. of the 22nd International Joint Conference on Artificial Intelligence (IJCAI-11) (2011)
47. Zheng, Y., Li, Q., Chen, Y., Xie, X., Ma, W.Y.: Understanding mobility based on gps data. In: Proc. of International Conference on Ubiquitous Computing (UbiComp), pp. 312–321 (2008)
48. Zheng, Y., Xie, X.: Learning travel recommendations from user-generated gps traces. ACM Transactions on Intelligent Systems and Technology (TIST) 2(1), 2 (2011)

Chapter 7
Trajectory Analysis for Driving

John Krumm

Abstract This chapter discusses the analysis and use of trajectories from vehicles on roads. It begins with techniques for creating a road map from GPS logs, which is a potentially less expensive way to make up-to-date road maps than traditional methods. Next is a discussion of map matching. This is a collection of techniques to infer which road a vehicle was on given noisy measurements of its location. Map matching is a prerequisite for the next two topics: location prediction and route learning. Location prediction works to anticipate where a vehicle is going, and it can be used to warn drivers of upcoming traffic situations as well as give advertising and alerts about future points of interest. Route learning consists of techniques for automatically creating good route suggestions based on the trajectories of one or more drivers.

7.1 Introduction

GPS trajectories from vehicles are useful in a variety of ways to make driving better. It is easy to gather trajectories using small GPS loggers, and it can be even easier with GPS-equipped mobile phones and in-car personal navigation devices. Figure 7.1 shows a typical GPS trajectory recorded from a vehicle. Such trajectories typically consist of a sequence of time-stamped latitude/longitude points.

Many of the techniques discussed in this chapter can be considered crowdsourcing, which take a large collection of GPS traces from different drivers to create something useful. The chapter starts with techniques for making road maps from GPS trajectories. This is an alternative to expensive, special purpose road surveys that can have trouble keeping up with road changes. The chapter next discusses map matching. Given a road map, this is the problem of assigning GPS points to particu-

John Krumm
Microsoft Research, Microsoft Corpration, Redmond, WA USA
e-mail: jckrumm@microsoft.com

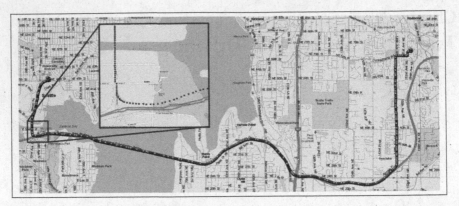

Fig. 7.1 The black dots show a GPS trajectory recorded in a vehicle, staring from the right side. The inset is a close-up of part of the trajectory, showing some of the individual points. In this case, the points were recorded every one second.

lar roads. Map matching is often an annoying, but sometimes necessary prerequisite to the chapter's next two topics: destination prediction and route learning. Destination prediction attempts to predict a driver's destination during a trip, with the goal of providing warnings and useful information in time for the driver to act. Route learning is a process that observes drivers' preferences based on GPS trajectories and uses these preferences to influences future route suggestions.

7.2 Making Road Maps from Trajectories

Digital road maps are one of the most important resources for assisting drivers, from planning a route on a desktop PC to real time route guidance in a car. Creating these maps is expensive, because it usually requires trained, dedicated personnel in specially equipped vehicles to drive the streets for the sole purpose of mapping. Technologists have explored other, less expensive methods of making digital road maps. One of the earliest attempts was to use aerial imagery, from airplanes and satellites, along with computer vision algorithms, to automatically find roads, such as the early work by Tavakoli and Rosenfeld [4].

Another inexpensive approach is manual crowdsourcing, as exemplified by the successful OpenStreetMap (OSM) project founded by Steve Coast [13]. OSM's registered users can update the map by editing and adding entities, including geographic features derived from aerial images, out-of-copyright maps, and uploaded GPS traces.

In this section, however, we explore the automated processing of GPS trajectories to create a map. Figure 7.2 shows a set of GPS trajectories that our research group has collected over the past few years in Seattle, WA USA. Given that the raw traces

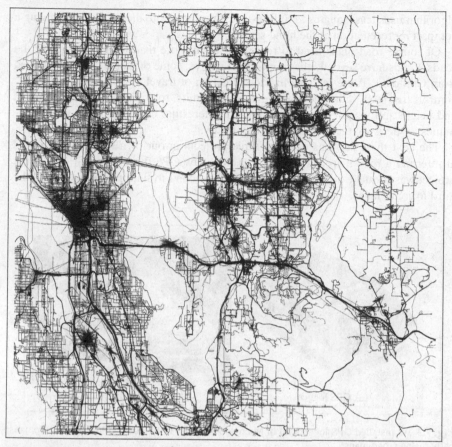

Fig. 7.2 These are GPS traces taken from volunteer drivers in Seattle, WA USA. Based on GPS traces like this, it seems plausible that one could build a road map. The dark, star-like clusters come from idling GPS loggers that occasionally recorded an outlier location.

already look like a road map, it seems plausible to automatically make a road map from such data.

One of the first attempts to make a road map from GPS traces came from Rogers et al. in 1999 [24]. In a series of papers, this idea was developed into an end-to-end system that starts with differential GPS data and concludes with a refinement of an existing map, including finding lanes and lane transitions through the intersections [9]-[25]. The process involves smoothing and filtering the GPS data, matching to an existing map, spline fitting for the road centerline, clustering to find lanes, and refinement of the intersection geometry.

Worrall and Nebot approached the problem of finding roads in a large mining site without using any previous maps [14]. Their algorithm starts by clustering GPS points with similar locations and headings. These clusters are in turn linked, and then the linked clusters are segmented into straight line segments and curves. The algorithm uses least squares to fit lines and circular arcs to these segments, giving a compact representation of the mine's roads.

Of course road maps need to represent much more than the centerlines of the roads. Roads have several other important features, some of which could be derived from GPS data. These features include direction of travel, number of lanes, traffic controls like stop lights and stop signs, speed limits, road names, turn restrictions, and height and weight restrictions. It is an interesting challenge to imagine techniques for deriving these features automatically.

The next three sections describe three projects in our own research lab aimed at inferring a road map and road details from raw GPS data. The goal of the first project was to make a routable road map, the second to find intersections, and the third to count lanes.

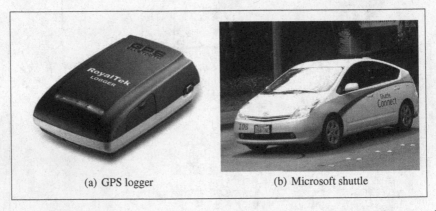

(a) GPS logger   (b) Microsoft shuttle

Fig. 7.3 We installed GPS loggers (a) in Microsoft shuttles (b) to record GPS data for automatically making road maps.

7.2.1 Routable Road Map

Our lab's first project in this area was an attempt to create a routable road map from recorded GPS data [10]. For this experiment, we installed GPS loggers on Microsoft shuttle vehicles operating on and between Microsoft corporate campuses in the Seattle, WA area, shown in Figure 7.3. The GPS loggers we used were Royal-Tek RBT-2300 models. These loggers are convenient because they can hold 400,000 GPS points (including latitude, longitude, altitude, speed, and time stamp), and they can be powered from the vehicle's cigarette lighter. For this experiment, we deployed loggers on 55 different Microsoft shuttles, recording GPS points every 1 second for approximately three weeks.

We split the GPS traces from each shuttle into discrete trips and dropped some of the GPS points for easier processing. Specifically, we split the GPS points into trips by looking for gaps of at least 10 seconds or 100 meters between temporally adjacent GPS points. To reduce the amount of data, we dropped points that were within 30 meters of a previous point, unless the point was part of a turn (change in direction over last three points greater than 10 degrees), when we allowed the points to be as close as 10 meters apart.

Our process for making a map proceeded in two steps. First we "clarified" the GPS traces in an effort to mitigate the effect of measurement noise. Second, we clustered the clarified traces into discrete roads represented by a graph with nodes and edges.

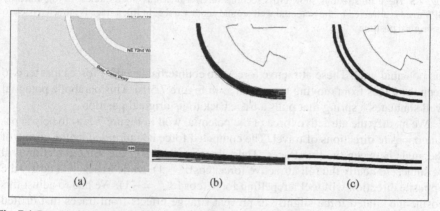

| (a) | (b) | (c) |

Fig. 7.4 Part (a) shows a road map that corresponds to the raw GPS traces in (b). After clarifying, we get the traces in (c) that are more compact and that show the roads' two directions of travel.

7.2.1.1 Clarifying GPS Traces

GPS measurements inevitably have noise, which can be reasonably modeled with a Gaussian distribution [13]. As an example of GPS noise, Figure 7.4 shows some of the GPS data we collected along with a section of a road map for the same area. The raw GPS traces are spread, making it more difficult to infer the location of the road and to differentiate the two directions of travel.

We developed a technique to clarify the GPS traces by imagining each trace as a special kind of electrostatically charged wire. Pairs of wires with the same direction of travel were attracted to each other over short distances, while pairs of wires with the opposite direction of travel repelled each other over short distances. The traces could move in response to these forces, but they were also anchored with imaginary springs to their original locations to prevent too much deviation.

These imaginary forces are simulated as energy potential wells, as shown in Figure 7.5. Figure 7.5(a) shows a cross sectional view of two traces. The inverted Gaussian potential well is centered on one trace, and it tends to pull the other trace toward it at the bottom of the potential well. The force is proportional to the derivative of

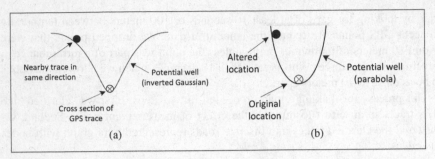

Fig. 7.5 These illustrations show cross sections of GPS traces as small circles. In (a), a nearby GPS trace is attracted to the potential well around another trace. In (b), a trace is attracted back to its original position.

the potential well. These attractive forces are counterbalanced by forces that tend to keep the traces from moving too much, as in Figure 7.5(b). This parabolic potential well simulates a spring that pulls a trace back to its original position.

We modify the attractive force of the potential well in Figure 7.5(a) to help separate opposite directions of travel. The computed force is multiplied by the cosine of the angle between the two traces. Thus, traces with nearly the same direction will be subject to nearly the full attractive force ($\cos 0° = 1$), while traces with nearly the opposite direction will feel a repelling force ($\cos 180° = -1$). We had to adjust this cosine-modulated force slightly to prevent strange effects from traces that drifted left past the opposite direction of travel. Details are in [10].

The two types of potential wells in Figure 7.5 are each governed by a small set of parameters giving their amplitude and width. One way to set these parameters is to experiment with actual traces to see which parameters work best. Instead, we wrote equations to simulate the behavior of the potential wells and traces and solved for the parameters numerically. We specifically looked at two scenarios that were relatively easy to capture with equations: adjacent lanes in the same direction and a road split, shown in Figure 7.6. For the adjacent lanes, we adjusted the potential well parameters so the traces would merge, and for the road split, we tried to maintain the location of the split in spite of tendency for traces in the two lanes to merge.

The result of the clarification step is shown in some examples in Figure 7.7. The procedure successfully consolidated traces going in the same direction and separated traces going in the opposite direction, helping to suppress the GPS noise.

7.2.1.2 Merging Traces

To build a routable road map, we need to convert the GPS traces into a road network represented by a graph of nodes and edges. After clarifying the GPS traces, the next step is to merge them into a graph. Figure 7.8 shows an example of merging. We begin with an empty graph and choose one clarified GPS trip as the first set of nodes and edges. Each GPS point is a node, and the connections between temporally

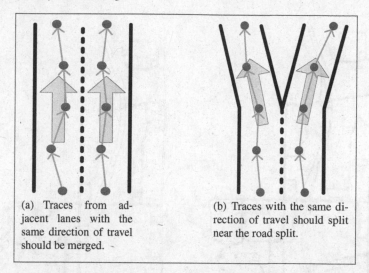

(a) Traces from adjacent lanes with the same direction of travel should be merged.

(b) Traces with the same direction of travel should split near the road split.

Fig. 7.6 We adjusted the parameters of our potential functions to achieve the desired effects in these two situations.

adjacent nodes are edges. To add more clarified traces, we merge new nodes with existing nodes. For a given candidate node to add, we look for the nearest nodes with the same direction of travel. If the nearest node is close enough, we merge the two nodes. If necessary, we also add an edge to preserve the connectivity of the candidate node with previous nodes in its trip. More details of this merging algorithm are in [10].

7.2.1.3 Routable Road Network

With the road network created, we can run a route-planning algorithm to compute actual driving routes. Some of these routes are shown in Figure 7.9, along with comparisons to routes computed with [TM] Maps. Our routes match well. Where there are deviations, one is due to the fact that a road had closed, which our graph represented correctly over the slightly out-of-date Bing[TM] version. Another deviation was due to the fact that we never observed a GPS trace on a road that was actually there. Overall, however, this technique of generating road maps shows that it is possible to create a routable road map from raw GPS traces without the expense of paid drivers in specialized vehicles.

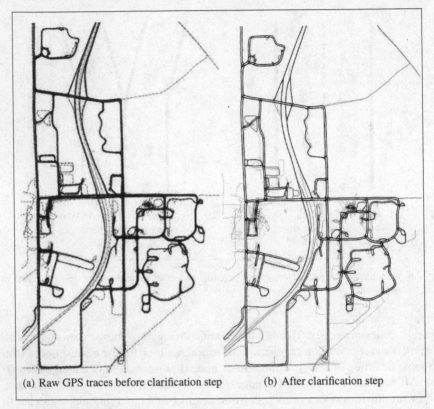

(a) Raw GPS traces before clarification step   (b) After clarification step

Fig. 7.7 The GPS traces in (a) were taken from shuttles around Microsoft in Redmond, WA USA. The clarified version of the traces is shown in (b). Here, the directions of travel have been separated, and traces in the same direction are pulled together.

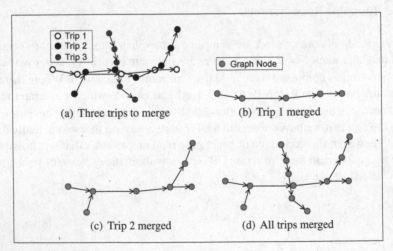

(a) Three trips to merge   (b) Trip 1 merged

(c) Trip 2 merged   (d) All trips merged

Fig. 7.8 This shows an illustration of our GPS trace merging algorithm to product a graph representation of the road network.

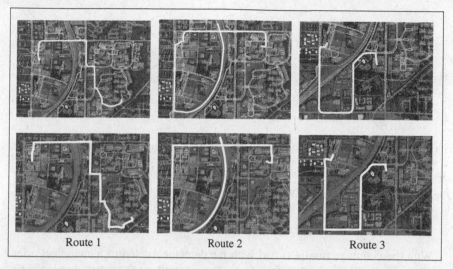

Fig. 7.9 The upper rows shows the road network we generated and routes planned on this network in white. The bottom row shows the same routes planned in Bing™ Maps.

7.2.2 Intersection Detection

While the work above on making a routable road network finds road intersections implicitly, we wanted an explicit way of finding intersections, because this is where roads are often most interesting.

7.2.2.1 Detecting Intersections

Figure 7.10(a) shows GPS traces on an intersection. It might be possible to create an algorithm that closely examines the geometry of GPS traces like this to infer that they represent an intersection, but it would be difficult to scale an algorithm like this to all types of intersections with varying densities of GPS traces. Instead, we created a shape detector that can be automatically trained to find intersections of many different types, as shown in Figure 7.10(b).

The shape detector works by virtually placing it at some latitude/longitude point and examining the underlying GPS traces. The detector is similar to a two-dimensional histogram in that each bin counts the number of traces passing through it, as shown in Figure 7.10(c). After completing the counts, we normalize them so they sum to one in an effort to make the detection less dependent on the absolute number of GPS traces at the intersection. We also rotate the shape detector to a canonical orientation. Specifically, we rotate so the bin with the maximum value is at an angle of zero. This makes the detector less sensitive to the absolute orientation of the intersection.

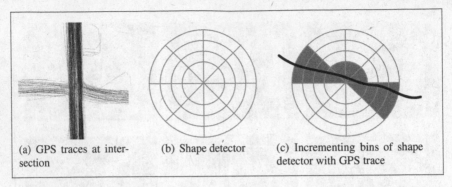

(a) GPS traces at inter-
section

(b) Shape detector

(c) Incrementing bins of shape
detector with GPS trace

Fig. 7.10 Part (a) shows typical GPS traces at a road intersection. The shape detector in (b) is used to find intersections. A GPS trace increments the bins it intersects in the shape detector, as in (c)

With the counts normalized and with the detector rotated to a canonical orientation, we extract the counts into a feature vector with each vector element corresponding to one bin of the shape detector.

Our goal is to apply a standard machine learning technique to the feature vectors.
In order to do this, we need training data consisting of positive and negative samples
of intersections. Since we have a ground truth map with the actual locations of
intersections in the region of our GPS data, we use it to find positive feature vectors
for training, and we take negative examples centered at GPS points that are at least
20 meters from any known intersection. Given these positive and negative training
samples, we use Adaboost [26] to learn a classifier, although several other classifiers
likely would have worked as well.

The algorithm outlined above reduces the problem of finding road intersections
to a classic detection problem. We can thus assess the performance of the algorith-
m with a receiver operating characteristic (ROC) curve, as shown in Figure 7.11.
This shows the tradeoff between correctly finding intersections (true positives) and
hallucinating intersections that are not actually there (false positives) as we adjust
the sensitivity of the Adaboost classifier. The ideal operating point is in the upper
left corner, where all the intersections are found and there are no false positives. We
optimized the geometry of our shape detector (e.g. number of circles, number of
angular slices) by finding the geometry that gave the best ROC curve.

7.2.2.2 Refining Intersections

Although our main goal is to find intersections, we find roads by looking for GPS
traces that directly connect the detected intersections. Some of the intersections and
roads we found are shown in Figure 7.12. Some of the intersections we did not find
are due to the fact that we did not have enough GPS traces passing through them.

With these roads, we can refine the locations of the intersections using the Iter-
ated Closest Point (ICP) algorithm [7]. In general, ICP is a technique to find the

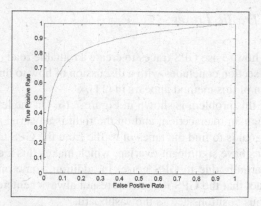

Fig. 7.11 This is the receiver operating characteristic (ROC) curve of our intersection detector. More correct detections (true positives) come at the expense of more false positives.

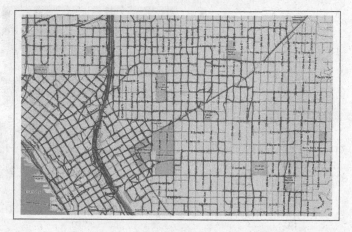

Fig. 7.12 These are some of the intersections found by our intersection detector and the roads filled between the detected intersections.

geometric transformation between measured points on a model curve or surface and the model curve or surface itself. In our case, the measured points are the GPS points near the intersection, and the model is the roads making up the intersection. We use ICP to find the geometric transformation between the intersection's roads and the GPS points, which tends to center the roads on the intersection, giving a more accurate measurement of the intersection's location. Before ICP, the mean distance between our detected intersections and their ground truth locations on a map was 7.2 meters. After ICP, the mean error was reduced to 4.6 meters.

More details about our intersection detection work can be found in [14].

7.2.3 Finding Traffic Lanes

Having discussed how to use GPS traces to create a routable road map and pinpoint intersections, this section concludes with a discussion of how to find traffic lanes. A detailed discussion of this method appears in [11].

An example of this problem is shown in Figure 7.13. On the left is a set of GPS traces going through an intersection, and on the right is an aerial view of the same intersection. Our goal is to find the lanes of traffic from the GPS traces. GPS traces from adjacent lanes have significant overlap, which makes this a difficult problem. This overlap is partially due to GPS noise. In addition, some of the GPS spread comes from the fact that the GPS loggers were not always centered in the vehicle, but placed at various positions across the dashboard.

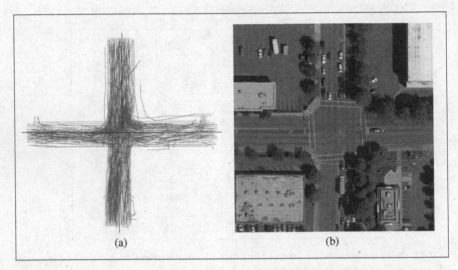

(a)   (b)

Fig. 7.13 The GPS traces in (a) are colored by their direction of travel. They came from the intersection in (b). The lanes in the GPS data are difficult to distinguish due to GPS noise.

If we model GPS noise as Gaussian as suggested in [13], this suggests that the spread of GPS traces across multiple lanes can be modeled as a mixture of Gaussians (GMM), as shown in Figure 7.14. Specifically, the GMM applies to the points where the GPS traces intersect a perpendicular line across the road. The mixture of Gaussians represents a continuous probability density expressed as a weighted sum of Gaussian distributions, as in Equation 7.1.

$$p(x) = \sum_{j=1}^{k} w_j \frac{1}{\sqrt{2\pi\sigma_j^2}} \exp\left(-\frac{(x-u_j)^2}{2\sigma_j^2}\right) \qquad (7.1)$$

Here, k gives the number of lanes across the road. The variables μ_j and σ_j give the center and standard deviation of the GPS traces in the jth lane, respectively. The w_j

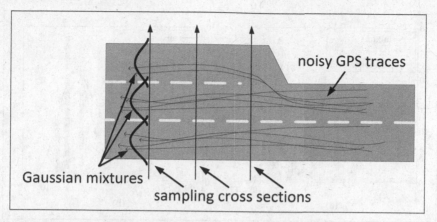

Fig. 7.14 We fit a Gaussian mixture to GPS traces to find lanes of traffic.

represents the fraction of GPS traces in each lane. While there are standard techniques to find k, $\mu_1 \ldots \mu_k$, and $\sigma_1 \ldots \sigma_k$, we found these did not work well. Instead, we developed a special expectation maximization (EM) algorithm to find the lane parameters. This algorithm is discussed in detail in [11]. The basic assumptions are:

- Lanes are evenly spread across the road, expressed as $\mu_j = \mu + (j-1)\triangle\mu$, $j = 1 \ldots k$.
- GPS traces have the same spread in each lane, i.e. $\sigma_j = \sigma$, $j = 1 \ldots k$.

In addition, we impose priors on $\triangle\mu$ and σ^2. To choose between different values of k (the number of lanes), we impose a penalty term that prefers lane widths of 5 meters. Figure 7.15 shows an example of how the fit improves with our restricted GMM. We tested this technique on roads approaching three intersections, with a separate model fit for each direction of travel. We looked at several cross sections along each road. Compared to a general GMM, our new technique achieved a lower percentage error in estimating the number of lanes as well as more consistent estimates of the lanes' spacing and spread.

7.3 Map Matching

It is common to convert GPS trajectories from a sequence of raw latitude/longitude coordinates to a sequence of roads. Knowledge of which road a vehicle is on is important for assessing traffic flow, inferring the vehicle's route, and predicting where the vehicle is going. The problem of converting a GPS sequence to a sequence of roads is called "map matching," and it is not as easy as it first appears. Figure 7.16 illustrates the problem, showing three points from a sequence and the multiple roads to which they could match.

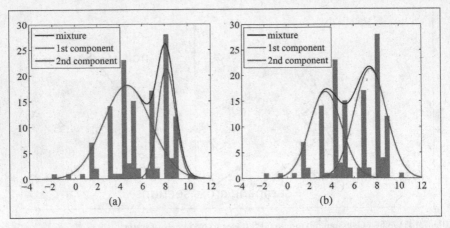

Fig. 7.15 Before imposing restrictions on our Gaussian mixture model, the resulting Gaussians are uneven, as in (a). After the restrictions, the two Gaussians have the same width as we expect of traffic lanes, as in (b).

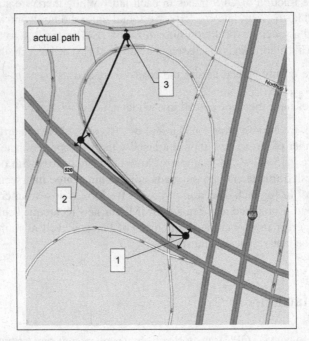

Fig. 7.16 Map matching is the process of finding which road corresponds to each GPS point. For these three GPS points, there are multiple nearby roads, but the true path is obvious considering continuity.

The first inclination for solving the map matching problem is to match the GPS point to the nearest road. This technique falls in the category of "geometric" algorithms for map matching, as explained in the survey articles in [6, 23]. The combination of GPS noise and multiple nearby roads, however, means that this technique often fails. In fact, matching to the nearest road generally serves only as a weak technique with which to demonstrate the superiority of more sophisticated algorithms.

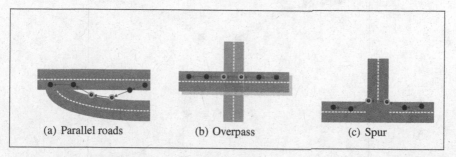

Fig. 7.17 These are three problematic scenarios for map matching. Simply matching to the nearest road will result in mistakes for the points with gray outlines.

Figure 7.17 shows three common scenarios where map matching can be difficult. In all three, multiple roads cause ambiguity in matching.

A step up in sophistication and robustness beyond geometric algorithms are so-call topological algorithms [6, 23] that pay attention to the connectivity of the road network. Clearly this would help solve the map matching problems illustrated in Figure 7.16 and Figure 7.17, where mistaken matches tend to violate the continuity of the drive. Representative of these algorithms are those that use the Fréchet distance to measure the fit between a GPS sequence and candidate road sequence [5, 8]. Here, candidate routes are assessed against the polyline created by connecting the GPS points with straight line segments. The candidate routes are created from the road network, and thus represent feasible driving paths which rule out sudden jumps to nearby roads. The Fréchet distance measures the dissimilarity between the GPS polyline and the candidate route. [5] describes the Fréchet distance as:

> Suppose a person is walking his dog, the person is walking on the one curve and the dog on the other. Both are allowed to control their speed but they are not allowed to go backwards. Then the Fréchet distance of the curves is the minimal length of a leash that is necessary for both to walk the curves from beginning to end.

In addition to noise in the GPS data, map matching algorithms must deal with occasional outliers. Also, the GPS sampling rate may be low. These problems can be addressed by probabilistic algorithms [23] that make explicit provisions for GPS noise and consider multiple possible paths through the road network to find the best one.

A new class of map matching algorithms has emerged recently that embrace both the topology of the road network and the noise and outliers in the GPS data, exemplified by [22]-[21]. These algorithms find a sequence of roads that simultaneously come close to the noisy GPS data and form a reasonable route through the road network. The description below concentrates on [22], from our lab, as a well-tested and representative sample of algorithms like this.

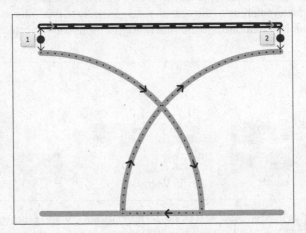

Fig. 7.18 GPS points 1 and 2 can match to either the black road or gray road. Matching to the black road is preferred, because the length of the resulting route is about the same as the distance between the GPS points.

7.3.1 Hidden Markov Model for Map Matching

There is a natural tradeoff in map matching between matching a GPS point to the nearest road and matching it to a road that makes sense in the context of other matched points. For example, Figure 7.16 shows three GPS points and nearby roads. None of the points is exactly on a road, due to GPS noise and possible geometric errors in the map. Thus, each GPS point has multiple candidate roads to which it could be matched, shown by the arrows at each point. However, it is obvious to us which route the driver took based on the continuity of the sequence of road matching candidates. There is a tradeoff between two criteria: the matched road should be close to the GPS point, but the sequence of roads should make a reasonable path. A hidden Markov model (HMM) makes this tradeoff explicit using probability distributions.

The first probability distribution for the HMM models the GPS observations and their possible matching roads. Specifically, this probability distribution models GPS noise, which can be represented by a Gaussian centered around the actual GPS measurement [13]. This means that roads that are farther away from the GPS point have a smaller probability of matching the GPS point. We measure the distance between the GPS point and the road as the great circle distance between the GPS point and the nearest point on the road. In Figure 7.16, these distances are the lengths of the lines with arrows. Based on ground truth data from manual map matching, we estimated the standard deviation of this Gaussian distribution to be 4.07 meters for our data, which is a reasonable value for GPS noise [22].

The other probability distribution in the HMM, the transition probabilities, helps enforce the continuity of the route in terms of the transitions between pairs of GPS points. Intuitively, we want to avoid matching to a nearby road that would require a convoluted route to get to the next matched road. For example, in Figure 7.16, if GPS point "2" matched to the highway rather than the curving ramp, the subsequent route to any of the roads near point "3" would be long and inefficient.

The structure of the HMM requires these transition probabilities be a function of two, temporally adjacent road match candidates. We achieve this by considering the distance between the GPS points (great circle distance) and the length of the route that it would take to traverse this distance starting and ending on the road match candidates.

Figure 7.18 shows an example. If the two GPS points match to the black road, then the route distance between the two matched points is approximately the same as the great circle distance between the GPS points. If they instead matched to the gray roads, the route distance is much longer. In our experiments, we found that the absolute value of the difference between the great circle distance and route distance for correct road matches follows an exponential probability distribution, i.e.

$$p(d) = \frac{1}{\beta} e^{-d/\beta} \tag{7.2}$$

where d is the absolute difference between the great circle distance and route distance. This distribution is maximum at $d = 0$, and decreases monotonically for $d > 0$. In the example in Figure 7.18, d would be small if the two GPS points matched to the black road, leading to a larger probability. Matching to the gray road, d would be much larger, along with a lower probability.

To implement the computation of these transition probabilities requires that we run a route planner between all candidate pairs of matches for all temporally adjacent GPS points. To make this faster, we ignore all road matches that are more than 200 meters away from the GPS measurement.

The transition probabilities above are computed by looking at the differences in length between the GPS pairs and resulting routes. In [20] we looked at using time differences instead, with similar results. Time differences, however, can vary due to traffic speeds.

For each GPS point, the HMM looks at all the possible road matches within 200 meters. The HMM algorithm finds the optimal sequence of matched roads that maximizes the product of the observation probabilities (Gaussian for GPS noise) and transition probabilities (exponential on distance differences). Figure 7.19 shows some examples of how our algorithm works in spite of the problems illustrated in Figure 7.17. In Figure 7.19, the roads highlighted in white show the correctly matched route segment. The roads highlighted in gray show the route the vehicle would have to drive if each GPS point matched to the nearest road. This often results in an unreasonable, circuitous route, necessary to pass through mistakenly matched roads.

Although the HMM is a robust technique for map matching, it can fail if the GPS point falls far from a road (e.g. a GPS outlier) or if the underlying map is wrong. For this reason, in [22] we show how to break the HMM at these points and recombine the matched route by removing offending GPS points.

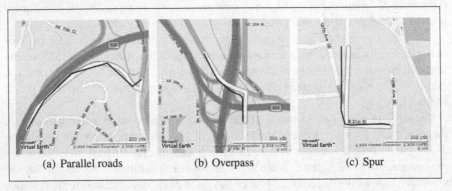

| (a) Parallel roads | (b) Overpass | (c) Spur |

Fig. 7.19 These three scenarios correspond to the three in Figure 7.17. The HMM map matching algorithm works correctly. The black lines connect the GPS points, the white road shows the correct match, and the gray path shows the path necessary to satisfy naively matching the nearest road to each GPS point.

7.4 Destination Prediction

Drivers could benefit from a prediction of their destination. They could be informed of upcoming traffic jams, road construction, speed limit changes, and other warnings. With prediction, drivers could also be alerted about gas stations or charging stations along their route. Retailers near the driver's destination might pay a premium to deliver ads to vehicles predicted to arrive nearby. People doing a local Web search on a mobile device could get results clustered around their destination.

Our lab has developed a general, probabilistic framework for destination prediction [17, 19]. It operates on a vehicle's partial trajectory to predict where the trajectory will end. The framework depends on a discretization of the map, such as a tiling of square cells, as in Figure 7.20(a). In this discretization, each GPS point is represented by cell that contains it. Although the remainder of this discussion will use the cell-based discretization, we have also experimented with alternative discretizations, such as representing each GPS point by its nearest road intersection (Figure 7.20(b)), or road. In any case, the GPS trail is converted to a sequence of discretized points $C = \{c_1, c_2, c_3, \ldots, c_n\}$. In general, this is a partial trajectory, and a new element is added to the end whenever the vehicle encounters a new cell.

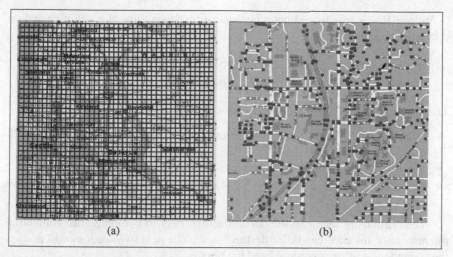

Fig. 7.20 For destination prediction, GPS points are discretized based on which grid cell they fall in (a), or which intersection they are close to (b).

Whenever a new element is added to the sequence C, our prediction is reinvoked to compute a probability for each cell on the map giving the probability that the current trip will end in that cell. Figure 7.21 shows an example of two partial trajectories starting near the center and going south. Cells with darker borders have higher probabilities.

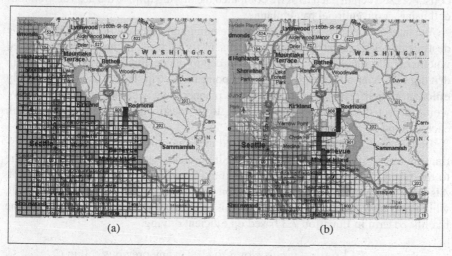

Fig. 7.21 As the trip progresses, we compute a destination probability for each cell. Cells with darker borders have higher destination probabilities.

We use Bayes rule to compute the destination probability of a cell c^* based on the partial trajectory C:

$$p(c^*|C) = \frac{p(C|c^*)p(c^*)}{\sum_{j=1}^{N} p(C|c_{(j)})p(c_{(j)})} \qquad (7.3)$$

The three parts of the right side of Equation 7.3 are:

- $p(C|c^*)$ - Likelihood of partial trajectory C given the candidate destination cell c^*.
- $p(c^*)$ - Prior probability of c^* being a destination.
- $\sum_{j=1}^{N} p(C|c_{(j)})p(c_{(j)})$ - normalizing factor that makes cell probabilities sum to one.

The normalizing factor in the denominator is easy to compute after computing the numerator for each candidate destination c^*. The next two sections will discuss the computation of the likelihood and the prior.

7.4.1 Destination Likelihood from Efficient Driving

Our destination likelihood term is based on our assumption that drivers take fairly efficient routes to their destination. In order to test this, we gathered GPS data from 118 volunteer drivers representing 4300 different trips [17]. We discretized each trip with the grid in Figure 7.20(a), so each trip was represented by a sequence of grid cells. For each cell in each trip, we computed the driving time to the last (destination) cell using a conventional route planner.

We measured efficiency by tracking the minimum computed driving time to the destination cell as each trip progressed. That is, for each cell in the trip, we computed whether or not the driving time from that cell to the destination was the minimum driving time over all the cells up to that point in the trip. If drivers were perfectly efficient in their driving, and if our route planner gave accurate driving times, we would expect the minimum driving time to the destination cell to decrease with each cell-to-cell transition.

In fact, we found that drivers decreased the minimum driving time to their destination for 62.5% of the cell-to-cell transitions. This seems low, but could be due to the fact that drivers may be sensitive to traffic, while our route planner was not. Also, we computed driving times based on the road nearest the center of each cell, which may also be inaccurate. Despite this, we can give a simple equation for the likelihood term in Equation 7.3 based on efficient driving.

$$p(C|c^*) = \prod_{i=2}^{n} \begin{cases} p & \text{if } c_i \text{ is closer to } c^* \text{ than any previous cell in } C \\ 1-p & \text{otherwise} \end{cases} \qquad (7.4)$$

Here $p=0.625$, based on our experiment. Also recall that $C = \{c_1, c_2, c_3, \ldots, c_n\}$, so n is the number of cells in the partial trajectory. By "closer" in Equation 7.4, we mean that c_i has the minimum driving time to c^* compared to all previous cells in the sequence. This likelihood term tends to reduce the probability of cells that the driver is driving away from, as illustrated in Figure 7.21. Equation 7.4 is invoked every time the driver enters a new cell, and it is evaluated for every cell c^* in the grid. As shown in Figure 7.21, the likelihood is generally reduced for cells that the driver appears to be passing up.

One challenge of implementing this likelihood based on driving times is the computation of driving times. For each cell in the trip $C = \{c_1, c_2, c_3, \ldots, c_n\}$, it is necessary to compute the driving time to each candidate cell c^* in the whole grid. This is the classic "single-source shortest path" problem where each cell in C serves as a starting point. In practice, we pre-compute the driving times between all pairs of cells in our grid. To save computation time and storage, we assume the driving time between two cells is independent of which cell is the start cell. Even with this assumption, for N cells in the grid, there are $N(N-1)/2$ distinct cell pairs. In the grid shown in Figure 7.20, there are $N = 41^2$ cells, which makes over 1.4 million distinct cell pairs. Using a convention route planner on a conventional PC, we can compute routes at a rate of about 60 per second, which translates to about 6.5 hours of computation time for our grid. We have recently switched to an algorithm called PHAST for much faster computation of "single-source shortest path" [12]. This is fast enough that we can compute driving times in real time without the need for pre-computation.

The next section discusses the computation of the prior probability, $p(c^*)$, which completes the terms in Equation 7.3.

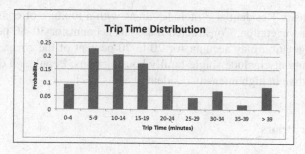

Fig. 7.22 We can use a distribution of trip times as a prior probability for destination prediction.

7.4.2 Destination Priors

We know that there are several other clues to a driver's destination in addition to the efficient driving likelihood discussed above. These other clues can be formulated as prior probabilities over the grid of cells, denoted as $p(c^*)$ in Equation 7.3. This section discusses some useful priors.

7.4.2.1 Driving Time

Our intuition is that most car trips are fairly short. For instance, if someone started a trip in Los Angeles, we might be suspicious if we saw a high probability destination prediction for New York City. This intuition is correct, according to the U.S. 2001 National Household Transportation Survey (NHTS) [3]. Figure 7.22 shows a probability distribution of trip times from the 2001 NHTS. Given a starting cell in the grid, we can use our computed trip times and the trip time distribution to compute a driving time prior for each cell.

7.4.2.2 Ground Cover

The U.S. Geological Survey classifies each 30m x 30m square of the U.S. into one of 21 different ground cover types, such as water, pasture, urban, and commercial [1]. An image of their classifications for the region around Seattle, Washington USA is shown in Figure 7.23(a). We looked up the ground cover for each trip in our GPS data to create a ground cover prior. In our experiment, we found that the two most popular ground cover types at destinations were:

- Commercial/Industrial/Transportation "Includes infrastructure (e.g. roads, rail-roads, etc.) and all highly developed areas not classified as High Intensity Residential." [2]
- Low Intensity Residential "Includes areas with a mixture of constructed materials and vegetation. Constructed materials account for 30–80 percent of the cover. Vegetation may account for 20 to 70 percent of the cover. These areas most commonly include single-family housing units. Population densities will be lower than in high intensity residential areas." [2]

7.4.2.3 Other Priors

Other destination priors include a driver's personal destinations, i.e. places that he or she goes frequently, such as their home and work place. In [19] we implemented this as an "open world" model, raising the probability of previously visited locations, but leaving some non-zero probability for new places.

Another prior could be a function of which types of businesses are present. We have found, not surprisingly, that restaurants are more common destinations than dentists. This fact could help raise the probability of certain cells in the map.

Of course, all of these priors could vary by the time of day and day of week, as we expect certain types of destinations to rise and fall in popularity as time progresses.

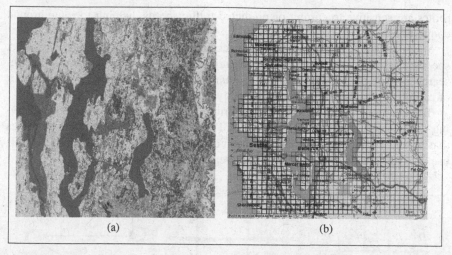

(a)   (b)

Fig. 7.23 (a) shows the USGS ground cover classes around the Seattle, Washington USA area. The resulting destination probabilities are shown in (b). Water and unpopulated areas are less popular destinations.

7.4.3 Route Prediction

While the focus of this section has been destination prediction, a related problem is route prediction, where the goal is predict a driver's entire route rather than just their stop location. In summary, here are three route prediction techniques we have explored:

Trip Observations: Observe several trips from a driver. When a new trip starts, attempt to find a good match with a previous trip, and use the matched trip as a route prediction [15].

Markov Model: For short sequences (1–10) of observed road segments for a particular driver, build a probabilistic model for which road segment comes next [16].

Turn Proportions: To predict which way a driver will turn at an intersection, look at which turns bring the driver closer to the most destinations. This tends to assign low probabilities to turns that result in dead ends or closed neighborhoods [18].

7.5 Learning Routes

It is easy to get driving directions from Web-based mapping sites and GPS-based navigation devices. When they were first available, these directions would give the fastest route assuming time-invariant driving speeds. More recently, the computed routes are sensitive to current or predicted traffic speeds, helping to route around congestion.

Such directions, however, still fail to account for route subtleties that may affect speeds (e.g. turn delays) or driving pleasure (e.g. complexity). To address this, researchers have built routing algorithms that are heavily influenced by the behavior of real drivers whose routes are presumably closer to optimal.

This section presents two routing algorithms that exemplify learning by observing real drivers. The first, T-Drive, uses GPS traces from taxi cabs to create routes that are measurably faster than those from conventional routers. The second, TRIP, uses an individual's driving behavior to create better, new routes.

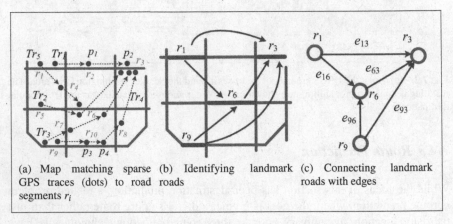

(a) Map matching sparse GPS traces (dots) to road segments r_i

(b) Identifying landmark roads

(c) Connecting landmark roads with edges

Fig. 7.24 These are the steps to creating a routing graph from taxi traces in T-Drive.

7.5.1 T-Drive: Learn from Taxis

T-Drive [28] looked at GPS traces from taxi drivers in Beijing, China to compute fast driving routes based on the assumption that taxi drivers are generally good at efficiently navigating through the city. This includes the fact that the best route may vary with the time of day as traffic conditions vary.

The research started with three months of GPS data from over 33,000 Beijing taxis. These taxis were already equipped with GPS sensors for the purpose of logging their locations for business intelligence. Unfortunately, the GPS points were separated in time by 2–5 minutes. This complicates the process of map matching, because the route taken between GPS samples is ambiguous. This led the same research team to create a new map matching algorithm specially aimed at sparsely sampled GPS traces [29]. This step of the process is illustrated in Figure 7.24(a).

It is unlikely that a newly requested start and end point will correspond exactly to a previous taxi route, especially a previous route at the same time of day as the request. Thus, T-Drive identified "landmark" road segments that contained a significant number of taxi traces. These landmarks are not necessarily connected, but they serve as start points, end points, and waypoints for computing new routes. Figure 7.24(b) shows an illustration of these landmark road segments.

Two landmark roads are connected to each other if there have been enough taxi traces connecting them. These connections, illustrated in Figure 7.24(c), serve as edges in a "landmark graph". This graph is similar to the traditional road network graph used for conventional route planning, but its nodes (landmark roads) represent popular taxi roads, and its edges (landmark edges) represent taxi routes between the landmark roads.

Based on GPS observations of the taxis, the travel times associated with the landmark edges vary. T-Drive introduces a clustering technique called Variance-Entropy Clustering that splits the observed driving times on these edges into time periods throughout the day. The clustering splits time to produce periods of stable travel time distributions for each edge. Thus, there may be one distribution of travel times for 7:15 a.m. to 7:50 a.m., followed by a different distribution for 7:50 a.m. to 8:10 a.m., etc.

In planning a route, the route planner uses the landmark graph as a conventional road graph, but with two changes. First, a conventional road graph has a single cost scalar associated with each edge, such as the travel time or length of the edge. Instead, T-Drive has travel time distributions for each edge. Thus, prior to each route computation, the user must specify a value of α, $0 \leq \alpha \leq 1$, which indicates how close they are to a taxi's speed and aggressiveness. This value is used to choose one travel time for each edge from the edge's travel time distribution.

The second difference is that the T-Drive graph has time-varying travel times. The clustering process, combined with a choice of α, gives different travel times in different time periods for each edge. Thus, the expected travel time of an edge near the end of the route may change as the driver executes the route. This is a so-called "time-dependent fastest path" problem for which there are standard solutions, one of which T-Drive uses.

The first test of the T-Drive router used GPS trajectories from 30 drivers to show that T-Drive gave accurate driving times compare to real driving times. The researchers found that $\alpha = 0.6$ produced a good match between the two.

Having established that T-Drive gave accurate driving times, it was compared against synthetic route queries computed on a graph with constant edge speeds and on a graph with real-time traffic information. For about 60% of the queries, T-Drive gave faster routes than the graph with constant travel times, and it also outperformed the graph with real-time traffic. This is because the traffic-sensitive router had to resort to constant speeds on roads where it had no data and because T-Drive could implicitly account for subtleties like pedestrians blocking the road and cars parking.

Finally, T-Drive was compared against real drivers taking T-Drive routes and routes computed from Google Maps[TM]. 81% of T-Drive's routes were faster, and saved 11.9% of driving time on average.

Recently, an extension of T-Drive system performs a self-adaptive driving direction service for a particular user. In this self-adaptive service, a mobile client (typically running on a GPS-phone) records a user's driving routes and gradually learns the user's driving behavior from the GPS logs [27]. Specifically, a user's driving behavior is represented by a vector of custom factors, $\boldsymbol{\alpha} = \{\alpha_1, \alpha_2, \ldots\}$, where α_i denotes how fast the user could drive on the landmark edge i as compared to taxi drivers. This is different from the original T-Drive system using only one custom factor for a user. The motivation of this work is that an individual's driving behavior varies in routes and driving experiences. For example, traveling on an unfamiliar route, a user has to pay attention to the road signs, hence drive relatively slowly. Meanwhile, after becoming experienced in driving a user is likely to drive faster than he/she did before.

7.5.2 *Learn from Yourself*

While T-Drive helps drivers imitate taxis to find the fastest route, Trip Router with Inividualized Preferences (TRIP) helps drivers find routes that match their implied preferences [27]. Using GPS data from 102 different drivers who made 2517 total trips, TRIP was designed to compute new routes that paid attention to both traffic speeds and individual driver's preferences for certain roads. TRIP was based on the fact that drivers do not always choose the fastest route to their destination, even if they are familiar with the area. This is illustrated in Figure 7.25, which shows four different routes for a driver's morning commute. The driver's usual route is significantly different from the shortest distance route, the fastest route computed based on traffic speeds, and the fastest route computed by Microsoft MapPoint®. In fact, by looking at the GPS trips in the TRIP study, drivers choose the fastest route only about 35% of their trips. These are drivers who are familiar with the area.

Interestingly, the work in TRIP was preceded by necessary work on map matching, similar to the work on T-Drive above, resulting in another HMM-based algorithm [20].

TRIP worked in two steps. The first was to assign time-varying road speeds to every road segment encountered by all the drivers. TRIP's road speeds were computed as the average speed seen on each road, separated into 15-minute time intervals to account for variations throughout the day. There was one set of speeds for weekdays and another set for weekends. If a road had no GPS data for a certain time interval, its speed was taken as the system-wide average of roads of the same type (e.g. arterial, highway, ramp) for that time interval. If there was no data available for a time interval, the road's speed was taken as its speed limit. Similar to T-Drive, these real time road speeds provide the basis for computing fast routes between any two points.

The efficient routes computed from measured road speeds served as a baseline for incorporating driving preferences that go beyond efficiency, which was TRIP's second step. TRIP examined all the recorded trips for each driver to compute each driver's "inefficiency ratio" r. The inefficiency ratio reflects the amount by which a driver's actual routes are less time-efficient than the computed fastest routes. For a given route, the numerator of the ratio is the duration of the fastest route between the endpoints, computed using TRIP's road speed estimates from above. The denominator is the driver's actual driving time. This r is almost always less than one, unless the driver has taken the fastest route or, occasionally, has found a route faster than that based on TRIP's speed estimates. The driver's average efficiency ratio, r, represents the amount he or she is implicitly discounting the time cost of their preferred roads.

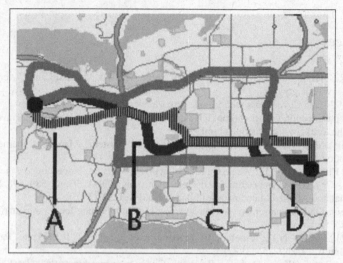

Fig. 7.25 This shows four routes for a drivers morning commute. (A) shows the drivers usual route, (B) is the shortest distance route, (C) is the fasted route based on our own GPS study, and (D) is the fastest route according to MapPoint®.

A driver's efficiency ratio is used to discount the cost of their preferred roads for computing routes. Specifically, TRIP computes the time cost of road segment i as t_i, based on GPS observations from all drivers as explained above. For planning a route, TRIP uses the driver's average efficiency ratio as follows to compute new costs c_i of every road segment:

$$c_i = \begin{cases} \bar{r}t_i & \text{if road } i \text{ previously driven by this driver} \\ t_i & \text{otherwise} \end{cases} \quad (7.5)$$

The effect of this discounting is a reduction in cost of roads the driver has already driven, meaning they are more likely to be chosen by the otherwise conventional route planner.

In tests, TRIP's computed route matched the driver's actual route in 46.6% of the test cases. This compares favorably with computing routes based on the estimated road speeds, which matched the test trips only 34.5% of the time. Using a traditional route planner based on static speed limits, the computed routes matched only 30% of the time.

7.6 Summary

This chapter discussed the use of GPS traces to first construct a road map, match new GPS traces to an existing road map, and then use matched traces to predict a driver's destination and to compute optimal routes. Each of these efforts depends on an easily gathered repository of GPS traces of drivers as they drive normally, which demonstrates the value of logged GPS data.

References

1. Multi-Resolution Land Characteristics Consortium (MRLC), National Land Cover Database. http://www.mrlc.gov/
2. NLCD 92 Land Cover Class Definitions. http://landcover.usgs.gov/classes.php
3. U.S. Department of Transportation, F.H.A. 2001 NHTS (National Household Travel Survey). http://nhts.ornl.gov/introduction.shtml#2001
4. Building and road extraction from aerial photographs. Systems, Man and Cybernetics, IEEE Transactions on 12(1), 84 –91 (1982)
5. Alt, H., Efrat, A., Rote, G., Wenk, C.: Matching planar maps. In: Proceedings of the fourteenth annual ACM-SIAM symposium on Discrete algorithms, SODA '03, pp. 589–598. Society for Industrial and Applied Mathematics, Philadelphia, PA, USA (2003)
6. Bernstein, D., Kornhauser, A.: An introduction to map matching for personal navigation assistants
7. Besl, P.J., McKay, N.D.: A method for registration of 3-d shapes. IEEE Trans. Pattern Anal. Mach. Intell. 14, 239–256 (1992)
8. Brakatsoulas, S., Pfoser, D., Salas, R., Wenk, C.: On map-matching vehicle tracking data. In: Proceedings of the 31st international conference on Very large data bases, VLDB '05, pp. 853–864. VLDB Endowment (2005)
9. Bruntrup, R., Edelkamp, S., Jabbar, S., Scholz, B.: Incremental map generation with gps traces. In: Intelligent Transportation Systems, 2005. Proceedings. 2005 IEEE, pp. 574 – 579 (2005)
10. Cao, L., Krumm, J.: From gps traces to a routable road map. In: Proceedings of the 17th ACM SIGSPATIAL International Conference on Advances in Geographic Information Systems, GIS '09, pp. 3–12. ACM, New York, NY, USA (2009)
11. Chen, Y., Krumm, J.: Probabilistic modeling of traffic lanes from gps traces. In: Proceedings of the 18th SIGSPATIAL International Conference on Advances in Geographic Information Systems, GIS '10, pp. 81–88. ACM, New York, NY, USA (2010)
12. Delling, D., Goldberg, A.V., Nowatzyk, A., Werneck, R.F.: Phast: Hardware-accelerated shortest path trees (2011)
13. van Diggelen, F.: Gnss accuracy: Lies, damn lies, and statistics
14. Fathi, A., Krumm, J.: Detecting road intersections from gps traces. In: Proceedings of the 6th international conference on Geographic information science, GIScience'10, pp. 56–69. Springer-Verlag, Berlin, Heidelberg (2010)
15. Froehlich, J., Krumm, J.: Route prediction from trip observations. In: Society of Automotive Engineers (SAE) 2008 World Congress
16. Krumm, J.: A markov model for driver turn prediction. In: Society of Automotive Engineers (SAE) 2008 World Congress
17. Krumm, J.: Real time destination prediction based on efficient routes. In: Society of Automotive Engineers (SAE) 2006 World Congress

18. Krumm, J.: Where will they turn: predicting turn proportions at intersections. Personal Ubiquitous Comput. **14**, 591–599 (2010)
19. Krumm, J., Horvitz, E.: Predestination: Inferring destinations from partial trajectories. In: Proceedings of the 8th International Conference on Ubiquitous Computing (UbiComp 2006)
20. Krumm, J., Letchner, J., Horvitz, E.: Map matching with travel time constraints. In: Society of Automotive Engineers (SAE) 2007 World Congress
21. Lou, Y., Zhang, C., Zheng, Y., Xie, X., Wang, W., Huang, Y.: Map-matching for low-sampling-rate gps trajectories. In: Proceedings of the 17th ACM SIGSPATIAL International Conference on Advances in Geographic Information Systems, GIS '09, pp. 352–361. ACM, New York, NY, USA (2009)
22. Newson, P., Krumm, J.: Hidden markov map matching through noise and sparseness. In: Proceedings of the 17th ACM SIGSPATIAL International Conference on Advances in Geographic Information Systems, GIS '09, pp. 336–343. ACM, New York, NY, USA (2009)
23. Quddus, M.A., Ochieng, W.Y., Noland, R.B.: Current map-matching algorithms for transport applications: State-of-the art and future research directions. Transportation Research Part C-emerging Technologies **15**, 312–328 (2007)
24. Rogers, S., Langley, P., Wilson, C.: Mining gps data to augment road models. In: Proceedings of the fifth ACM SIGKDD international conference on Knowledge discovery and data mining, KDD '99, pp. 104–113. ACM, New York, NY, USA (1999)
25. Schroedl, S., Wagstaff, K., Rogers, S., Langley, P., Wilson, C.: Mining gps traces for map refinement. Data Min. Knowl. Discov. **9**, 59–87 (2004)
26. Viola, P., Jones, M.: Rapid object detection using a boosted cascade of simple features. In: Computer Vision and Pattern Recognition, 2001. CVPR 2001. Proceedings of the 2001 IEEE Computer Society Conference on, vol. 1, pp. I–511 – I–518 vol.1 (2001)
27. Yuan, J., Zheng, Y., Xie, X., Sun, G.: Driving with knowledge from the physical world (to appear). In: Proceedings of the 17th ACM SIGKDD International Conference on Knowledge Discovery and Data Mining, KDD '11
28. Yuan, J., Zheng, Y., Zhang, C., Xie, W., Xie, X., Sun, G., Huang, Y.: T-drive: driving directions based on taxi trajectories. In: Proceedings of the 18th SIGSPATIAL International Conference on Advances in Geographic Information Systems, GIS '10, pp. 99–108. ACM, New York, NY, USA (2010)
29. Yuan, J., Zheng, Y., Zhang, C., Xie, X., Sun, G.Z.: An interactive-voting based map matching algorithm. In: Proceedings of the 2010 Eleventh International Conference on Mobile Data Management, MDM '10, pp. 43–52. IEEE Computer Society, Washington, DC, USA (2010)

Chapter 8
Location-Based Social Networks: Users

Yu Zheng

Abstract In this chapter, we introduce and define the meaning of location-based social network (LBSN) and discuss the research philosophy behind LBSNs from the perspective of users and locations. Under the circumstances of trajectory-centric LBSN, we then explore two fundamental research points concerned with understanding users in terms of their locations. One is modeling the location history of an individual using the individual's trajectory data. The other is estimating the similarity between two different people according to their location histories. The inferred similarity represents the strength of connection between two users in a location-based social network, and can enable friend recommendations and community discovery. The general approaches for evaluating these applications are also presented.

8.1 Introduction

8.1.1 Concepts and Definitions of LBSNs

A social network is a social structure made up of individuals connected by one or more specific types of interdependency, such as friendship, common interests, and shared knowledge. Generally, a social networking service builds on and reflects the real-life social networks among people through online platforms such as a website, providing ways for users to share ideas, activities, events, and interests over the Internet.

The increasing availability of location-acquisition technology (for example GPS and Wi-Fi) empowers people to add a location dimension to existing online social networks in a variety of ways. For example, users can upload location-tagged photos

Yu Zheng
Microsoft Research Asia, China
e-mail: yuzheng@microsoft.com

to a social networking service such as Flickr [2], comment on an event at the exact place where the event is happening (for instance, in Twitter [6]), share their present location on a website (such as Foursquare [3]) for organizing a group activity in the real world, record travel routes with GPS trajectories to share travel experiences in an online community (for example GeoLife [60, 57, 53, 61]), or log jogging and bicycle trails for sports analysis and experience sharing (as in Bikely [1] and [15]).

Here, a location can be represented in absolute (latitude-longitude coordinates), relative (100 meters north of the Space Needle), and symbolic (home, office, or shopping mall) form. Also, the location embedded into a social network can be a stand-alone instant location of an individual, like in a bar at 9pm, or a location history accumulated over a certain period, such as a GPS trajectory: "a cinema→a restaurant→a park→a bar."

The dimension of location brings social networks back to reality, bridging the gap between the physical world and online social networking services. For example, a user with a mobile phone can leave her comments with respect to a restaurant in an online social site (after finishing dinner) so that the people from her social structure can reference her comments when they later visit the restaurant. In this example, users create their own location-related stories in the physical world and browse other people's information as well. An online social site becomes a platform for facilitating the sharing of people's experiences.

Furthermore, people in an existing social network can expand their social structure with the new interdependency derived from their locations. As location is one of the most important components of user context, extensive knowledge about an individual's interests and behavior can be learned from her locations. For instance, people who enjoy the same restaurant can connect with each other. Individuals constantly hiking the same mountain can be put in contact with each other to share their travel experiences. Sometimes, two individuals who do not share the same absolute location can still be linked as long as their locations are indicative of a similar interest, such as beaches or lakes.

These kinds of location-embedded and location-driven social structures are known as location-based social networks, formally defined as follows:

A location-based social network (LBSN) does not only mean adding a location to an existing social network so that people in the social structure can share location-embedded information, but also consists of the new social structure made up of individuals connected by the interdependency derived from their locations in the physical world as well as their location-tagged media content, such as photos, video, and texts. Here, the physical location consists of the instant location of an individual at a given timestamp and the location history that an individual has accumulated in a certain period. Further, the interdependency includes not only that two persons co-occur in the same physical location or share similar location histories but also the knowledge, e.g., common interests, behavior, and activities, inferred from an individual's location (history) and location-tagged data.

In a location-based social network, people can not only track and share the location-related information of an individual via either mobile devices or desktop computers, but also leverage collaborative social knowledge learned from user-generated and location-related content, such as GPS trajectories and geo-tagged photos. One example is determining this summer's most popular restaurant by mining people's geo-tagged comments. Another example could be identifying the most popular travel routes in a city based on a large number of users' geo-tagged photos. Consequently, LBSNs enable many novel applications that change the way we live, such as physical location (or activity) recommendation systems [65, 63, 59, 50, 51, 58, 10] and travel planning [45, 46], while offering many new research opportunities for social network analysis (like user modeling in the physical world and connection strength analysis) [28, 39, 16, 20, 21, 19, 25, 44], spatio-temporal data mining [29, 47, 49, 42, 64], ubiquitous computing [55, 54, 52, 56, 48, 62], and spatio-temporal databases [35, 13, 12, 37, 14, 18].

8.1.2 Location-Based Social Networking Services

Existing applications providing location-based social networking services can be broadly categorized into three folds: geo-tagged-media-based, point-location-driven and trajectory-centric.

- *Geo-tagged-media-based.* Quite a few geo-tagging services enable users to add a location label to media content such as text, photos, and videos generated in the physical world. The tagging can occur instantly when the medium is generated, or after a user has returned home. In this way, people can browse their content at the exact location where it was created (on a digital map or in the physical world using a mobile phone). Users can also comment on the media and expand their social structures using the interdependency derived from the geo-tagged content (for example, in favor of the same photo taken at a location). Representative websites of such location-based social networking services include Flickr, Panoramio, and Geo-twitter. Though a location dimension has been added to these social networks, the focus of such services is still on the media content. That is, location is used only as a feature to organize and enrich media content while the major interdependency between users is based on the media itself.

- *Point-location-driven.* Applications like Foursquare and Google Latitude encourage people to share their current locations, such as a restaurant or a museum. In Foursquare, points and badges are awarded for "checking in" at venues. The individual with the most number of "check-ins" at a venue is crowned "Mayor." With the real-time location of users, an individual can discover friends (from her social network) around her physical location so as to enable certain social activities in the physical world, e.g., inviting people to have dinner or go shopping. Meanwhile, users can add "tips" to venues that other users can read, which serve as suggestions for things to do, see, or eat at the location. With this kind of service, a venue (point location) is the main element determining the interdependency connecting users, while user-generated content such as tips and badges feature a point location.

- *Trajectory-centric.* In a trajectory-centric social networking service, such as Bikely, SportsDo, and Microsoft GeoLife, users pay attention to both point locations (passed by a trajectory) and the detailed route connecting these point locations. These services do not only tell users basic information, such as distance, duration, and velocity, about a particular trajectory, but also show a user's experiences represented by tags, tips, and photos for the trajectory. In short, these services provide "how and what" information in addition to "where and when." In this way, other people can reference a user's travel/sports experience by browsing or replaying the trajectory on a digital map, and follow the trajectory in the real world with a GPS-phone.

Table 8.1 provides a brief comparison among these three services. The major differences between the point-location-driven and the trajectory-centric LBSN lie in two aspects. One is that a trajectory offers richer information than a point location, such as how to reach a location, the temporal duration that a user stayed in a location, the time length for travelling between two locations, and the physical/traffic conditions of a route. As a result, we are more likely to accurately understand an individuals behavior and interests in a trajectory-centric LBSN. The other is that in a point-location-driven LBSN users usually share their real-time location while the trajectory-centric more likely delivers historical locations as users typically prefer to upload a trajectory after a trip has finished (though it can be operated in a continuously uploading manner). This property could compromise some scenarios based on the real-time location of a user, however, it reduces to some extent the privacy issues in a location-based social network. In other words, when people see a users trajectory the user is no longer there.

Table 8.1 Comparison of different location-based social networking services

LBSN Services	Focus	Real-time	Information
Geo-tagged-media-based	Media	Normal	Poor
Point-location-driven	Point location	Instant	Normal
Trajectory-centric	Trajectory	Relatively Slow	Rich

Actually, the location data generated in the first two LBSN services can be converted into the form of a trajectory which might be used by the third category of LBSN service. For example, if we sequentially connect the point locations of the geo-tagged photos taken by a user over several days, a sparse trajectory can be formulated. Likewise, the check-in records of an individual ordered by time can be regarded as a low-sampling-rate trajectory. However, due to the sparseness, i.e., the distance and time interval between two consecutive points in a trajectory could be very big, the uncertainty existing in a single trajectory from the first two services is increased. Aiming to put these trajectories into trajectory-centric LBSN services, we need to use them in a collective and collaborative way.

The following sections will pay closer attention to trajectory data, which is the most complex data structure to be found in the three LBSN services, and provides the richest information. If it is handled well, other data sources become easier to deal with. Moreover, as mentioned above, location data can be converted into a trajectory on many occasions. Consequently, some methodologies designed for trajectory data can be employed by the first two LBSN services.

8.1.3 Research Philosophy of LBSN

User and location are two major subjects closely associated with each other in a location-based social network. As illustrated in Fig. 8.1, users visit some locations in the physical world, leaving their location histories and generating location-tagged media content. If we sequentially connect these locations in terms of time, a trajectory will be formulated for each user. Based on these trajectories, we can build three graphs: a location-location graph, a user-location graph, and a user-user graph.

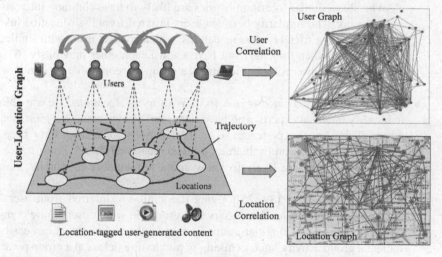

Fig. 8.1 Research philosophy of a location-based social network

In the location-location graph (demonstrated in the bottom-right of Fig. 8.1), a node (a point on the graph) is a location and a directed edge (a line on the graph) between two locations indicates that some users have consecutively traversed these two locations during a trip. The weight associated with an edge represents the correlation between the two locations connected by the edge.

In the user-location graph (depicted in the left part of Fig. 8.1), there are two types of nodes: users and locations. An edge starting from a user and ending at a location indicates that the user has visited this location, and the weight of the edge can indicate the number of visits.

In the user-user graph (shown in the top-right of Fig. 8.1), a node is a user and an edge between two nodes consists of two folds. One is the original connection between two users in an existing social network like Twitter. The other is the new interdependency derived from their locations, e.g., two users have visited the same location, or similar types of places, in the real world over a certain number of visits. The latter information, initially inferred from a user's locations, can be transferred to the former through a recommendation mechanism. In other words, we can recommend users to an individual based on the inferred interdependency. Once the individual accepts the recommendation, the relationship switches from the second category to the first.

Using these graphs, we can understand users and locations respectively, and explore the relationship between them. Though the research topics are listed individually from the perspective of users and locations as follows, these two subjects have a mutually reinforcing relationship that cannot be studied alone:

1) Understanding users: Here, we aim to understand users based upon their locations.

   • *Estimate user similarity* [28, 16]: An individual's location history in the real world implies, to some extent, her interests and behaviors. Accordingly, people who share similar location histories are likely to have common interests and behavior. The similarity between users inferred from their location histories can enable friend recommendations, which connect users with similar interests even when they may not have known each other previously [63], and community discovery that identifies a group of people sharing common interests.

   • *Finding local experts in a region* [65]: With users' locations, we are able to identify the local experts who have richer knowledge about a region than others. Their travel experiences, e.g., the locations where they have been, are more accountable and valuable for travel recommendation. For instance, local experts are more likely to know about high-quality restaurants than some tourists.

   • *Community discovery* [39, 25]: Using the similarity inferred from users' locations, we can cluster these users into groups in which users share common interests like visiting museums. Consequently, an individual can easily initiate a group activity, such as hiking or purchasing tickets at a group price, by sending an invitation to the appropriate users in a social site.

2) Understanding locations: Here, we focus on understanding locations based upon user information.

   • *Generic travel recommendations*:

     – Mining the most interesting locations [65]: Finding the most interesting locations in a city as well as the travel sequences among these locations is a general task that a tourist wants to fulfill when traveling to an unfamiliar city. Location-based social networks provide us with the opportunity to identify such information by mining a large number of users' location histories (represented by trajectories). Refer to Section 9.2.1.

- Itinerary planning [45, 46]: Sometimes, a user needs a sophisticated itinerary conditioned by the user's travel duration and departure place. The itinerary could include not only stand-alone locations but also detailed routes connecting these locations and a proper schedule, e.g., the typical time of day that most people reach the location and the appropriate time length that a tourist should stay there. Planning a trip in terms of the collective knowledge learned from many people's trajectories is an interesting research topic. Refer to Section 9.2.2.
- Location-activity recommender [51]: This recommender provides a user with two types of recommendations: 1) The most popular activities that can be performed in a given location and 2) the most popular locations for conducting a given activity, such as shopping. These two categories of recommendations can be mined from a large number of users' trajectories and location-tagged comments. Refer to Section 9.2.3.

- *Personalized travel recommendations*
  - User-based collaborative filtering [63]: In this scenario, the similarity between each pair of users (introduced above) is incorporated into a collaborative filtering model to conduct a personalized location recommendation system, which offers locations matching an individual's preferences. The general idea behind collaborative filtering [23, 30] is that similar users vote in a similar manner on similar items. Thus, if similarity is determined between users and items, predictions can be made about a user's potential ratings of those items. For instance, if we know user A and B are very similar (in terms of their location histories), we can recommend the locations where user A has already been to user B and vice versa. Refer to Section 9.3.2 for details.
  - Location-based collaborative filtering [59, 58]: User-based collaborative filtering is able to accurately model an individual's behavior. However, it suffers from the increasing scale of users (in a real system) since the model needs to calculate the similarity between each pair of users. To address this issue, location-based collaborative filtering is proposed. This model regards a physical location as an item and computes the correlation between locations based on the location histories of the users visiting these locations. Given the limited geographical space (i.e., the number of locations is limited), this location-based model is more practical for a real system. The main challenge to the location-based model is how to embody an individual's behavior which is the advantage of a user-based model. Refer to Section 9.3.3 for details.

- *Events discovery from social media* [29, 27]
  Quite a few projects aim to detect anomalous events, such as concerts, traffic accidents, sales promotions, and festivals, using media (such as geo-tagged photos and tweets) posted by a large number of users in a location-based social network. Intuitively, people witnessing such an event would post a

considerable amount of media (e.g., tweets) in the location where the event occurs. By grouping and mining media that co-occurs in particular locations, we can get a sense of geo-social events automatically.

Actually, the problems that traditional social networks have exist in location-based social networks, and become more challenging due to the following reasons:

- The graph representing a location-based social network is heterogeneous, consisting of at least two types of nodes (user and location) and three kinds of links (user-user, location-location, and user-location). Or, we can say there are at least three tightly associated graphs modeling a LBSN (as mentioned previously). If it is a trajectory-centric LBSN, trajectories can be regarded as another kind of node in the social network; so do geotagged videos and photos. Location is not only an additional dimension of the user, but also an important object in a LBSN. Under the circumstances, determining the connecting strength between two users in a LBSN needs to involve the information from the other graphs, such as the linking structure of user-location and location-location, besides that of users (refer to Fig. 8.1).

- Location-based social networks are constantly evolving at a faster pace than traditional social networks, in both social structure and properties of nodes and links. Though academic social networks are also heterogeneous with authors, conferences, and papers, its evolves at a much slower speed than LBSNs do. For example, it is much easier to add a new location to a LBSN (by check-in) than launching a new conference or publishing a paper. That is, the number of nodes in a LBSN increases faster than an academic social network. Also, it is common for users to visit locations (e.g., restaurants and shopping malls) they have never been before. However, researchers will not constantly attend new conferences. Thus, the linking structure of a LBSN evolves much faster than an academic social network. Furthermore, the properties of nodes and links in a LBSN evolve more quickly than in an academic social network. A user can become a travel expert in a city after visiting many interesting locations over several months, while a researcher needs years before becoming an expert in a research area.

- A location has unique features beyond that of other objects in a social network. Besides general linking relationship between locations, the hierarchical and sequential properties of locations are unique. A location can be as small as a restaurant or as big as a city. Locations with different granularities formulate hierarchies between them. For example, a restaurant belongs to a neighborhood, and the neighborhood pertains to a city. Further, the city will belong to a county and a country, and so on. Using different granularities, we will obtain different location graphs even given the same trajectory data. This hierarchical property does not hold in an academic social network as a conference never belongs to others. Regarding the sequential property, each link between two locations is associated with temporal and directional information. Moreover, these links can construct a sequence carrying a particular semantic meaning, e.g., a popular travel route.

There are other important research points in location-based social networks. For example, from the perspective of data management, streaming databases and indexing user-generated location data are vital. Also, user privacy in location-based social networks deserves to be further studied. As these topics have been discussed extensively in other chapters, they will not be covered here.

So far, there is no dedicated conference for researchers and professionals to share the research into LBSNs. While people submit LBSN-related papers to a number of conferences such as WWW, Ubicomp, and ACM GIS, ACM SIGSPATIAL Workshop on Location-Based Social Networks provides a dedicated international forum for LBSN researchers and practitioners from academia and industry to share their ideas, research results, and experiences. This workshop was launched in 2009 and has been in conjunction with ACM SIGSPATIAL GIS conference from 2009 to 2011.

8.2 Modeling Human Location History

8.2.1 Overview

To carry out the above-mentioned research, it is first necessary to model the location history of an individual from raw sensor data, such as GPS readings. The methods presented in most literature [8, 24] solely pay attention to detecting significant places from the sensor data, without considering the social computing among different users. That is, they do not study how to compare different users' location histories when modeling the location data of multiple persons. Since 2008, a series of publications [65, 63, 59, 10, 28, 39] proposed a systematical solution for this problem, following the paradigm of "sensor data→ geospatial locations (significant places)→semantic meanings (e.g., restaurants)." Beyond the related methods, this solution has the following two advantages: 1) Modeling the location history of an individual and that of many users simultaneously, thereby making different users' location histories comparable and computable; 2) Modeling an individual's travel in geospatial and semantic spaces respectively, allowing deeper understanding of the individual's behavior and interests.

Given these advantages, this solution will be further introduced in later sections. This paradigm is further illustrated using Fig. 8.2 as an example, in which two users visited some locations and created two trajectories, Tr_1 and Tr_2, respectively.

Directly measuring these two users' location histories based on the GPS readings (denoted by points) is difficult for two reasons. First, the raw sensor readings of these two users are different even if they were visiting the same location, such as A and C. This is caused by the intrinsic positioning error of a location-acquisition technology and the randomness of people's movement (e.g., people exit a building from different gates). Second, defining a proper distance threshold (e.g. 100 meters) is often arbitrary in determining whether two readings belong to the same location. If we regard two points with a Euclidian distance smaller than 100 meters as readings from the same location, why do those points having a distance of 101 meters not also pertain to the same location? To address this issue, we need to convert a user's location history from sensor readings into a sequence of comparable locations in the geographic spaces, for instance, $Tr_1 : A \rightarrow C$, and $Tr_2 : A \rightarrow B \rightarrow C \rightarrow D$. Note that the focus is on the significant places like A and B where an individual carried out some meaningful behavior (reflecting her interests), such as shopping and watching a movie, instead of some points generated when an individual passes by a location like a crossroad without taking any essential action. This process will be discussed in more detail in Section 8.2.2.

However, knowing an individual's movement in the geographic spaces is not enough to understand the individual's interests. The semantic meaning of a physical location, e.g., a shopping center, will bring richer knowledge and context to explore a user's behavior. Given this reason, a user's trajectory is further converted from "$Tr_1 : A \rightarrow C$" to "a lake$\rightarrow$ a shopping center," thereby modeling the user's location history in terms of semantic spaces. Refer to Section 8.2.3 for details.

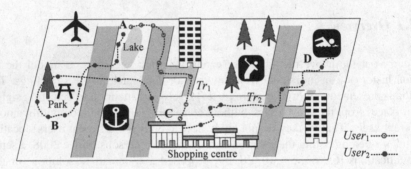

Fig. 8.2 Modeling the location history of a user from sensor data

8.2.2 Geospatial Model Representing User Location History

In this section, a framework is proposed, called a hierarchical graph, to uniformly model each individual's location history in the geospatial spaces [63, 28]. The framework consists of the following three steps, which are further illustrated in Fig. 8.3 and respectively detailed in later sections.

1) Detect significant places: The stay points are determined, each of which denotes a geographic region where an individual stayed for a certain duration, from the trajectory data. As compared to a raw sensor reading, each stay point carries a particular semantic meaning, such as the shopping malls and restaurants visited by an individual. Refer to Section 8.2.2.1 for details.

2) Formulate a shared framework: All users' stay points are placed together into a dataset. Using a density-based clustering algorithm, this dataset is recursively clustered into several clusters in a divisive manner. Thus, similar stay points from various users are assigned to the same cluster, and the clusters on different layers represent locations (geographical regions) of different granularities. This structure of clusters, referred to as a hierarchical framework, provides various users with a uniform framework to formulate their own graphs. Refer to the middle box shown at the bottom of Fig. 8.3.

3) Construct a personal location history: By projecting the individual location history onto the shared hierarchical framework, each user can build a personal directed-graph, in which a graph node is the cluster containing the user's stay points and a graph edge stands for the user's traveling sequence between these clusters (geographic regions). To simplify the problem, we do not differentiate between the diverse paths that a user created between two places (clusters).

In later sections, GPS logs are used as an exemplary trajectory to illustrate the methodology. Of course, this solution can be applied to other trajectory data sources, such as geo-tagged photos.

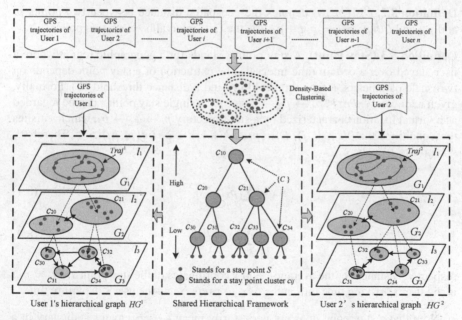

Fig. 8.3 Framework for modeling users location history in geographical spaces

8.2.2.1 Detecting Significant Places

A significant place denotes the location where an individual carried out some meaningful behavior (reflecting her interests), such as shopping, watching a movie, or
visiting a museum. These significant places allow a better understanding of an
individual's interests, thereby accurately computing the interdependency between
different users. At the same time, other sensor readings outside of these significant
places can be skipped, saving the computational load in a real system. Literature
that introduces a method for detecting significant places includes [10, 28, 8, 24].
Basically, they share the idea of using a spatial and temporal constraint to delineate
a location from a sequence of GPS coordinates. One representative method [28, 48]
is selected and introduced below. Before going into detail, it is necessary to first
define some terms that will be used in Chapters 8 and 9.

Definition 8.1 (GPS Trajectory). A GPS trajectory $Traj$ is a sequence of time-
stamped points, $Tra = p_0 \rightarrow p_1 \rightarrow \cdots \rightarrow p_k$, where $p_i = (x,y,t),(i = 0,1,\ldots,k)$;
(x,y) are latitude and longitude respectively, and t is a timestamp. $\forall 0 \leq i \leq k, p(i+1).t > p_i.t$.

Definition 8.2. $Dist(p_i,p_j)$ denotes the geospatial distance between two points p_i
and p_j, and $Int(p_i,p_j) = |p_i.t - p_j.t|$ is the time interval between two points.

Definition 8.3 (Stay Point). A stay point s stands for a geographic region where a
user stayed over a certain time interval. The extraction of a stay point depends on
two scale parameters, a time threshold (τ) and a distance threshold (δ). Formally,
given a trajectory, $Traj : p_1 \rightarrow p_2 \rightarrow \cdots \rightarrow p_n$, a single stay point s can be regarded
as a virtual location characterized by a sub-trajectory $p_i \rightarrow \cdots \rightarrow p_j$, which satisfies
the conditions that $\forall k \in [i,j)$, $Dist(p_k,p(k+1)) < \delta$, $Int(p_i,p_j) > \tau$. Therefore,
$s = (x,y,t_a,t_l)$, where

$$s.x = \sum_{k=i}^{j} p_k.x/|s|, \tag{8.1}$$

$$s.y = \sum_{k=i}^{j} p_k.y/|s|, \tag{8.2}$$

respectively stands for the average x and y coordinates of the stay point s; $s.t_a = p_i.t$
is the user's arriving time on s and $s.t_l = p_j.t$ represents the user's departure time.

Note that a stay point does not necessarily mean a user remains stationary in a
location. Also, we do not expect to include the circumstance when an individual
is stuck in a traffic jam or waiting for a traffic signal. Instead, we aim to detect
the significant stays reflecting the semantic meanings of an individual's behavior
and interests, which usually occur in the following two situations. One is when
people enter a building and lose satellite signal over a time interval before coming
back outdoors. Figure 8.4 A) shows an example in which an individual visited a
shopping mall and stayed inside for a period of time. The other situation is when a

user exceeds a time limit at a certain geospatial area (outdoors). For instance, people strolling along a nice beach (refer to Fig. 8.4 B)), or being attracted by a landmark (See Fig. 8.4 C)) could generate a stay point.

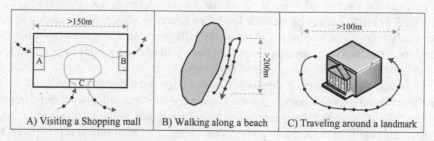

A) Visiting a Shopping mall	B) Walking along a beach	C) Traveling around a landmark

Fig. 8.4 Some examples of stay points

Figure 8.5 demonstrates the algorithm for stay point detection, using a trajectory ($p_1 \to p_2 \to \cdots \to p_7$). Overall, the stay point detection algorithm includes two operations: checking spatio-temporal constraint and expanding. As depicted in Fig. 8.5 B), p_1 and p_2 cannot formulate a stay point as $Dist(p_1, p_2)$ exceeds the corresponding threshold δ. Then, we move to p_2 and find that $Dist(p_2, p_3) < \delta$ and $Dist(p_2, p_4) < \delta$ while $Dist(p_2, p_5) > \delta$ (see Fig. 8.5 C)). If the time interval between p_2 and p_4 is larger than time threshold τ, the three points form a small cluster representing a stay point. However, they might not be the entire set of the points in this stay. Accordingly, we try to expand the stay point by continuously checking the distance between p_4 and the remaining points (p_5, p_6, p_7) in the trajectory. As depicted in Fig. 8.5 D), p_5 and p_6 are added into this stay point since they also meet the spatio-temporal constraints. Finally, we detect ($p_2 \to p_3 \to p_4 \to p_5 \to p_6$) as a stay point because we cannot expand the cluster any further. That is, all the points in the cluster have a distance farther than δ to p_7.

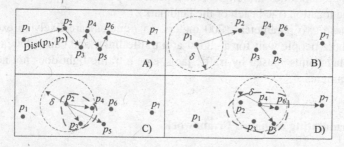

Fig. 8.5 An example of stay pints detection

At this point, some people might ask why not use an already existing clustering algorithm like DBSCAN to determine the stay points. There are two reasons for not doing so. On the one hand, as depicted in Fig. 8.4 A), if stay points are detected by directly clustering raw GPS points, most significant places like shopping malls and restaurants will remain undetected. This is caused by the fact that GPS devices lose satellite signal indoors, i.e., few GPS points will be generated at those places. However, some places like intersections passed by many people will be identified as stay points. On the other hand, if we use an interpolation operation (to fill the lost GPS points), the computational load for clustering such a big dataset will be extremely heavy. For instance, a 2-hour stay will generate 720 points if we use 5-seconds as the interpolating frequency. The workload of clustering is impractical for a real system with an increasing number of users.

However, the selection of thresholds δ and τ for the algorithm is still not easy and depends on people's commonsense knowledge. For example, in the experiment of [28, 63], if an individual spent more than 15 minutes within a distance of 200 meters, the region is detected as a stay point. Although the aim is to represent each stay of a user as precisely as possible, we have to use a proper geo-region to specify an individual's stay for a number of reasons.

First, a strict region size, such as 20×20 meters, might be more capable of accurately identifying a business like a Starbucks visited by a user; however, it would cause many stays to remain undetected. As demonstrated in Fig. 8.4 A), a user could enter a shopping mall from Gate A while leaving the mall from Gate B (see the blue line). Given that a shopping mall could cover a 150×150 meter geo-region, the distance between the last GPS point before entering the mall and the first point after coming out from the mall could be larger than 150 meters; i.e., the user's stay at this shopping mall cannot be detected using a very small region constraint like 20 meters. Moreover, even if a user leaves the shopping mall from the same gate they entered, in most cases, the distance between the last GPS point before entering and the first point after coming out could be larger than 100 meters. Typically, GPS devices need some time to re-locate themselves after returning outdoors.

Second, a very small region constraint could cause the stays of people to be over-detected. As shown in Fig. 8.4 B) and C), multiple trivial stay points could be detected in one location when people stroll along a beach or wander around a landmark. The data does not align with a person's perceptions that she has only accessed one location (the beach or the landmark).

Third, these two parameters (200 meters, 15 minutes) are likely to exclude a situation where people wait for a signal at a traffic light, and can reduce to some extent the stay points caused by traffic jams, e.g., a traffic light does not normally last for 10 minutes.

8.2.2.2 Formulating a Shared Framework

After detecting the stay points from an individual's GPS trajectories, we can model the individual's location history with a sequence of stay points, which is defined as follows:

Definition 8.4 (Location histroy). Generally, location history is a record of locations that an entity visited in geographical spaces over an interval of time. In this book, an individual's location history (LocH) is represented by a sequence of stay points (s) they visited with the corresponding arrival and departure times.

$$LocH = (s_1 \xrightarrow{\Delta t_1} s_2 \xrightarrow{\Delta t_2} \cdots \xrightarrow{\Delta t_{n-1}} s_n) \qquad (8.3)$$

where $s_i \in S$ and $\Delta t_i = s_{i+1}.t_a - s_i.t_l$.

However, different people's location histories represented by a sequence of stay points are still inconsistent and incomparable even if they visited the same place. Besides the intrinsic positioning error of GPS sensors (as explained in Section 8.3.1), people accessing a location in a variety of ways, such as different directions, entrances, and exits, could also generate a variety of stay points in the location. Figure 8.4 A) gives an example where two users visit one building from two different gates and generate two different stay points.

To uniformly model each individual's location history, a shared framework that is formulated by hierarchically clustering all users' stay points is proposed, as formally defined in Definition 8.5.

Definition 8.5 (Shared Hierarchical Framework F**).** F is a collection of stay point-based clusters C with a hierarchy structure L. $F = (C, L)$, where $L = l_1, l_2, \ldots, l_n$ denotes the collection of layers of the hierarchy. $C = \{c_{ij} | 1 \le i \le |L|, 0 \le j \le |C_i|\}$, where c_{ij} denotes the jth cluster of stay points on layer $l_i (l_i \in L)$, and C_i is the collection of clusters on layer l_i.

Figure 8.3 illustrates the process for formulating a shared hierarchical framework. The stay points from different users are placed into one dataset, and recursively clustered into several clusters in a divisive manner using a density-based clustering algorithm, such as DBSCAN or OPTICS [7]. As compared to an agglomerative method like K-Means, these density-based approaches are capable of detecting clusters with irregular structures, which may stand for a set of nearby restaurants, a beach, or a shopping street.

As a result, the similar stay points from different users are assigned to the same cluster, and the clusters on different layers denote locations of different granularities. From the top to the bottom of the hierarchy, the geospatial scale of these clusters decreases while the granularity of the locations (corresponding to the clusters) becomes finer. For example, cluster c_{20} on the second layer is divided into two clusters c_{30} and c_{31} on the third layer. So, c_{30} and c_{31} have a smaller size in geographical spaces while they are with a finer granularity than c_{20}. This hierarchical feature is useful for differentiating people with different degrees of similarity. Intuitively, people sharing the same location histories on a deeper layer might be more correlated than on a higher layer. For instance, people visiting the same museum are more likely to be similar than those visiting the same city.

Overall, the shared hierarchical framework provides different users with a uniform foundation to re-formulate their own location history, which looks like a hierarchical graph. Meanwhile, this shared framework is the model representing the location histories of a myriad of users in a LBSN.

8.2.2.3 Constructing Personal Location History

A user's personal hierarchical graph can be constructed by substituting each stay point in the user's original location history (refer to Definition 8.4) with the cluster (from different layers of the shared framework) the stay point pertains to. For example, as illustrated in Fig. 8.3, User 2's location history is originally represented as

$$LocH = s_1 \xrightarrow{\Delta t_1} s_2 \xrightarrow{\Delta t_2} s_3 \xrightarrow{\Delta t_3} s_4 \xrightarrow{\Delta t_4} s_5 \xrightarrow{\Delta t_5} s_6 \xrightarrow{\Delta t_6} s_7 \xrightarrow{\Delta t_7} s_8. \tag{8.4}$$

After projecting these stay points onto the third layer of the shared framework, User 2's location history can be transferred to,

$$LocH = c_{31} \xrightarrow{\Delta t_1} c_{34} \xrightarrow{\Delta t_2} c_{33} \xrightarrow{\Delta t_3} c_{32} \xrightarrow{\Delta t_4} c_{31} \xrightarrow{\Delta t_5} c_{32} \xrightarrow{\Delta t_6} c_{32} \xrightarrow{\Delta t_7} c_{31}. \tag{8.5}$$

Where c_{ij} is the jth cluster on the ith layer. For instance, s_1, s_5, s_6, and s_8 belong to cluster c_{31}. Further, we merge the same cluster (like c_{31}) continuously appearing in a user's location history.

$$LocH = c_{31} \xrightarrow{\Delta t_1} c_{34} \xrightarrow{\Delta t_2} c_{33} \xrightarrow{\Delta t_3} c_{32} \xrightarrow{\Delta t_4} c_{31} \xrightarrow{\Delta t_6} c_{32} \xrightarrow{\Delta t_7} c_{31}. \tag{8.6}$$

This transformation from a stay point to a cluster ID is performed on each layer of the shared framework. As a result, User 2's location history is denoted as a set of sequences of clusters. Since a user could visit a cluster multiple instances at different times, the presentation of a user's location (in sequences) looks more like a hierarchical graph. Generally speaking, the personal hierarchical graph is the integration of two structures: a shared hierarchical framework F and a graph G on each layer of the F. The tree expresses the parent-children (or ascendant-descendant) relationships of the nodes pertaining to different levels, and the graphs specify the peer relationships among the nodes on the same level. Refer to the bottom part of Fig. 8.3 for two examples.

8.2.3 Semantic Model Representing User Location History

In this section, an individual's stay in the physical world is provided with some semantic meanings, e.g., "museum→ cinema → restaurant," aiming to transfer human location history from the geographical spaces into semantic spaces. The semantic meaning of a location reveals the interests of an individual better than its original geo-position, and enables detection of similar users without any overlapping of geographic spaces, e.g., people living in different cities.

Expanding the method introduced in Section 8.2.2, [39] proposed a solution that is comprised of the following three steps: 1) stay point representation in semantic spaces, 2) the formulation of a shared semantic framework, and 3) the construction of personal location histories. The major difference between this method and that designed for geographical spaces lies in the first step. The three steps are detailed in the following subsections.

8.2.3.1 Stay Point Representation

This step aims to represent a stay point (detected in Section 8.2.2.1) with the semantic meaning (e.g., a restaurant) of the location where the stay occurred. However, it is almost impossible to identify the exact point of interest (POI) an individual has visited given a stay point, because of the GPS positioning error and the crowded distribution of POIs in a city. In practice, as shown in Fig. 8.6, a GPS reading usually has a 10-meter or more error in its real position. Accordingly, there could be multiple POIs of different categories involved in this distance. Unfortunately, the nearest POI to the center of a stay point may not be the actual place that an individual visited. What is worse, many POIs, like restaurants, shopping malls, and cinemas, often overlap in the same building.

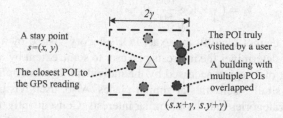

Fig. 8.6 Challenges in discovery of the semantic meaning of a stay point

Due to the challenge mentioned above, it is necessary to first expand a stay point to a stay region covering the POI that a user has visited. For example, as depicted in Fig. 8.6, a stay point s is expanded to a region $[s.x - \gamma, s.x + \gamma] \times [s.y - \gamma, s.y + \gamma]$ where γ is a parameter formulating a bounding box. The value of γ is related to the threshold δ for detecting a stay point.

After that, a feature vector is constructed for each stay region according to the POIs located in a region (defined in Definition 8.6). Here, TF-IDF (term frequency-inverse document frequency) [32, 33], a statistical measurement used to evaluate how important a word is to a document in a collection or corpus, is employed. The importance increases proportionally to the number of times a word appears in the document but is offset by the frequency of the word in the corpus.

Similarly, the method proposed in [39] regards categories of POIs as words and treats stay regions as documents. Intuitively, if POIs of a category occur in a region many times, this POI category is important in representing this region. Furthermore, if a POI category (e.g., "museum" and "natural parks") occurs rarely in other regions, the category is more representative for the region (in which it is located) beyond a common POI category, e.g., "restaurant," which appears in many places. Thus, both the occurrence frequency of a POI category in a region (similar to TF) and the inverse location frequency (equivalent to IDF) of this category have been considered in [39]. Combining these two factors, the feature vector is defined as follows:

Definition 8.6 (Feature Vector). The feature of a stay region r in a collection of regions R is $f_r = <w_1, w_2 \ldots, w_K>$, where K is the number of unique POI categories in a POI database and w_i is the weight of POI category i in the region r. The value of w_i is calculated as Eq. 8.7:

$$w_i = \frac{n_i}{N} \times \log \frac{|R|}{|\{\text{Regions containing i}\}|} \qquad (8.7)$$

Suppose that s_1 contains two restaurants and one museum, and s_2 only has four restaurants. The total number of stay regions created by all the users is 100, in which 50 have restaurants and two contain museums. So, the feature vectors of s_1 and s_2 are f_1 and f_2 respectively:

$$f_1 = (\frac{2}{3} \times \log \frac{100}{50}, \frac{1}{3} \times \log \frac{100}{2}, \ldots),$$
$$f_2 = (\frac{4}{4} \times \log \frac{100}{50}, 0, \ldots).$$

Although we still cannot identify the exact POI category visited by an individual, this feature vector determines the interests of a user to some extent by extracting the semantic meaning of a region accessed by the individual. For example, people are likely to conduct similar activities at similar places. Also, users visiting locations with similar POI categories may have similar interests. Consequently, the representation of a stay point carries advanced semantic information (beyond its geographical position), contributing to a broad range of applications in LBSNs, such as the calculating of similarity between two users in terms of their location histories and activity inferences.

8.2.3.2 Building a Semantic Location History

Step 2: Formulating a shared semantic framework: The second step clusters the stay regions into groups according to their feature vectors. The stay regions in the same cluster can be regarded as locations of similar type with similar semantic meanings. However, a flat clustering is insufficient to differentiate similar users of different extents. Intrinsically, we are more capable of discriminating similar users given categories with a finer granularity. For example, "restaurant" helps identify users who like dining out, while "Indian restaurant" and "Japanese restaurant" enable us to differentiate people interested in different types of food.

Considering this factor, the feature vectors are hierarchically clustered in a divisive manner, building a tree-structured sematic location hierarchy. This is similar to generating a shared framework in the geographical spaces (refer to Section 8.2.2.2). As shown in the middle part of Fig. 8.7, feature vectors of all users are placed into

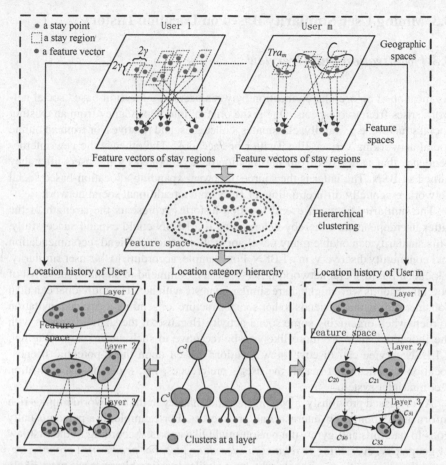

Fig. 8.7 Modeling human location history in semantic spaces

one cluster and this cluster is treated as the root (i.e., cluster at layer 1). Each cluster c at layer j ($j \geq 0$) is split into a set of sub-clusters by using a flat clustering algorithm. The resulting sub-clusters of c are considered c's child nodes at layer $j+1$. This procedure repeats a given number of times, leading to a tree-structured hierarchy where clusters at a lower layer have a finer granularity.

Step 3: Construct personal location history: In the third step, a location history is constructed for each user based on the semantic location hierarchy and the user's stay points. Originally, a user's location history in the geographic spaces is represented by a sequence of stay points with the travel time between each two consecutive stay points. Then, on each layer of the semantic location hierarchy, a stay point is substituted with the semantic location that the stay point's feature vector pertains to. After this projection, different users' location histories become comparable in the semantic spaces.

8.3 Mining User Similarity Based on Location History

8.3.1 Motivation and Overview

As mentioned before, the connection between users in a location-based social network arises from two aspects. One is the original interdependency from an existing social structure, e.g., family, classmates, colleagues, and relatives, or from an online social networking service like Twitter or Facebook. The other is the new interdependency that is derived from the location data generated by the users after they joined a LBSN. The latter is the source of power expanding a location-based social network, essentially differentiating a LBSN from a traditional social network.

The similarity between users' location histories represents the strength of the latter interdependency, thereby determining if a LBSN could expand successfully. This similarity can enable many novel applications, such as friend recommendation and community discovery, in a LBSN. For example, according to this user similarity, a location-based social networking service can recommend to an individual a list of potential friends who might share similar interests with her. The individual can then consider adding these friends to her social structure, or sending a targeted invitation to them when organizing some social activity. Because of the shared interests with the individual, they are more likely to be receptive to such an invitation. Further, a LBSN service can discover new locations (based upon these potential friends' location histories) that match the user's preferences, i.e., a personalized location recommender system.

As discussed previously, a person's location history in the real world implies rich information about their interests and preferences. For example, if a person usually goes to stadiums and gyms, the person might like sports. According to the first law of geography, *everything is related to everything else, but near things are more related than distant things*, people who have similar location histories are more likely to share similar interests and preferences. The more location histories they share, the more correlated these two users would be. Note that the location history mentioned here includes its representation in both geographical and semantic spaces. This claim even makes more sense in the semantic spaces as compared to geographical spaces. That is, people accessing locations with similar semantic meanings like a cinema are more likely to be similar.

[28] is the first publication proposing a framework to estimate the similarity between users in terms of their location histories, followed by a series of similar work [39, 16, 25, 44]. In this framework, the similarity between each pair of users is calculated according to two steps. First, find a set of similar subsequences shared by two users on each layer of their hierarchical graphs. Here a similar sequence stands for two individuals who have visited the same sequence of places for similar time intervals. Second, given the similar sequences, calculate a similarity score for the pair of users involving the following three factors:

- Sequential property of users' movements: This framework takes into account not only the locations they accessed, but also the sequence in which these locations were visited. The longer the similar sequences shared by two users' location histories are, the more related these two users might be.

- Hierarchical property of geographic spaces: This framework mines user similarity by exploring movements on different scales of geographic (or, semantic) spaces. Users who share similar location histories on a space of finer granularities might be more correlated. For example, people accessing the same building could be more similar then those visiting the same city. In this example, a building belongs to a lower layer of the geographic hierarchy than the city. This claim also holds in semantic spaces. For instance, two users sharing an interest in dining at Chinese restaurants might be more similar than others who generally like dinning in any restaurant. Here, the Chinese restaurant is a subset of restaurants, thereby having a finer granularity.

- Popularity of different locations: Analogous to inverse document frequency (IDF) [34], the proposed framework considers the visited popularity of a location when measuring the similarity between users. Two users who access a location visited by a few people might be more correlated than others who share a location history accessed by many people. For example, a myriad of people have visited the Great Wall, a well-known landmark in Beijing. It might not mean all these people are similar to one another. If two users visited a small museum, however, they might indeed share some similar preferences.

The input of this framework is the location histories (i.e., two hierarchical graphs) of two users in geographical or semantic spaces, and the output is a similarity score indicating how similar these two users are.

8.3.2 Detecting Similar Sequences

In this step, the sub-sequences shared by two users at each layer of their hierarchical graph are determined. Intuitively, users sharing the habit of "cinema→ restaurant→ shopping" are more similar to each other than those visiting these three places separately or in a different order. Therefore, the simple method counting the number of items shared by two sequences will lose a great deal of information about an individual's behavior and preferences. To address this issue, we must consider both the order of visitation and the travel time between two locations when detecting similar sub-sequences. Under the circumstances, *Travel Match* and *Maximum Travel Match* are defined as follows:

Notation: Given sequence $Seq = (c_1 \xrightarrow{\Delta t_1} c_2 \xrightarrow{\Delta t_2} \cdots \xrightarrow{\Delta t_{m-1}} c_m)$, we denote the i-th item of Seq as $Seq[i]$ (e.g., $Seq[1]=c_1$) and represent its subsequence as $Seq[a_1, a_2, \ldots, a_k]$ where $1 \le a_1 < a_2 < \cdots \le m$, for instance, $Seq[1,3,6,7]= c_1 \to c_3 \to c_6 \to c_7$.

Definition 8.7 (Travel Match). Given a temporal constraint factor $\rho \in [0,1]$ and two sub-sequences $Seq_1[a_1, a_2, \ldots, a_k]$ and $Seq_2[b_1, b_2, \ldots, b_k]$ from two sequences

Seq_1 and Seq_2 respectively, these two sub-sequences formulate a k-length travel match if they hold the following two conditions:

1. $\forall i \in [1,k]$, $a_i = b_i$, and

2. $\forall i \in [1,k)$, $\frac{|\triangle t_i = \triangle t_i'|}{\max(\triangle t_i, \triangle t_i')} \le p$, where $\triangle t_i$ is the travel time between a_i and a_{i+1}, and $\triangle t_i'$ denotes that between b_i and b_{i+1}.

This travel match is represented by $(a_1, b_1) \to (a_2, b_2) \to \cdots \to (a_k, b_k)$.

Definition 8.8 (Maximum Travel Match). A travel match $(a_1, b_1) \to (a_2, b_2) \to \cdots \to (a_k, b_k)$ between two sequences Seq_1 and Seq_2 is a maximum travel match if,

1. No left increment: $\nexists a_0 < a_1, b_0 < b_1$, s.t.,
$(a_0, b_0) \to (a_1, b_1) \to (a_2, b_2) \to \cdots \to (a_k, b_k)$;
2. No right increment: $\nexists a_{k+1} > a_1, b_{k+1} > b_k$, s.t.,
$(a_1, b_1) \to (a_2, b_2) \to \cdots \to (a_k, b_k) \to (a_{k+1}, b_{k+1})$;
3. No internal increment: $\forall i \in [1,k], \nexists a_i < a_{i'} < a_{i+1}$ and $b_i < b_{i'} < b_{i+1}$, s.t.,
$(a_1, b_1) \to (a_2, b_2) \to \cdots \to (a_i, b_i) \to (a_{i'}, b_{i'}) \to (a_{i+1}, b_{i+1}) \to \cdots \to (a_k, b_k)$

Essentially, a travel match is a common sequence of locations visited by two users in a similar amount of time, and a maximum travel match is a travel match that is not contained in any other travel matches. Note that 1) the locations in a travel match do not have to be consecutive in the user's original location history, and 2) what we need to detect for the calculating of user similarity are the maximum travel matches. Additionally, the location in a travel match can be a cluster of stay points in the geographical spaces, or a cluster in semantic spaces.

Figure 8.8 demonstrates an example of a maximum travel match between two sequences Seq_1 and Seq_2. Here, a node stands for a location and the letter in a node represents the ID of the location. The numbers on the top of the box denotes the index of a node in a sequence, e.g., location A is the first node in both Seq_1 and Seq_2. The number appearing on a solid edge means the travel time between two consecutive nodes, and the number shown on a dashed edge denotes the duration that a user stayed in a location.

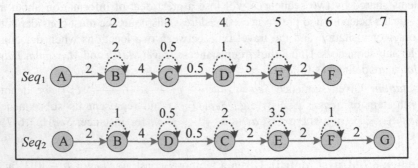

Fig. 8.8 An example of finding maximal travel match

Let $\rho = 0.2$ in this example. First, $(1,1) \rightarrow (2,2)$, i.e., $A \rightarrow B$, is a travel match, because the travel times $(A \rightarrow B)$ in Seq_1 and Seq_2 are identical, $|2 - 2|/2 = 0$. Then, we find that $(2,2) \rightarrow (3,4)$, i.e., $B \rightarrow C$, also satisfies the conditions defined in Definition 8.7. Though B and C are not directly connected in Seq_2, the travel time between these two locations is $4 + 0.5 + 0.5 = 5$, which is very similar to that of Seq_1. In short, $|5 - 4|/5 = 0.2$. However, both $A \rightarrow B$ and $B \rightarrow C$ are not the maximum travel match in this example as they are contained in $A \rightarrow B \rightarrow C$, i.e., $(1,1) \rightarrow (2,2) \rightarrow (3,4)$. Later, $C \rightarrow E$ and $C \rightarrow F$ cannot formulate travel matches due to the difference between corresponding travel times. Using the same approach, we find $(1,1) \rightarrow (2,2) \rightarrow (4,3) \rightarrow (5,5) \rightarrow (6,6)$, i.e., $A \rightarrow B \rightarrow D \rightarrow E \rightarrow F$, is another maximum travel match. Overall, we detect two maximum travel matches, $A \rightarrow B \rightarrow C$ and $A \rightarrow B \rightarrow D \rightarrow E \rightarrow F$ from Seq_1 and Seq_2.

Some well-known sequence matching algorithms, such as longest common subsequences (LCSS) searching [36]] and dynamic time wrapping (DTW) [43], cannot satisfy the need to discover the maximum travel matches as they do not incorporate the travel time between two locations in the matching process. Due to this reason, a method has been proposed in [39] for detecting the maximum travel matches from two sequences. This method consists of two steps, summarized as follows:

The first step detects the 1-length travel matches between two sequences and identifies a precedence relation between these 1-length matches. For example, A in Fig. 8.8, i.e., $(1,1)$, is a 1-length travel match between Seq_1 and Seq_2, and A is a precedence of B. Then, the 1-length matches and their precedence relation are transferred into a precedence graph G, where a node is a 1-length match and an edge corresponds to the precedence relation between 1-length matches.

The second step searches graph G for the maximum length path which has been proved equivalent to the maximum matches.

Following the case illustrated in Fig. 8.8, Fig. 8.9 shows an example of building graph G based on Seq_1 and Seq_2. As demonstrated in Fig. 8.9 A), the identical items in two sequences are first detected by putting these two sequences into a matching matrix. The numbers that stand on the top and left of the matrix denote the index of an item in a sequence. For example, A_{11} means that A is the first item in both sequences. In Fig. 8.9 B), each node corresponds to a trivial match, and an edge between two nodes stands for a precedent relation between two trivial matches. The number in a node indicates its order being added to the graph. For instance, F_{66} is the first node being added to graph G.

After the graph building process, precedence graph G is a directed acyclic graph in which a path represents a travel match (between two sequences). More specifically, if $(a_1, b_1) \rightarrow (a_2, b_2) \rightarrow \cdots \rightarrow (a_k, b_k)$ is a path in G, $Seq_1[a_1, a_2, \ldots, a_k]$ and $Seq_2[b_1, b_2, \ldots, b_k]$ form a travel match, and vice versa. Meanwhile, path P in G corresponds to a maximum travel match if the first node of P has zero in-degree and the last node has zero out-degree. For instance, path $A_{11} \rightarrow B_{22} \rightarrow C_{34}$ in Fig. 8.9 b) corresponds to the maximum travel match $(1,1) \rightarrow (2,2) \rightarrow (3,4)$ in Fig. 8.8.

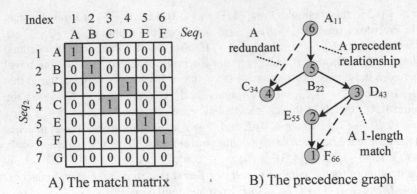

A) The match matrix   ·   B) The precedence graph

Fig. 8.9 The precedence graph for Seq_1 and Seq_2

8.3.3 Calculating Similarity Scores

After detecting the maximum travel matches from two users' location sequences, a similarity score can be calculated for the two users according to the following three factors: visited popularity of a location, sequential properties, and hierarchical properties, which were introduced in the beginning of Section 8.3.2 and are formally defined in Eq. 8.8, 8.9, 8.10, and 8.11.

$$SimUser(LocH_1, LocH_2) = \sum_{l=1}^{L} f_w(l) \times SimSq(Seq_1^l, Seq_2^l); \qquad (8.8)$$

$$SimSq(Seq_1, Seq_2) = \frac{\sum_{j=1}^{m} simTM(t_j)}{|Seq_1| \times |Seq_2|}, \qquad (8.9)$$

$$SimTM(s) = g_w(k) \times \sum_{i=1}^{k} vp(c_i); \qquad (8.10)$$

$$vp(c) = \log \frac{N}{n}, \qquad (8.11)$$

where N is the total number of users in the dataset and n is the number of users visiting location c.

Given two users' location histories $LocH_1$ and $LocH_2$, the similarity between them can be computed by summing up the similarity score at each layer of the hierarchical graph (refer to Definition 8.5 for details) in a weighted way. A function $f_w(l)$ is employed to assign a bigger weight to the similarity of sequences occurring at a lower layer, e.g., $f_w(l) = 2^{l-1}$, where l is the depth of a layer in the hierarchy.

Then, the similarity between two sequences Seq_1 and Seq_2 at a layer, $SimSq(Seq_1, Seq_2)$, is represented by the sum of the similarity score, $simTM(t_j)$, of each maximum travel match between Seq_1 and Seq_2. Here, m is the total number of maximum matches. Meanwhile, $SimSq(Seq_1, Seq_2)$ is normalized by the production of the lengths of the two sequences, since a longer sequence has a higher probability of having long matches. That is, a user with a longer history of data is more likely to be similar to others (than a user having a shorter period of data) without performing the normalization.

Further, the similarity score of a maximum travel match t, $simTM(t)$, is calculated by summing up the vp (visited popularity) of each location c contained in t. At the same time, the $simTM(t)$ is weighted in terms of the length k of t, e.g., $g_w(k) = 2^{k-1}$. The insight leading to Eq. 8.10 and 8.11 is based on two aspects. First, the longer the similar sequences shared by two users' location histories, the more related these two users are likely to be (this is known as the sequential property). Second, users who have accessed a location visited by a few people might be more correlated than others who share a location history accessed by many people (the visited popularity of a location). According to the experimental results, it was discovered that the number of shared sub-sequences exponentially decreases with the increase of the length of the sub-sequence. So, in the implementation, it's preferable to use an exponential weight function, assigning a higher weight to the longer sequences.

Note that this framework can be applied to the location history modeled either in geographical or semantic spaces. Specifically, when applying this framework to the location history in geographical spaces, a location is a cluster of stay points as depicted in Fig. 8.3, while a location is replaced by a group of semantic features in the semantic spaces illustrated in Fig. 8.7.

8.4 Friend Recommendation and Community Discovery

8.4.1 Methodology

With the user similarity calculated above, we can hierarchically cluster users into groups in a divisive manner by using some clustering algorithms like K-mean. Consequently, as depicted in Fig. 8.10, we can build a user cluster hierarchy, where a cluster denotes a group of users sharing some similar interests and different layers represent different levels of similarity. The clusters shown on a higher layer could stand for big communities in which people share some high-level interests, such as sports. The clusters occurring on the lower layers denote people sharing some narrower interests, like hiking (the layer of the hierarchy can be determined based on the needs of applications). Meanwhile, we can find one representative user (the center) for each cluster according to the similarity scores between each pair of users. For instance, the individual with the minimal distance to other users in the cluster (the individual pertains to) can be selected as the representative user of the cluster.

This user hierarch brings us two types of advantages:

1) Fast retrieval of similar users: Instead of checking all the users, we can retrieve the top k similar users for an individual by only ranking the users from the same cluster (the individual belongs to) in terms of similarity score. This retrieval process

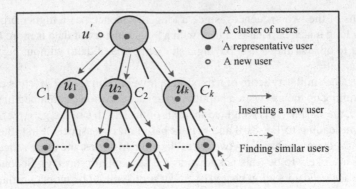

Fig. 8.10 Finding similar users and inserting new users in hierarchical user clusters

can start from the bottom layer of the hierarchy, as depicted by the blue dash arrow in Fig. 8.10. If the number of users is less than k in the bottom-layer cluster, we can further check the parent node (cluster) of this cluster until finding a cluster with more than k users.

2) Insert new users: When a new user u' enters the system, it is not necessary to compute the similarity score between u' and each user in the system. This process is very time consuming and will become more difficult as the number of users increases. Instead, we only need to insert this user into the most appropriate clusters on each layer of the hierarchy by computing the similarity between u' and the representative user in a cluster. For example, as demonstrated by the red solid arrows in Fig. 8.10, we first compute the similarity between u' and $(u_1, u_2, \ldots, u_k)$ who are representative users in each cluster. If u_2 is the most similar user to u' out of the k users, we insert u' into u_2's cluster C_2. Then, we further check the children clusters of C_2 and insert u' into the clusters whose representative user is the most similar to u'. This process is performed iteratively until reaching the bottom layer of the hierarchy.

In practice, we do not need to re-build this hierarchy unless the number of newly inserted users exceeds a certain threshold. That is, in most cases we can find similar users for a person very efficiently.

Evaluating the applications in a location-based social network, such as friend recommendation and community discovery, is a non-trivial research topic due to the following challenges: data, ground truth, and metrics.

8.4.2 Public Datasets for the Evaluation

The biggest challenge of the evaluation comes from the data, consisting of location data such as GPS trajectories and the social structure, of many users. To collect the data, a research group typically needs to deploy a location-based social networking service and encourage enough people to use this service in a certain period, e.g., 3 months. Without an online LBSN service, they could assign some location-acquisition devices like GPS loggers to a group of users and collect the data offline. Both ways are very time-consuming and resource-intensive, thereby becoming a major barrier to many professionals stepping into this field.

In recent years, a few real-world datasets created by some pioneers were made available on the Internet for free download, for example, "the reality mining dataset" [5] from MIT media laboratory and "GeoLife GPS Trajectories" [4] from Microsoft research. The reality mining dataset was collected by one hundred human subjects with a Bluetooth-enabled mobile phone over the course of nine months, representing 500,000 hours of data on users' location, communication, and device usage behavior.

The GeoLife GPS Trajectories was collected by 170 users with a GPS logger or GPS-phone (see Fig 8.11) over a period of four years (from April 2007 to the date when this book was published). This dataset is still growing and upgrading with an annual release. The latest version (released in July, 2011) is comprised of 17,085 GPS trajectories with a total distance over 1,000,000km and an effective duration over 48,000 hours. 95 percent of these trajectories are logged in a dense representation, e.g., every 2~5 seconds or every 5~10 meters per point. Figure 8.12 shows the distribution of this dataset in the urban area of Beijing, where the figures associated with the colored bar indicate the number of GPS points in a location.

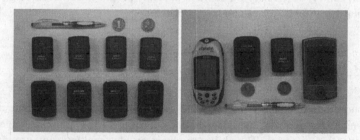

Fig. 8.11 GPS devices used for collecting data in GeoLife Project

This dataset recorded a broad range of users' outdoor movements, including not only daily routines like going to work but also some entertainment and sports activities, such as shopping, sightseeing, dining, hiking, and cycling. A part of these trajectories has a label of transportation modes including driving, riding a bike, taking a bus, and walking. These datasets provide professionals with a good resource to evaluate their early research into LBSN, significantly boosting the LBSN community. Detailed information can be found at the website [4].

The advent of some commercial LBSN services like Foursquare brings new opportunities to carry out evaluations using large-scale and real-world data. For example, Foursquare released an API set allowing LBSN researchers to crawl the publically available check-in records generated by users. The collected data includes the venue where a user checked in and a corresponding timestamp as well as the tips

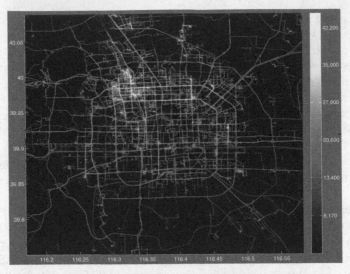

Fig. 8.12 The distribution of GeoLife dataset in the urban area of Beijing

that the user left in the venue. At the same time, the social structure of a user can be obtained by using this API. A great deal of research based on such data has been published [16, 41, 40], verifying some hypothesis proposed in LBSN, e.g., people with similar location histories can be correlated.

8.4.3 Methods for Obtaining Ground Truth

The second challenge stems from ground truth. For example, to evaluate a friend recommendation, we need to rank people according to the similarity inferred in terms of location histories. The ability to obtain an idea ranking (i.e., ground truth) is important. Generally speaking, there are two ways of generating ground truth. One is performing a questionnaire-style user study. The other is to extract ground truth from an individual's social structure [17, 38, 9, 31].

The former approach usually provides the users who collect location data for the research with a questionnaire inquiring about their interests. In the GeoLife project, for example, each user answered the questions shown in Fig. 8.13 A) by giving a rank (1~4) to denote different degrees of desire for an activity. A user's answer, e.g., Fig. 8.13 B), is regarded as an interest vector, in which each entry is the user's rank to a corresponding question. In the example, the user's interest vector is $< 3, 2, 1, 4, 3, 1, 3, 1, 2, 3, 1, 1 >$. A cosine similarity between two users' interest vectors can be calculated and used to rank a group of people for an individual. As a result, the top k people can be retrieved as a ground truth. Due to the intensive human effort, this kind of approach can only be applied to a small scale of people such as when using the Reality Mining Dataset and GeoLife GPS Trajectories.

Where do you like to go in weekends? Please rank from 1(dislike) to 4(favorite).	Example response
1. Shopping	3
2. Theatre	2
3. Karaoke	1
4. Go out for dinner	4
5. Outdoor sports, e.g., hiking	3
6. Indoor sports, e.g., gym and bowling	1
7. Natural parks	3
8. Exhibition, museum	1
9. Stay home; not go to any places	2
10. Go to office; over-time working	3
11. Visit parents, relatives, or friends	1
12. Campus	1
A)	B)

Fig. 8.13 A questionnaire A) and an example of answers B)

The latter approach uses the closeness between two users inferred from the connections in their social structure as the ground truth. For example, random walk theory [17] can be used to analyze the closeness of two nodes (i.e., friendship strength in a social network context) using the resistance distance, which is the random walk steps for the electrons traveling from one node to the other, in a social graph. However, the random walk theory completely overlooks semantic information contained in social networks. As a result, recently, more advanced research have been proposed to analyze the closeness between two users considering: 1) the similarity of their profile [38, 11], e.g., demographics like age, gender, and hometown, and 2) interaction activities [9, 22, 31], e.g., commenting, tagging, and group communication patterns.

8.4.4 Metrics for the Evaluation

The third challenge is the metric used to measure the effectiveness of inferred user similarity, given the data and ground truth. As mentioned before, user similarity is a metric specifying to what extent two users are similar to each other, instead of a binary value indicating whether two users are similar or not. The goal is to rank a group of people for an individual according to similarity and recommend the top k people as potential friends to the individual. Accordingly, it is natural to look at user similarity as an information retrieval problem.

Given an individual, the top k similar users to the individual can be retrieved according to their similarity scores (inferred by the approach mentioned previously). An idea rank can be formulated from the ground truth (which was obtained by using one of the methods mentioned in Section 8.4.3). Based on these two ranking lists, *MAP* (Mean Average Precision) and *nDCG* (Normalized Discounted Cumulated

Gain) are calculated for retrieval. After testing all users, a mean value of *MAP* and *nDCG* is computed respectively.

More specifically, when generating the ground truth for an individual, users are divided into groups according to their similarity scores to the individual. As demonstrated in Fig. 8.14, users are ranked in terms of the similarity scores to the individual, and then split into 5 classes: 0~4. The users in class 4 have a higher similarity score than those in a lower class. The split can be driven by evenly partitioning the similarity scores or by a uniform division of users. The number of classes is determined by application which can assign each cluster a semantic meaning as in the example shown in Fig. 8.14. Afterwards, the numeric value of a (ground truth) similarity is replaced by the class ID it pertains to.

In the testing phase, the top k users are retrieved for an individual according to our method (based on location history). Then, a ranking list e.g., $G = (U_3, U_2, \ldots, U_5)$ can be obtained. By replacing these user IDs with the corresponding class IDs, another ranking list, e.g., $G = (4, 3, 2, 3, , 0)$, is formulated. Now, a score for this ranking list can be calculated in terms of *nDCG*. *nDCG* is used to compute the relative-

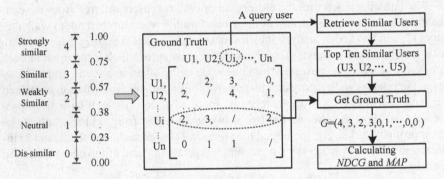

Fig. 8.14 Evaluation metric for user similarity detection

to-the-ideal performance of information retrieval techniques [26]. The discounted cumulative gain of G is computed as follows: (In our experiments, $b = 2$.)

$$DCG[i] = \begin{cases} G[1], & \text{if } i = 1 \\ DCG[i-1] + G[i], & \text{if } i < b \\ DCG[i-1] + \frac{G[i]}{\log_b i}, & \text{if } i \geq b \end{cases} \quad (8.12)$$

Given the ideal discounted cumulative gain DCG', then *nDCG* at i-th position can be computed as $nDCG[i] = DCG[i]/DCG'[i]$. According to Eq. 8.12, $nDCG[3]$ of $G = (4, 2, 3, 3, 0)$ can be calculated as follows:

$$DCG[1] = G[1] = 4;$$
$$DCG[2] = DCG[1] + G[2] = 4 + 2 = 6;$$
$$DCG[3] = DCG[2] + (G[3])/(\log_2 3) = 6 + 1.893 = 7.893;$$

However, the idea ranking should be $G' = (4, 3, 3, 2, 0)$. According to the same method, the $DCG'[3] = 8.893$. As a result,

$$nDCG[3] = \frac{DCG[3]}{DCG'[3]} = \frac{7.893}{8.893} = 0.888.$$

8.5 Summary

This chapter defined a location-based social network and discussed a research philosophy from the perspective of user and location. Three categories of location-based social networking services were classified in terms of the location data powering a service and a user's preferences in the service. Then, research focusing on understanding users in a location-based social network was gradually explored from modeling the location history of an individual to estimating the similarity between different users, and then moving to high-level applications, such as friend recommendation and community discovery. Some possible methods for evaluation of these applications were discussed, and a number of publically available datasets have been listed as well. All these efforts are enabled by the unprecedented wealth of user-generated trajectories.

References

1. Bikely. http://www.bikely.com
2. Flickr. http://www.flickr.com
3. Foursquare. https://foursquare.com
4. GeoLife GPS Trajectories. http://research.microsoft.com/en-us/downloads/b16d359d-d164-469e-9fd4-daa38f2b2e13/default.aspx
5. The Reality Mining Dataset. http://reality.media.mit.edu/dataset.php
6. Twitter. http://twitter.com
7. Ankerst, M., Breunig, M.M., Kriegel, H.P., Sander, J.: Optics: ordering points to identify the clustering structure. In: Proceedings of the 1999 ACM SIGMOD international conference on Management of data, SIGMOD '99, pp. 49–60. ACM, New York, NY, USA (1999)
8. Ashbrook, D., Starner, T.: Using gps to learn significant locations and predict movement across multiple users. Personal Ubiquitous Comput. **7**, 275–286 (2003)
9. Backstrom, L., Leskovec, J.: Supervised random walks: predicting and recommending links in social networks. In: Proceedings of the fourth ACM international conference on Web search and data mining, WSDM '11, pp. 635–644. ACM, New York, NY, USA (2011)
10. Cao, X., Cong, G., Jensen, C.S.: Mining significant semantic locations from gps data. Proc. VLDB Endow. **3**, 1009–1020 (2010)
11. Chen, J., Geyer, W., Dugan, C., Muller, M., Guy, I.: Make new friends, but keep the old: recommending people on social networking sites. In: Proceedings of the 27th international conference on Human factors in computing systems, CHI '09, pp. 201–210. ACM, New York, NY, USA (2009)

12. Chen, Y., Jiang, K., Zheng, Y., Li, C., Yu, N.: Trajectory simplification method for location-based social networking services. In: Proceedings of the 2009 International Workshop on Location Based Social Networks, LBSN '09, pp. 33–40. ACM, New York, NY, USA (2009)
13. Chen, Z., Shen, H.T., Zhou, X., Zheng, Y., Xie, X.: Searching trajectories by locations: an efficiency study. In: Proceedings of the 2010 international conference on Management of data, SIGMOD '10, pp. 255–266. ACM, New York, NY, USA (2010)
14. Chow, C.Y., Bao, J., Mokbel, M.F.: Towards location-based social networking services. In: Proceedings of the 2nd ACM SIGSPATIAL International Workshop on Location Based Social Networks, LBSN '10, pp. 31–38. ACM, New York, NY, USA (2010)
15. Counts, S., Smith, M.: Where were we: communities for sharing space-time trails. In: Proceedings of the 15th annual ACM international symposium on Advances in geographic information systems, GIS '07, pp. 10:1–10:8. ACM, New York, NY, USA (2007)
16. Cranshaw, J., Toch, E., Hong, J., Kittur, A., Sadeh, N.: Bridging the gap between physical location and online social networks. In: Proceedings of the 12th ACM international conference on Ubiquitous computing, Ubicomp '10, pp. 119–128. ACM, New York, NY, USA (2010)
17. Doyle, P.G., Snell, J.L.: Random walks and electric networks (1984)
18. Doytsher, Y., Galon, B., Kanza, Y.: Querying geo-social data by bridging spatial networks and social networks. In: Proceedings of the 2nd ACM SIGSPATIAL International Workshop on Location Based Social Networks, LBSN '10, pp. 39–46. ACM, New York, NY, USA (2010)
19. Eagle, N., de Montjoye, Y.A., Bettencourt, L.M.A.: Community computing: Comparisons between rural and urban societies using mobile phone data. In: Proceedings of the 2009 International Conference on Computational Science and Engineering - Volume 04, pp. 144–150. IEEE Computer Society, Washington, DC, USA (2009)
20. Eagle, N., Pentland, A., Lazer, D.: Inferring social network structure using mobile phone data. Proceedings of the National Academy of Sciences (PNAS) **106**, 15,274–15,278 (2007)
21. Eagle, N., (Sandy) Pentland, A.: Reality mining: sensing complex social systems. Personal Ubiquitous Comput. **10**, 255–268 (2006)
22. Gilbert, E., Karahalios, K.: Predicting tie strength with social media. In: Proceedings of the 27th international conference on Human factors in computing systems, CHI '09, pp. 211–220. ACM, New York, NY, USA (2009)
23. Goldberg, D., Nichols, D., Oki, B.M., Terry, D.: Using collaborative filtering to weave an information tapestry. Commun. ACM **35**, 61–70 (1992)
24. Hariharan, R., Toyama, K.: Project lachesis: Parsing and modeling location histories. In: Proceedings of the 3rd International Conference on Geographic Information Science, pp. 106–124 (2004)
25. Hung, C.C., Chang, C.W., Peng, W.C.: Mining trajectory profiles for discovering user communities. In: Proceedings of the 2009 International Workshop on Location Based Social Networks, LBSN '09, pp. 1–8. ACM, New York, NY, USA (2009)
26. Järvelin, K., Kekäläinen, J.: Cumulated gain-based evaluation of ir techniques. ACM Trans. Inf. Syst. **20**, 422–446 (2002)
27. Lee, R., Sumiya, K.: Measuring geographical regularities of crowd behaviors for twitter-based geo-social event detection. In: Proceedings of the 2nd ACM SIGSPATIAL International Workshop on Location Based Social Networks, LBSN '10, pp. 1–10. ACM, New York, NY, USA (2010)
28. Li, Q., Zheng, Y., Xie, X., Chen, Y., Liu, W., Ma, W.Y.: Mining user similarity based on location history. In: Proceedings of the 16th ACM SIGSPATIAL international conference on Advances in geographic information systems, GIS '08, pp. 34:1–34:10. ACM, New York, NY, USA (2008)
29. Liu, W., Zheng, Y., Chawla, S., Yuan, J., Xie, X.: Discovering spatio-temporal causal interactions in traffic data streams. In: The 17th ACM SIGKDD international conference on Knowledge Discovery and Data mining, KDD '11. ACM, New York, NY, USA (2011)
30. Nakamura, A., Abe, N.: Collaborative filtering using weighted majority prediction algorithms. In: Proceedings of the Fifteenth International Conference on Machine Learning, ICML '98, pp. 395–403. Morgan Kaufmann Publishers Inc., San Francisco, CA, USA (1998)

31. Roth, M., Ben-David, A., Deutscher, D., Flysher, G., Horn, I., Leichtberg, A., Leiser, N., Matias, Y., Merom, R.: Suggesting friends using the implicit social graph. In: Proceedings of the 16th ACM SIGKDD international conference on Knowledge discovery and data mining, KDD '10, pp. 233–242. ACM, New York, NY, USA (2010)
32. Salton, G., Buckley, C.: Term-weighting approaches in automatic text retrieval. Inf. Process. Manage. 24, 513–523 (1988)
33. Salton, G., Fox, E.A., Wu, H.: Extended boolean information retrieval. Commun. ACM 26, 1022–1036 (1983)
34. Sparck Jones, K.: A statistical interpretation of term specificity and its application in retrieval, pp. 132–142. Taylor Graham Publishing, London, UK, UK (1988)
35. Tang, L.A., Zheng, Y., Xie, X., Yuan, J., Yu, X., Han, J.: Retrieving k-nearest neighboring trajectories by a set of point locations. In: The 12th Symposium on Spatial and Temporal Databases (2011)
36. Vlachos, M., Gunopoulos, D., Kollios, G.: Discovering similar multidimensional trajectories. In: Proceedings of the 18th International Conference on Data Engineering, ICDE '02, pp. 673–684. IEEE Computer Society, Washington, DC, USA (2002)
37. Wang, L., Zheng, Y., Xie, X., Ma, W.Y.: A flexible spatio-temporal indexing scheme for large-scale gps track retrieval. In: Proceedings of the The Ninth International Conference on Mobile Data Management, pp. 1–8. IEEE Computer Society, Washington, DC, USA (2008)
38. Xiang, R., Neville, J., Rogati, M.: Modeling relationship strength in online social networks. In: Proceedings of the 19th international conference on World wide web, WWW '10, pp. 981–990. ACM, New York, NY, USA (2010)
39. Xiao, X., Zheng, Y., Luo, Q., Xie, X.: Finding similar users using category-based location history. In: Proceedings of the 18th SIGSPATIAL International Conference on Advances in Geographic Information Systems, GIS '10, pp. 442–445. ACM, New York, NY, USA (2010)
40. Ye, M., Shou, D., Lee, W.C., Yin, P., Janowicz, K.: On the semantic annotation of places in location-based social networks. In: The 17th ACM SIGKDD international conference on Knowledge Discovery and Data mining, KDD '11. ACM, New York, NY, USA (2011)
41. Ye, M., Yin, P., Lee, D.L., Lee, W.C.: Exploiting geographical influence for collaborative point-of-interests recommendation. In: The 34th international ACM SIGIR conference on Research and development in information retrieval, SIGIR '11. ACM, New York, NY, USA (2011)
42. Ye, Y., Zheng, Y., Chen, Y., Feng, J., Xie, X.: Mining individual life pattern based on location history. In: Proceedings of the 2009 Tenth International Conference on Mobile Data Management: Systems, Services and Middleware, MDM '09, pp. 1–10. IEEE Computer Society, Washington, DC, USA (2009)
43. Yi, B.K., Jagadish, H.V., Faloutsos, C.: Efficient retrieval of similar time sequences under time warping. In: Proceedings of the Fourteenth International Conference on Data Engineering, ICDE '98, pp. 201–208. IEEE Computer Society, Washington, DC, USA (1998)
44. Ying, J.J.C., Lu, E.H.C., Lee, W.C., Weng, T.C., Tseng, V.S.: Mining user similarity from semantic trajectories. In: Proceedings of the 2nd ACM SIGSPATIAL International Workshop on Location Based Social Networks, LBSN '10, pp. 19–26. ACM, New York, NY, USA (2010)
45. Yoon, H., Zheng, Y., Xie, X., Woo, W.: Smart itinerary based on user-generated gps trajectories. In: Proceedings of the 7th international conference on Ubiquitous intelligence and computing, UIC'10, pp. 19–34. Springer-Verlag, Berlin, Heidelberg (2010)
46. Yoon, H., Zheng, Y., Xie, X., Woo, W.: Social itinerary recommendation from user-generated digital trails. Personal and Ubiquitous Computing (2011)
47. Yuan, J., Zheng, Y., Xie, X., Sun, G.: Driving with knowledge from the physical world. In: The 17th ACM SIGKDD international conference on Knowledge Discovery and Data mining, KDD '11. ACM, New York, NY, USA (2011)
48. Yuan, J., Zheng, Y., Xie, X., Sun, G.: Where to find the next passenger. In: Proceedings of the 13th ACM international conference on Ubiquitous computing, Ubicomp '11. ACM, New York, NY, USA (2011)

49. Yuan, J., Zheng, Y., Zhang, C., Xie, W., Xie, X., Sun, G., Huang, Y.: T-drive: driving directions based on taxi trajectories. In: Proceedings of the 18th SIGSPATIAL International Conference on Advances in Geographic Information Systems, GIS '10, pp. 99–108. ACM, New York, NY, USA (2010)

50. Zheng, V.W., Cao, B., Zheng, Y., Xie, X., Yang, Q.: Collaborative filtering meets mobile recommendation: A user-centered approach. In: Proceedings of AAAI conference on Artificial Intelligence (AAAI 2010), pp. 236–241. ACM, New York, NY, USA (2010)

51. Zheng, V.W., Zheng, Y., Xie, X., Yang, Q.: Collaborative location and activity recommendations with gps history data. In: Proceedings of the 19th international conference on World wide web, WWW '10, pp. 1029–1038. ACM, New York, NY, USA (2010)

52. Zheng, Y., Chen, Y., Li, Q., Xie, X., Ma, W.Y.: Understanding transportation modes based on gps data for web applications. ACM Trans. Web 4, 1:1–1:36 (2010)

53. Zheng, Y., Chen, Y., Xie, X., Ma, W.Y.: Geolife2.0: A location-based social networking service. In: Proceedings of the 2009 Tenth International Conference on Mobile Data Management: Systems, Services and Middleware, MDM '09, pp. 357–358. IEEE Computer Society (2009)

54. Zheng, Y., Li, Q., Chen, Y., Xie, X., Ma, W.Y.: Understanding mobility based on gps data. In: Proceedings of the 10th international conference on Ubiquitous computing, UbiComp '08, pp. 312–321. ACM, New York, NY, USA (2008)

55. Zheng, Y., Liu, L., Wang, L., Xie, X.: Learning transportation mode from raw gps data for geographic applications on the web. In: Proceeding of the 17th international conference on World Wide Web, WWW '08, pp. 247–256. ACM, New York, NY, USA (2008)

56. Zheng, Y., Liu, Y., Xie, X.: Urban computing with taxicabs. In: Proceedings of the 13th ACM international conference on Ubiquitous computing, Ubicomp '11. ACM, New York, NY, USA (2011)

57. Zheng, Y., Wang, L., Zhang, R., Xie, X., Ma, W.Y.: Geolife: Managing and understanding your past life over maps. In: Proceedings of the The Ninth International Conference on Mobile Data Management, pp. 211–212. IEEE Computer Society, Washington, DC, USA (2008)

58. Zheng, Y., Xie, X.: Learning location correlation from gps trajectories. In: Proceedings of the 2010 Eleventh International Conference on Mobile Data Management, MDM '10, pp. 27–32. IEEE Computer Society, Washington, DC, USA (2010)

59. Zheng, Y., Xie, X.: Learning travel recommendations from user-generated gps traces. ACM Trans. Intell. Syst. Technol. 2, 2:1–2:29 (2011)

60. Zheng, Y., Xie, X., Ma, W.Y.: Geolife: A collaborative social networking service among user, location and trajectory. IEEE Data Eng. Bull. 33(2), 32–39 (2010)

61. Zheng, Y., Xie, X., Zhang, R., Ma, W.Y.: Searching your life on web maps. In: SIGIR Workshop on Mobile Information Retrieval (2008)

62. Zheng, Y., Yuan, J., Xie, W., Xie, X., Sun, G.: Drive smartly as a taxi driver. In: Proceedings of the 2010 Symposia and Workshops on Ubiquitous, Autonomic and Trusted Computing, UIC-ATC '10, pp. 484–486. IEEE Computer Society, Washington, DC, USA (2010)

63. Zheng, Y., Zhang, L., Ma, Z., Xie, X., Ma, W.Y.: Recommending friends and locations based on individual location history. ACM Trans. Web 5, 5:1–5:44 (2011)

64. Zheng, Y., Zhang, L., Xie, X., Ma, W.Y.: Mining correlation between locations using human location history. In: Proceedings of the 17th ACM SIGSPATIAL International Conference on Advances in Geographic Information Systems, GIS '09, pp. 472–475. ACM, New York, NY, USA (2009)

65. Zheng, Y., Zhang, L., Xie, X., Ma, W.Y.: Mining interesting locations and travel sequences from gps trajectories. In: Proceedings of the 18th international conference on World wide web, WWW '09, pp. 791–800. ACM, New York, NY, USA (2009)

Chapter 9
Location-Based Social Networks: Locations

Yu Zheng and Xing Xie

Abstract While chapter 8 studies the research philosophy behind a location-based social network (LBSN) from the point of view of users, this chapter gradually explores the research into LBSNs from the perspective of locations. A series of research topics are presented, with respect to mining the collective social knowledge from many users' GPS trajectories to facilitate travel. On the one hand, the generic travel recommendations provide a user with the most interesting locations, travel sequences, and travel experts in a region, as well as an effective itinerary conditioned by a user's starting location and an available time length. On the other hand, the personalized travel recommendations find the locations matching an individual's interests, which can be learned from the individual's historical data.

9.1 Introduction

The increasing availability of location-acquisition technologies and Internet access in mobile devices is fostering a variety of location-based services generating a myriad of spatio-temporal data, especially in the form of trajectories [45, 42, 41, 6]. These trajectories reflect the behavior and interests of users, thereby enabling us to better understand an individual and the similarity between different individuals [20, 34, 7, 15, 36]. Research and applications were introduced in Chapter 8 in which the users are the focus and locations are employed as enhanced information for better understanding them. Instead, this chapter discusses the research topics that aim at understanding locations based upon the collective social knowledge of users (e.g., the knowledge contained in their GPS trajectories) starting with generic travel recommendations [48, 44, 37, 38, 40] and then looking at personalized recommendations [46, 43, 39, 13, 33].

Yu Zheng · Xing Xie
Microsoft Research Asia, China
e-mail: {yuzheng, xing.xie}@microsoft.com

Regardless of an individual's preferences, the generic travel recommender systems mine a vast number of trajectories (generated by multiple users) and provide an individual with travel recommendations following a paradigm of "trajectories → interesting locations → popular travel sequences → itinerary planning → activities recommendation." Specifically, these recommender systems first infer the most interesting locations in a region from the given trajectories, and then detect the popular travel sequences among these locations [48]. An interesting location is defined as a culturally important place, such as Tiananmen Square in Beijing or the Statue of Liberty in New York (i.e., popular tourist destinations), and commonly frequented public areas, such as shopping malls/streets, restaurants, cinemas, and bars. With these interesting locations and travel sequences, an ideal itinerary can be planned for a user according to her departure location, destination, and available time [37, 38]. Finally, the generic travel recommendations provide users with some popular activities, e.g., dinning and shopping, that could be performed in a location [40]. All these recommendations mentioned above facilitate a user to travel to an unfamiliar place and plan a journey with minimal effort.

However, the personalized recommender systems learn an individual's interests from her personal location data (e.g., GPS trajectories) and suggest locations to the individual matching her preferences. Specifically, the personalized recommender uses the times that a particular individual has visited a location as her implicit ratings on that location, and estimates an individual's interests in unvisited places by considering her location history and those of other users [46, 44]. As a result, some locations with high ratings that might match the user's tastes can be recommended.

Two collaborative filtering (CF) models are individually used to infer a user's ratings of these unvisited locations. First, the personalized location recommendation is equipped with a user-based CF model, which employs user similarity introduced in Section 8.3 as a distance function between different users [46]. This model is able to capture people's mobility, such as the sequential and hierarchical properties of human movement in the physical world, while suffering from poor scalability caused by the heavy computation of user similarity. This user-based CF model is detailed in Section 9.3.2. Second, to address the problem of scalability, a location-based CF model is proposed [44]. This model uses the correlation between locations mined from many users' GPS traces [43] as a distance measure between two different locations. The location-based CF model is slightly less effective than the user-based one while being much more efficient. Refer to Section 9.3.3 for details.

9.2 Generic Travel Recommendations

This section describes the generic travel recommendation following the paradigm of "trajectories → interesting locations → popular travel sequences → itinerary planning → activities recommendation." Specifically, Section 9.2.1 introduces the detection of interest locations and travel sequences [48]. Section 9.2.2 then presents itinerary recommendation [37, 38]. Finally, a location-activity recommender [40] is discussed in Section 9.2.3.

9.2.1 Mining Interesting Locations and Travel Sequences

9.2.1.1 Background

Traveling to an unfamiliar city or region, people usually like to know the most inter-esting locations and the most popular travel sequences. In fact, this kind of informa-tion evolves as time goes by and varies in quite a few factors, such as time of day, day of the week, and the seasons. For example, the Forbidden City was the most popular tourist attraction in the urban area of Beijing before 2008. However, it has recently been replaced by the Olympic Park of Beijing. Locals particularly enjoy the Olympic Park on weekend evenings during the summer. Other popular destina-tions include Houhai Bar Street for sightseeing in the daytime and drinking in the evening, or some newly-built movie theatres offering half-price tickets every Tues-day night. Note that the interesting places do not only include tourist attractions but also restaurants and shopping malls popular among residents. Consequently, trav-el agencies and travel books cannot always provide the latest and most effective recommendations that a user needs.

In order to deal with this dilemma, it is necessary to gather travel recommen-dations automatically and in a timely manner from social media such as the vast amount of GPS trajectories generated by the large number of users travelling in a city. GPS trajectories can be formulated in terms of users' geo-tagged photos and check-in records, or obtained from some trajectory-sharing social networking ser-vices like GeoLife [45, 42, 41]. A number of studies [23, 2] have introduced the methods for extracting trips from geo-tagged photos, and some professionals have explored the idea of mining generic travel recommendations from GPS trajectories [48, 5, 44]. Particularly, one paper [48] first proposed a learning model to infer the most interesting locations in a city as well as the popular travel sequences among these locations, followed by a few expanded studies reported in [5, 44]. The mined locations and travel sequences are used to enable a generic travel recommender il-lustrated in Fig. 9.1 and Fig. 9.2.

Figure 9.1 illustrates the user interface of a generic travel recommender run-ning on desktop computers. The right column shows the top five most interesting locations and the five most experienced users in the region (specified by the present view of the map). The top five most popular travel sequences within this region are also displayed on the map. By zooming in/out and panning, a user can retrieve such results within any region. In addition, the photos taken in an interesting location will be presented on the bottom of the window after a user clicks the icon representing the location on the map.

As shown in Fig. 9.2, a user with a GPS-phone can find the top five most interest-ing locations as well as the five most popular sequences near her present geographic position (denoted as the red star). Additionally, when the user reaches a location, the recommender system will provide her with a further suggestion by presenting the top three most popular sequences starting from this location.

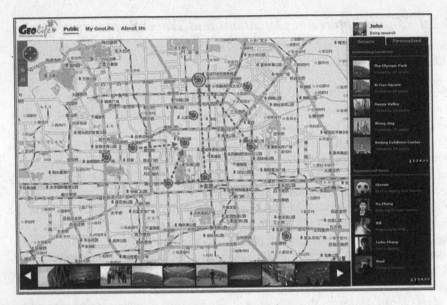

Fig. 9.1 The user interface of a generic location recommender

Fig. 9.2 Location recommendations on a GPS-phone

However, we will be faced with some challenges when conducting the generic recommendations. The first is to determine the interest level of a location. Intrinsically, the interest level of a location does not only depend on the number of users visiting this location but also on these users' travel experiences (knowledge). Intuitively, different people have different degrees of knowledge about a geospatial region. During a journey, the users with more travel experience of a region would be more likely to visit interesting locations in that region. For example, the residents of Beijing are more capable than overseas tourists of finding high quality restaurants and shopping malls in Beijing. If we do not consider the travel knowledge of a user, the hot spots like railway stations and airports will be most recommended. Second, an individual's travel experience and interest level of a lo-cation are relative values (i.e., it is not reasonable to judge whether or not a location is interesting), and are region-related (i.e., conditioned by the given geospatial region). An individual who has visited many places in New York might have no idea about Beijing. Likewise, the most interesting restaurant in a district of a city might not be the most interesting one in the whole city (as restaurants from other districts might outperform it).

9.2.1.2 Methodology for Mining Interesting Locations

To address the above challenges, the location histories of users are first modeled with a tree-based hierarchical graph (TBHG) according to the following two steps demonstrated in Fig. 9.3.

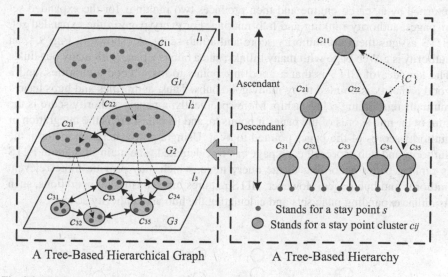

A Tree-Based Hierarchical Graph   A Tree-Based Hierarchy

Fig. 9.3 Building a tree-based hierarchical graph

1) Formulate a shared hierarchical framework F: This step is the same as that presented in Section 8.2.2.2. That is, the stay points detected from users' GPS logs are put into a dataset, and then hierarchically clustered into geospatial regions using a density-based clustering algorithm in a divisive manner. As a consequence, the similar stay points from various users would be assigned to the same clusters on different levels. Here, a stay point stands for a location where a user stayed for a certain period of time, formally defined in Definition 8.3.

2) Build location graphs on each layer: Based on shared framework F and users' location histories, the clusters on the same level are connected with directed edges. If two consecutive stay points from one trip are individually contained in two clusters, a link is generated between the two clusters in a chronological direction according to the time serial of the two stay points. Note that different from the third step of modeling an individual's location history (introduced in Section 8.2.1), this step feeds all users' location histories (sequences of stay points) into the shared framework. Therefore, this tree-based hierarchical graph models the location history of all users in a location-based social networking service.

Then, a HITS(Hypertext Induced Topic Search)-based inference model is proposed with the TBHG. This inference model regards an individual's access to a location as a directed link from the user to that location. This model infers two values, the interest level of a location and a user's travel experience, by taking into account 1) the mutuallly reinforcing relationship between the two values and 2) the geo-regional conditions. See details in the following paragraphs.

Concept of HITS model: HITS stands for hypertext induced topic search [17], which is a search-query-dependent ranking algorithm for Web information retrieval. When the user enters a search query, HITS first expands the list of relevant pages returned by a search engine and then produces two rankings for the expanded set of pages, authority ranking and hub ranking. For every page in the expanded set, HITS assigns them an authority score and a hub score. As shown in Fig. 9.4, an authority is a Web page with many inlinks, and a hub is a page with many out-links. The key idea of HITS is that a good hub points to many good authorities, and a good authority is pointed to by many good hubs. Thus, authorities and hubs have a mutually reinforcing relationship. More specifically, a page's authority score is the sum of the hub scores of the pages it points to, and its hub score is the integration of authority scores of the pages pointed to by it. Using a power iteration method, the authority and hub scores of each page can be calculated. The main strength of HITS is ranking pages according to the query topic, which may provide more relevant authority and hub pages. However, HITS requires time consuming operations, such as online expanding page sets and calculating the hub and authority scores.

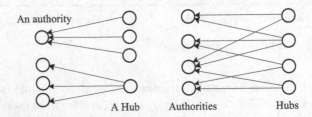

Fig. 9.4 The basic concept of HITS model

Mutually reinforcing relationship: Using the third level of the TBHG shown in Fig. 9.3 as an example, Fig. 9.5 illustrates the main idea of the HITS-based inference model. Here, a location is a cluster of stay points, like c_{31} and c_{32}. This model regards an individual's visit to a location as an implicitly directed link from the individual to that location. For instance, cluster c_{31} contains two stay points respectively detected from u_1's and u_2's GPS traces, i.e., both u_1 and u_2 have visited this location. Thus, two directed links are generated respectively to point to c_{31} from u_1 and u_2. Similar to HITS, in this model, a hub is a user who has accessed many places, and an authority is a location which has been visited by many users. Intuitively, a user with rich travel experience (knowledge) in a region is able to visit many interesting places in that region, and a very interesting place in that region could be accessed by many users with rich travel experiences. Therefore, users' travel experiences (hub scores) and the interest level of locations (authority scores) have a mutually reinforcing relationship. More specifically, a user's travel experience is represented by the sum of the interest values of the locations that the user has been to, and the interest value of a location is denoted by the sum of the experiences of users who have visited this location. For simplicity's sake, in the remainder of this chapter, a user with rich travel experience (i.e., relatively high hub score) in a region is called an experienced user of that region and a location that attracts people's profound interests (relatively high authority score) is denoted as an interesting location.

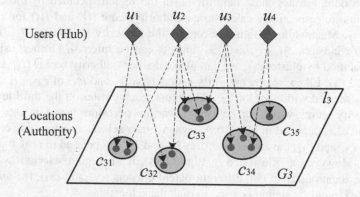

Fig. 9.5 The HITS-based inference model

Region-related: Intrinsically, a user's travel experience is region-related, i.e., a user who has a great deal of travel knowledge of a city might have no idea about another city. Also, an individual, who has visited many places in a particular part of a city might know little about another part of the city (especially if the city is very large, like New York). This concept is aligned with the query-dependent property of HITS. Thus, specifying a geospatial region (a topic query) and formulating a dataset that contains the locations in this region are needed for conducting the HITS-based inference model. However, an online data selection strategy (i.e., specifying a region based on an individual's input) will generate a great deal of resource-consuming operations, thereby diminishing the feasibility of our system. Therefore, a smart data selection strategy should be considered to fit this region-related feature of the HITS-based inference model.

Strategy for Data Selection: Actually, on a TBHG, the shape of a graph node (cluster of stay points) provides an implicit region for its descendent nodes. These regions covered by the clusters on different levels of the hierarchy might stand for various semantic meanings, such as a city, a district, or a community. Therefore, the interest of every location can be calculated in advance using the regions specified by their ascendant clusters. In other words, a location might have multiple authority scores based on the different regions it falls in. Also, a user might have multiple hub scores conditioned by the regions of different clusters.

Definition 9.1 (Location Interest). In this system, the interest of a location (c_{ij}) is represented by a collection of authority scores $I_{ij} = \{I_{ij}^1, I_{ij}^2, \ldots, I_{ij}^l\}$. Here, I_{ij}^l denotes the authority score of cluster c_{ij} conditioned by its ascendant nodes on level l, where $1 \leq l < i$.

Definition 9.2 (User Travel Experience). In our system, a user's (e.g., u_k) travel experience is represented by a set of hub scores $e^k = \{e_{ij}^k \mid 1 \leq i < |L|, 1 \leq j \leq |C_i|\}$ (refer to Definition 8.5), where e_{ij}^k denotes u_k's hub score conditioned by region c_{ij}.

Figure 9.6 demonstrates these definitions. In the region specified by cluster c_{11}, it is possible to respectively calculate an authority score (I_{21}^1 and I_{22}^1) for clusters c_{21} and c_{22}. Meanwhile, within this region, the authority scores (I_{31}^1, I_{32}^1, I_{33}^1, I_{34}^1 and I_{35}^1) for clusters c_{31}, c_{32}, c_{33}, c_{34} and c_{35} can be inferred. Further, using the region specified by cluster c_{21}, we can also calculate authority scores (I_{31}^2 and I_{32}^2) for c_{31} and c_{32}. Likewise, the authority scores (I_{33}^2, I_{34}^2 and I_{35}^2) of c_{33}, c_{34} and c_{35} can be re-inferred within region c_{22}. Therefore, each cluster on the third level has two authority scores, which can be used on various occasions based on user inputs. For instance, as depicted in Fig. 9.6 A), when a user selects a region only covering locations c_{31} and c_{32}, the authority scores I_{31}^2 and I_{32}^2 can be used to rank these two locations. However, as illustrated in Fig. 9.6 B), if the region selected by a user covers the locations from two different parent clusters (c_{21} and c_{22}), the authority values I_{32}^1, I_{33}^1 and I_{34}^1 should be used to rank these locations.

A strategy that allows for multiple hub scores for a user and multiple authority scores for a location has two advantages. First, it is able to leverage the main strength of HITS to rank locations and users within the context of geospatial regions (query topics). Second, these hub and authority scores can be calculated offline, thereby ensuring the efficiency of a recommender system while allowing users to specify any region on a map.

Inference: Given the locations pertaining to the same ascendant cluster, we are able to build an adjacent matrix M between users and locations based on the users' visits to these locations. In this matrix, item v_{ij}^k stands for the times that u_k (a user) has visited to cluster c_{ij} (the jth cluster on the ith level). Such matrixes can be built offline for each non-leaf node. For instance, the matrix M formulated for the example shown in Fig. 9.5 can be represented as follows, where all five clusters pertain to c_{11}:

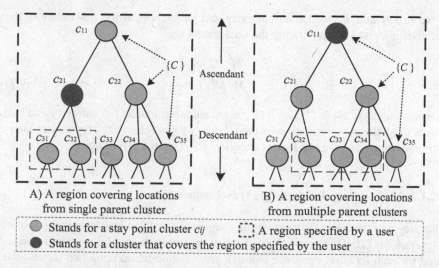

A) A region covering locations   B) A region covering locations
   from single parent cluster   from multiple parent clusters

◯ Stands for a stay point cluster c_{ij} ⬚ A region specified by a user
⬤ Stands for a cluster that covers the region specified by the user

Fig. 9.6 Some cases demonstrating the data selection strategy

$$
M = \begin{matrix} & \begin{matrix} c_{31} & c_{32} & c_{33} & c_{34} & c_{35} \end{matrix} \\ \begin{matrix} u_1 \\ u_2 \\ u_3 \\ u_4 \end{matrix} & \begin{bmatrix} 1 & 1 & 0 & 0 & 0 \\ 1 & 1 & 2 & 0 & 0 \\ 0 & 0 & 1 & 2 & 0 \\ 0 & 0 & 0 & 1 & 1 \end{bmatrix} \end{matrix} \tag{9.1}
$$

Then, the mutually reinforcing relationship of user travel experience e_{ij}^k and location interest I_{ij}^l is represented as follows:

$$
I_{ij}^l = \sum_{u_k \in U} e_{lq}^k \times v_{ij}^k; \tag{9.2}
$$

$$
e_{lq}^k = \sum_{c_{ij} \in c_{lq}} v_{ij}^k \times I_{ij}^l; \tag{9.3}
$$

where c_{lq} is c_{ij}'s ascendant node on the lth level, $1 \le l < i$. For instance, as shown in Fig. 9.6, c_{31}'s ascendant node on the first level of the hierarchy is c_{11}, and its ascendant node on the second level is c_{21}. Thus, if $l = 2$, c_{lq} stands for c_{21} and $(c_{31}, c_{32}) \in c_{21}$. Also, if $l = 1$, c_{lq} denotes c_{11}, and $(c_{31}, c_{32}, \ldots, c_{35}) \in c_{11}$.

Writing them in matrix form, we use J to denote the column vector with all the authority scores, and use E to denote the column vector with all the hub scores. Conditioned by the region of cluster c_{11}, $J = (I_{31}^1, I_{32}^1, \ldots, I_{35}^1)$, and $E = (e_{11}^1, e_{11}^2, \ldots, e_{11}^4)$.

$$
J = M^J \cdot E \tag{9.4}
$$

$$
E = M \cdot J \tag{9.5}
$$

If we use J_n and E_n to denote authority and hub scores at the nth iteration, the iterative processes for generating the final results are

$$J_n = M^J \cdot M \cdot J_{n-1} \tag{9.6}$$

$$E_n = M \cdot M^J \cdot E_{n-1} \tag{9.7}$$

Starting with $J_0 = E_0 = (1,1,\ldots,1)$, we are able to calculate the authority and hub scores using the power iteration method. Later, we can retrieve the top n most interesting locations and the top k most experienced users in a given region.

9.2.1.3 Methodology for detecting travel sequences

This step detects the top k most popular sequences of locations from the graph on a layer of the TBHG (refer to Fig. 9.3 for an example). A popularity score is calculated for each location sequence within a given region based upon two factors: the travel experiences of the users taking this sequence and the interests of the locations contained in the sequence. Since there would be multiple paths starting from the same location, the interest value of this location should be distributed to these paths according to the probability that users would take a path.

Figure 9.7 demonstrates the calculation of the popularity score for a 2-length sequence (i.e., a sequence containing two locations), $A \to C$. In this figure, the graph nodes $(A,B,C,D,$ and $E)$ stand for locations, and the graph edges denote people's transition sequences among them. The number associated with an edge represents how many times that users have taken the sequence. Eq. 9.8 computes the popularity score of sequence $A \to C$, consisting of contributions from the following three parts:

- The authority score of location $A(I_A)$ weighted by the probability of people leaving by this sequence (Out_{AC}). Clearly, there are seven (5+2) links pointing to other nodes from node A, and five out of seven of these links point directly to node C. So, $Out_{AC} = 5/7$, i.e., only five sevenths of location A's authority (I_A) should be propagated to sequence $A \to C$, and the rest of I_A should be distributed to $A \to B$.
- The authority score of location $C(I_C)$ weighted by the probability of people's entering by this sequence (In_{AC}).
- The hub scores of the users (U_{AC}) who have taken this sequence.

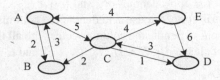

Fig. 9.7 Demonstration of mining popular travel sequences from a graph

$$S_{AC} = \sum_{u_k \in U_{AC}} (I_A \cdot Out_{AC} + I_C \cdot In_{AC} + e^k)$$

$$= |U_{AC}| \cdot (I_A \cdot Out_{AC} + I_C \cdot In_{AC}) + \sum_{u_k \in U_{AC}} e^k$$

$$= 5 \times \left(\frac{5}{7} \times I_A + \frac{5}{8} I_C\right) + \sum_{u_k \in U_{AC}} e^k. \tag{9.8}$$

Following this method, the popularity score of sequence $C \to D$ is calculated as follows:

$$S_{CD} = 1 \times \left(\frac{1}{7} \times I_C + \frac{1}{7} I_D\right) + \sum_{u_k \in U_{CD}} e^k. \tag{9.9}$$

Thus, the popularity score of sequence $A \to C \to D$ equals:

$$S_{ACD} = S_{AC} + S_{CD} \tag{9.10}$$

The detection of popular travel sequences starts with computing the popularity score for each 2-length sequence, and then searches for 3-lenth sequences based on these 2-length sequences. Though searching for the top k n-length most popular sequences in a graph is time consuming, there are a few optimization methods using some upper bound to filter unnecessary search spaces. Moreover, the size of a location graph is usually small as the number of interesting locations in a city is limited. Meanwhile, as people do not normally travel to too many places during a trip, it is not necessary to provide a user with very long travel sequences. Sometimes, a sequence with three locations is more useful than longer ones.

9.2.2 Itinerary Recommendation

9.2.2.1 Background

The interesting locations and travel sequences mentioned above can facilitate travel to an unfamiliar place. However, people are still faced with particular challenges when planning their trips.

First, while there are many location candidates that can be considered, a traveler usually wants to maximize her travel experience, i.e., visit as many interesting locations as possible in a comfortable manner without wasting too much time traveling between locations. To achieve this task, the typical duration that people stay in a location and the average travel time between two locations should be considered. Second, an effective itinerary needs to adapt to a traveler's present location and available time length as well as her destination. Some popular travel routes can be available in a book but may not be feasible for a particular individual in the real world because the attractions contained in these routes might be too far away, or the individual does not have enough time for the trip.

What a traveler needs is an effective itinerary, which can be adapted to the traveler's requests (consisting of a starting location, destination, and available time) and includes the information of not only a travel route passing some interesting locations but also the typical duration spent in each location and the general travel time between locations.

Itinerary recommendation has been studied in quite a few research projects. Some recommender systems [9, 3, 16] need a user's intervention when generating an itinerary for the user. For example, [9] presented an interactive travel itinerary planning system where a user defines which places to visit and avoid. Similarly, [3] reported on an interactive system where a user specifies general constraints, such as time and attractions to be included in the itinerary. The advantage of such interactive recommender systems is that the more a user knows about traveling in the area, the better the itinerary is. However, this assumption is not practical for most novice travelers who lack prior knowledge of a region.

To alleviate the human intervention and prior knowledge needed, [37, 38, 18, 8] presented relatively automated recommenders. Particularly, [37, 38] proposed a social itinerary recommendation service that generates an effective itinerary based on a user's query and social knowledge learned from user-generated GPS trajectories, with a major application scenario described as follows. Imagine that a researcher is attending a conference in Beijing. At the end of the conference, she has 8 hours to spend before catching her flight. She is a first time visitor to the city and has no idea how to plan an effective travel route, and is thereby relying on a social itinerary recommendation service. She is starting from her current location, which is automatically recognized with a GPS-enabled phone. She marks the Beijing Capital International Airport in the map as her destination, inputs 8 hours for travel duration, and sends the query. As a result, she receives an itinerary recommendation visualized on the map which shows interesting locations to be visited, a recommended amount of time to stay in each location, and an estimate of the time needed to travel between any two locations. With the information at hand, she obtains a good idea of where she might go and is able to manage her time effectively.

9.2.2.2 Methodology for Itinerary Recommendation

As illustrated in Fig. 9.8, the framework for this itinerary recommender consists of an online component and an offline component.

Offline component: This component data mines the collective social intelligence from a database of GPS trajectories in terms of the following two steps:

The first step detects stay points from each GPS trajectory and clusters these stay points into locations. A location graph can be formulated according to the method described in Section 9.2.1.2 (the bottom layer of the hierarchy shown in Fig. 9.3 is used here as a demonstration). Remember that each stay point has properties pertaining to arrival and departure times, which indicate the length of time stayed in

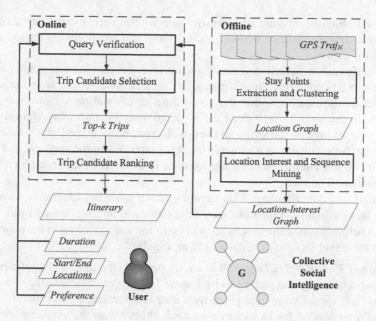

Fig. 9.8 Framework of the itinerary recommender

a location. So, the typical stay duration in each location and general travel duration between two locations (in a location graph) can be calculated, for example, using the median of all people's stay times in a location. These values associated with each location are used to estimate the duration of an itinerary. This clustering operation picks out the locations accessed by a significant number of people, ensuring the accuracy of the estimated travel and stay times. This operation also contributes to the second step by reducing the sparseness of the connections between users and locations.

The second step infers the interest value of each location in the location graph using the approach introduced in Section 9.2.1.2 (see Fig. 9.5), and calculates the popularity score of each 2-length travel sequence in terms of the method presented in Section 9.2.1.3. As a result, the output of this component is a location-interest graph, in which a node is a location associated with an interest value and a typical stay duration and an edge denotes people's transitions (between locations) and the general travel duration. The offline component will not be detailed further since they have been introduced in previous sections and in Chapter 8.

Online component: This component accepts a user-generated query (consisting of a starting location, a destination, and an available duration), and returns an effective itinerary comprised of a sequence of locations with a stay time in each location and travel times between two consecutive locations. This component can be decomposed into three steps, introduced as follows:

1) Query Verification: This step checks the feasibility of a query according to spatial and temporal constraints. In some cases, a user might set a short time length

with a far destination, making all itineraries impossible. Such queries can be filtered out by checking the distance between the start and end location with respect to the duration (of a query).

2) Trip Candidate Selection: This step searches a location-interest graph for candidate itineraries satisfying a user's query, i.e., each path has to start from the source and reach the destination within the given time length and pass some interesting locations on the way. Though there are some advanced path-finding algorithms, a straightforward method retrieving all the possible paths in a brute-force manner will work given the small size of a location-interest graph. When the start and end point of a query do not fall into any existing location in the graph, the nearest location to these points will be used.

3) Trip Candidate Ranking: This step first ranks candidate itineraries according to three factors: elapsed time ratio, stay time ratio, and interest density ratio, as introduced below. Then, these itineraries will be re-ranked according to the popularity score of the travel sequences pertaining to an itinerary.

- *Elapsed Time Ratio (ETR): ETR* is a ratio between the time length of a recommended itinerary and that given by a user. The bigger value this factor has, the more substantially the time given by a user is leveraged by an itinerary. If the total time needed for an itinerary is much shorter than the available time, the remaining time is wasted.
- *Stay Time Ratio (STR):* This factor considers how the available time is spent by calculating a ratio between the time that a user could stay in a location and that for traveling between locations. Intuitively, travelers prefer to spend more time in interesting locations rather than traveling to them. Therefore, an itinerary with a bigger *STR* is considered a better choice, i.e., a user can spend a longer time visiting actual places.
- *Interest Density Ratio (IDR): IDR* is the sum of the interest values of the locations contained in an itinerary. The general assumption is that visitors like to visit as many highly interesting locations as possible on a trip. Therefore, the bigger the IDR value, the better the itinerary. In the implementation, the IDR of an itinerary should be normalized to [0, 1] and divided by the maximum IDR in the candidate itineraries.

As shown in Fig. 9.9, a good itinerary candidate is a point located in the upper-right corner of this cube, i.e., simultaneously having larger *ETR*, *STR*, and *IDR* values.

To rank candidate itineraries, a Euclidian distance in these three dimensions is calculated as Eq. 9.11:

$$ED = \sqrt{\alpha_1 \cdot (ETR)^2 + \alpha_2 \cdot (STR)^2 + \alpha_3 \cdot (IDR)^2} \qquad (9.11)$$

- *Popular Travel Sequence:* This factor represents how popular a recommended itinerary is (according to people's travel history), by summing up the popularity scores of the travel sequences contained in the itinerary (the popularity score of a travel sequence is computed in Section 9.2.1.3). This factor uses collective social knowledge to further differentiate the itineraries returned by the first

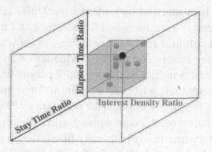

Fig. 9.9 Idea itinerary candidates

ranking step and guarantees the feasibility of an itinerary. As a result, the top k itineraries will be recommended to a user.

9.2.3 Location-Activity Recommendation

Besides the need for an itinerary, people usually have two types of questions in mind when traveling. They wonder where to go for sightseeing and food, and they wonder what there is to do at a particular location. The first question corresponds to location recommendation given a particular activity query, which might include restaurants, shopping, movies/shows, sports/exercise, and sightseeing. The second question corresponds to activity recommendation given a particular location query.

This section introduces a location-activity recommender system [40] which answers the above questions by mining a myriad of social media, such as tips-tagged trajectories or check-in sequences. Regarding the first question, this system provides a user with a list of interesting locations, e.g., the Forbidden City and the Great Wall, which are the top k candidate locations for conducting a given activity. With respect to the second question, if a user is visiting the Olympic Park of Beijing, the recommender suggests that the user can also try some exercise activities and nice restaurants nearby. This recommender integrates location recommendation and activity recommendation into one knowledge-mining process, since locations and activities are closely related in nature.

9.2.3.1 Data Modeling

Location-activity matrix: As mentioned before, to better share experiences, an individual can add comments or tips to a point location in a trajectory. For example, in Foursquare a user can leave some tips or a to-do-list in a venue so that her friends are able to view these tips when they arrive at the venue. Sometimes, these tips and to-do-lists clearly specify a user's activity in a location, enabling us to study the correlation between user activities and a location, for instance, what kinds of activities can be performed in a location, and how often a particular activity is conducted in the location. Consequently, a location-activity matrix can be built, in which rows stand for locations and columns represent activities, as shown in the middle part of Fig. 9.10. An entry in the matrix denotes the frequency of an activity performed in a location. For example, if 5 users had dinner and 7 people watched a movie in this location in a week, the frequency of activity "dining" and "watching movies" is 5 and 7 respectively. This frequency denotes the popularity of an activity in a location and indicates the correlation between an activity, and a location.

If this location-activity matrix is completely filled, the above-mentioned recommendations can be easily achieved. Specifically, when conducting the location recommendation given an activity, we can rank and retrieve the top k locations with a relatively high frequency from the column that corresponds to that activity. Likewise, when performing activity recommendation for a location, the top k activities can be retrieved from the row corresponding to the location.

However, the location-activity matrix is incomplete and very sparse. Intuitively, people will not leave tips and to-do-lists in every restaurant and shopping mall. In short, many venues will not have labels of user activities. To address this issue, the information from another two matrices, respectively shown in the left and right part of Fig. 9.10, can be leveraged. One is a location-feature matrix; the other is an activity-activity matrix.

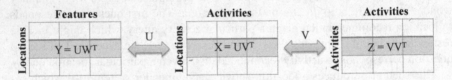

Fig. 9.10 The collaborative location-activity learning model

Location-feature matrix: In this matrix, a row stands for a location, and a column denotes a category (referred to as feature in this section), such as restaurants, cafes, and bars, illustrated in the left part of Fig. 9.10. Usually, a location might include multiple points of interest (POI) pertaining to different categories. For example, a mall would include different types of shops, movie theaters, and cafes. Further, a movie theater could have a few bars and restaurants inside. At the same time, a single venue could belong to multiple categories. For instance, some bars can also be regarded as a restaurant or a cafe. The motivation for building this location-feature matrix lies in the insight that people could carry out similar activities in similar locations.

Specifically, this matrix is built based on a POI database. Each POI in this database is associated with a set of properties, including name, address, GPS coordinates, and category. Given a location, the number of POIs pertaining to each category (and falling in this location) can be counted. Note that a location can be represented by a point or a small region [48, 46] like a cluster of stay points mentioned in Section 8.2.1.1, depending on the data source from different applications.

Suppose there are 4 restaurants, 2 bars and 5 shops in a location, a feature vector $v = < \ldots, 4, 2, 5, \ldots >$ is formulated for the location. To further differentiate the representativeness of each category in a location, a TF-IDF (term frequency-inverse document frequency [26, 27]) value is calculated for each category according to Eq. 8.7. Intuitively, if POIs of a category occur in a region many times, this POI category is important in representing this region. Furthermore, if a POI category (e.g., "museum" or "natural parks") occurs rarely in other regions, the category is more representative for the region in which it is located than a common POI category (e.g., "restaurant") that appears in many places. As a result, each item in a location-feature matrix is a TF-IDF value of a category in a location.

Activity-activity matrix: The activity-activity matrix, demonstrated in the right part of Fig. 9.10, models the correlation between two different activities, which contributes to the inferences of user activities that can be performed at a location. In other words, if a user performs some activity at a location, how likely would she perform another activity? One possible way to calculate this correlation is based upon user-generated tips and to-do-lists. In case the user-generated data is not large enough, the results returned by a search engine (like Google and Bing) can be used to compute the correlation as the correlation between two activities should be generally reflected by the World Wide Web. Specifically, we can send a pair of activities like "*shopping*" and "*food*" as a query to a search engine and count the returned results. The bigger this count is, the more these two activities are correlated. For example, the count of results returned for "*shopping*" and "*food*" is much larger than that of "*sports*" and "*food*," indicating that the former pair of activities is more related than the latter. Later, these counts are normalized into $[0, 1]$, representing the correlation between different pairs of activities.

9.2.3.2 Collaborative Inference

The data modeling has allowed for the compilation of location-activity, location-feature, and activity-activity matrices. The objective is to fill the missing entries in the location-activity matrix with the information learned from the other two matrices. A collaborative filtering (CF) approach based on collective matrix factorization [31] can be employed to train a location-activity recommender, using these matrices as inputs. Specifically, to infer the value of each missing entry an objective function is defined according to Eq. 9.12, which is iteratively minimized using a gradient descent method. Based on the filled location-activity matrix, it is possible to rank and retrieve the top k locations/activities as recommendations to users.

As shown in Fig. 9.10, a location-activity matrix $X_{m \times n}$ can be decomposed into a product of two matrices $U_{m \times k}$ and $V_{n \times k}$ (the superscript "T" for $V_{n \times k}^T$ denotes the matrix transpose), where m is the number of locations and n stands for the number of activities. k is the number of latent factors (topics), usually $k < n$. In the implementation, k was set to 3 as there are three topics: location, activity, and feature. Likewise, location-feature matrix $Y_{m \times l}$ is decomposed as a product of matrices $U_{m \times k}$ and $W_{l \times k}$, and activity-activity matrix $Z_{n \times n}$ is decomposed as a self-product of $V_{n \times k}$.

So, this location-activity matrix shares the location information with the location-feature matrix via $U_{m \times k}$, and shares the activity knowledge with the activity-activity matrix via $V_{n \times k}$. In short, the inference model propagates the information among $X_{m \times n}$, $Y_{m \times l}$ and $Z_{n \times n}$ by the low-rank matrices $U_{m \times k}$ and $V_{n \times k}$. Finally, an objective function is formulated as Eq. 9.13:

$$L(U,V,W) = \frac{1}{2}\|I \circ (X - UV^T)\|_F^2 + \frac{\lambda_1}{2}\|Y - UW^T\|_F^2 +$$
$$\frac{\lambda_2}{2}\|Z - VV^T\|_F^2 + \frac{\lambda_3}{2}\left(\|U\|_F^2 + \|V\|_F^2 + \|W\|_F^2\right), \quad (9.12)$$

where $\|\cdot\|_F$ denotes the Frobenius norm, and I is an indicator matrix with its entry $I_{ij} = 0$ if X_{ij} is missing, $I_{ij}=1$ otherwise. The operator "$\circ$" denotes the entry-wise product. The first three terms in the objective function control the loss in matrix factorization, and the last term controls the regularization over the factorized matrices so as to prevent over-fitting. λ_1, λ_2 and λ_3 are three parameters respectively weighting the contributions of location features, activity correlations, and the regularization term. These parameters can be learned using a training dataset.

In the objective function, the first term $(X - UV^T)$ measures the prediction loss of the location-activity matrix. The second term $(Y - UW^T)$ measures the prediction loss of the location-feature matrix. Minimizing it enforces the location latent factor U to be good as well in representing the location features. In other words, it helps to propagate the information of location features Y to the prediction of X. The third term $(Z - VV^T)$ measures the prediction loss of the activity-activity correlations. Minimizing it enforces the activity latent factor V to be good as well in representing the activity correlations. In other words, it helps to propagate the information of activity correlations Z to the prediction of X.

In general, this objective function is not jointly convex to all the variables, $U_{m \times k}$, $V_{n \times k}$, and $W_{l \times k}$. Also, there is no closed-form solution for minimizing the objective function. As a result, a numerical method such as the gradient descent is employed to determine the local optimal solutions. Specifically, the gradient (denoted as ∇) for each variable is represented as follows. Using the gradient descent, a converged $X_{m \times n}$ is returned with all the entries filled:

$$\nabla_U L = \left[I \circ (UV^T - X)\right]V + \lambda_1 (UW^T - Y)W + \lambda_3 U, \quad (9.13)$$

$$\nabla_V L = \left[I \circ (UV^T - X)\right]^T U + 2\lambda_2 (VV^T - Z)V + \lambda_3 V, \quad (9.14)$$

$$\nabla_W L = \lambda_1 (UW^T - Y)^T U + \lambda_3 W. \quad (9.15)$$

9.3 Personalized Travel Recommendations

While a generic travel recommender system can provide users with a variety of locations regardless of their personal interests, a personalized recommender offers locations matching an individual's preferences, which are learned from the individual's location history [44, 46]. Specifically, a personalized recommender uses a particular individual's number of visits to a location as their implicit rating of that location, and predicts the user's interest in an unvisited location in terms of their location history and those of other users. A matrix between users and locations, like the M shown in Eq. 9.16, is formulated, where rows stand for users and columns denote users' ratings of locations (represented by the times that a user has been to a location). One approach for building this matrix with user-generated GPS trajectory has been introduced in Section 8.2.1.

$$
M = \begin{array}{c} \\ u_1 \\ u_2 \\ u_3 \\ u_4 \end{array}
\begin{array}{ccccc}
c_{31} & c_{32} & c_{33} & c_{34} & c_{35} \\
\left[\begin{array}{ccccc}
1 & 1 & 2 & 0 & 4 \\
1 & 1 & 3 & 0 & 2 \\
0 & 0 & 1 & 2 & 0 \\
0 & 0 & 0 & 1 & 1
\end{array}\right]
\end{array}
\tag{9.16}
$$

Based on user-location matrix M, a collaborative filtering model can be employed to infer a user's ratings of some unvisited location. Later, by ranking and retrieving the top k unvisited locations in terms of the inferred values from the row (in M) corresponding to a particular user, we can provide the user with a personalized recommendation.

In the following sections, the general idea of a CF model is first introduced and then two types of CF-based models that have been used in previous literature to create a personalized location recommender system. One is a user-based location recommender [46]; the other is an item-based one [44].

9.3.1 Collaborative Filtering

Collaborative filtering is a well-known model widely used in recommender systems. The general idea behind collaborative filtering [11, 24] is that similar users make ratings in a similar manner for similar items. Thus, if similarity is determined between users and items, a potential prediction can be made as to the rating of a user with regards to future items. According to [4], algorithms for collaborative recommendations can be grouped into two general classes: memory-based (or heuristic-based) and model-based.

Memory-based: Memory-based algorithms are essentially heuristics that make ratings predictions based on the entire collection of previously rated items by users [1]. That is, the value of the unknown rating for a user and an item is usually computed as an aggregate of the ratings of some other (usually, the N most similar) users for the same item. There are two classes of memory-based collaborative filtering: user-based [29, 25] and item-based techniques [19, 28].

1) User-based techniques are derived from similarity measures between users. The similarity between two users (A and B) is essentially a distance measure and is used as a weight. In other words, when predicting user A's rating of an item, the more similar user A and B are, the more weight user B's rating of the item will carry. In most approaches, the similarity between two users is based on their ratings of items that both users have rated, using the Pearson correlation or the Cosine similarity measures. Spertus et al. [32] present an extensive empirical comparison of six distinct measures of similarity for recommending online communities to members of the Orkut social network. As a result, under the circumstances of the above-mentioned approach, they found that the Cosine similarity measure showed the best empirical results ahead of other measures, such as log odds and point-wise mutual information.

2) Item-based techniques predict the ratings of one item based on the ratings of another item. Examples of binary item-based collaborative filtering include Amazon's item-to-item algorithm [22], which computes the Cosine similarity between binary vectors representing the purchases in a user-item matrix. Slope One [19] is the simplest form of non-trivial item-based collaborative filtering. Its simplicity makes it especially easy to implement it efficiently while its accuracy is often on par with more complicated and computationally expensive algorithms. This algorithm is detailed in Section 9.3.2.2.

Model-based: In contrast to memory-based methods, model-based algorithms [10, 12] use the collection of ratings to form a model, which is then used to predict ratings. For example, Breese et al. [4] proposed a probabilistic approach to collaborative filtering. It is assumed that rating values are integers between 0 and n, and the probability expression is the probability that a user will give a particular rating to an item given that user's ratings of previously rated items. Hofmann et al. [12] proposed a collaborative filtering method in a machine learning framework where various machine learning techniques (such as artificial neural networks) coupled with feature extraction techniques can be used.

9.3.2 Location Recommenders Using User-Based CF

This section presents a personalized location recommender system [46] using a user-based CF model. Given a user-location matrix like M shown in Eq. 9.16, this location recommender operates according to the following three steps:

1) Infer the similarity between users: This personalized location recommender estimates the user similarity between two users in terms of their location histories (detailed in Section 8.3), instead of using traditional similarity measures, such as the Cosine similarity or the Pearson correlation. Typically, in a user-based CF model, the similarity between two users is based upon the ratings provided by both users. For example, the similarity between u_1 and u_2 (shown in Eq. 9.16) can be represent-

ed by the Pearson correlation between the ratings $< 1,1,2,4 >$ and $< 1,1,3,2 >$. Eq. 9.17 details the computation of the Pearson correlation.

$$sim(u_p, u_p) = \frac{\sum_{i \in S(R_p) \cap S(R_q)} (r_{pi} - \overline{R_p}) \cdot (R_{qi} - \overline{R_q})}{\sqrt{\sum_{i \in S(R_p) \cap S(R_q)} (r_{pi} - \overline{R_p})^2 \cdot \sum_{i \in S(R_p) \cap S(R_q)} (R_{qi} - \overline{R_q})^2}} \qquad (9.17)$$

Notations: The ratings from a user u_p are represented as an array $R_p =< r_{p0}, r_{p1}, \ldots, r_{pn} >$, where r_{pi} is u_p's implicit ratings (the occurrences) in a location i. $S(R_p)$ is the subset of R_p, $\forall r_{pi} \in S(R_p)$, $r_{pi} \neq 0$, i.e., the set of items (locations) that has been rated (visited) by u_p. The average rating in R_p is denoted as $\overline{R_p}$. In this example, $R_1 =< 1,1,2,0,4 >$, $R_2 =< 1,1,3,0,2 >$, $S(R_1) =< 1,1,2,4 >$, and $S(R_2) =< 1,1,3,2 >$.

This rating-based similarity measure well represents the similarity between two users when the items rated by users are relatively independent, such as books, videos, or music. However, dealing with locations (especially, when these locations are derived from user-generated trajectories), this approach loses the information of people's mobility (e.g., the sequential property) and the hierarchical property of locations. As we mentioned in Section 8.3, users accessing the same locations (A, B, C) could be similar to each other. However, they would be more similar if they visited these locations in the same sequence like $A \rightarrow B \rightarrow C$. Also, people who visited the same building could be more similar to one another than those who traveled to the same city.

Two studies based on a real GPS trajectory dataset generated by 109 users over a period of 2 years, one using the geographical model [21], the other using the semantic model [35], have shown that the user similarity based on location history outperforms the Cosine similarity and the Pearson correlation.

2) Location selection: For a user, this step selects some locations that have not been visited by the user but have been accessed by other users. Note that the inferred rating of a location would not be very accurate if the location has only been accessed by a few users. At the same time, using the personalized recommender, a user needs to have some location data accumulated in the system.

3) Rating inference: Given the user-location matrix, user p's interest (r_{pi}) in a location i can be predicted according to the following three Equations, which is a common implementation of user-based collaborative filtering. All the notations used here have the same meanings with that of Eq. 9.17:

$$r_{pi} = \overline{R_p} + d \sum_{u_q \in U'} sim(u_p, u_q) \times (r_{qi} - \overline{R_q}); \tag{9.18}$$

$$d = \frac{1}{|U'|} \sum_{u_q \in U'} sim(u_p, u_q); \tag{9.19}$$

$$\overline{R_p} = \frac{1}{|S(R_P)|} \sum_{i \in S(R_p)} r_{pi}. \tag{9.20}$$

As shown in Eq. 9.18, the similarity between users u_p and u_q, $sim(u_p, u_q)$, is essentially a distance measure and is used as a weight. That is, the more similar u_p and u_q are, the more weight r_{qi} will carry in the prediction of r_{pi}. Here, $sim(u_p, u_q)$ is calculated according to the method introduced in Section 8.3.3. However, different people may visit places a varying number of times (e.g., one user might visit a park twice while another person may access the same park four times, although both of them are equally interested in the park), i.e., they use the rating scale differently. Therefore, an adjusted weighted sum is used here.

First, instead of using the absolute values of ratings, the deviations from the average rating of the corresponding user are used, i.e., $r_{qi} - \overline{R_q}$, where $\overline{R_q}$ denotes the average rating of u_q. Second, a normalizing factor d is involved, calculated in terms of Eq. 9.19 where U' is the collection of users who are the most similar to u_q. Third, u_p's rating scale is considered by calculating the average rating $(\overline{R_p})$ of u_p as Eq. 9.20, where $S(R_p)$ represents the collection of locations accessed by u_p.

Actually, these equations illustrate a well-known method [1, 24], which has been used widely in many recommendation systems. Therefore, it is not necessary to explain them in more detail.

9.3.3 Location Recommenders Using Item-Based CF

The user-similarity-based CF model accurately reflects the sequential and hierarchical properties of locations, providing an individual with effective location recommendations. However, this model has a relatively poor scalability as it needs to compute the similarity between each pair of users. Though the approximated method introduced in 8.4.1 can be used to alleviate this problem to some extent, the constantly increasing number of users in a real system leads to a huge computational burden. To address this issue, a location recommender using item-based collaborative filtering was proposed in [44], which is comprised of the following two steps: 1) Mining the correlation between locations, and 2) rating inference.

9.3.3.1 Mining the Correlation between Locations

There are a variety of approaches to determine the correlations between locations, for example, according to the distance between them (i.e., in geographical spaces)

[14], or in terms of the category of POIs located in a location (i.e., in category spaces) [30]. However, this section focuses on introducing the correlation between locations in the spaces of user behavior, specifically, to what extent two locations are correlated in people's minds [43, 47].

Typically, people might visit a few locations during a trip, such as going to a few shopping malls, traveling to a bunch of landmarks on a sightseeing tour, or going to a cinema after a restaurant. These locations might be similar or dissimilar, or nearby or far away from one another; but they are correlated from the perspective of human behavior. For example, a cinema and a restaurant are not similar in terms of the business categories they pertain to. However, in a user's mind, these places would be correlated as many people visit both these places during a trip. As another example, when shopping for something important like a wedding ring, an individual will visit similar shops selling jewelry sequentially. In short, these shops visited by this individual might be correlated. However, these similar shops might be far away from each other, i.e., they might not be co-located in geographical spaces. Therefore, this kind of correlation between locations can only be inferred from a large number of users' location history in a collective way.

The correlation between locations mentioned above can enable many valuable services, such as location recommender systems, mobile tour guides, sales promotions and bus route design. For instance, as shown in Fig. 9.11 A), a new shopping mall was built in location A recently. The mall operator is intending to set up some billboards or advertisements in other places to attract more attention, thereby promoting sales at this mall. Knowing that locations B, C and E have a much higher correlation with location A in contrast to locations D and F (according to a large number of users' location histories), the operator is more likely to maximize the promotion effect with minimal investment by putting the billboards or promotion information in locations B, C and E. Another example can be demonstrated using Fig. 9.11 B). If a museum and a landmark are highly correlated to a lake in terms of people's location histories, the museum and landmark can be recommended to tourists when they travel to the lake. Otherwise, people would miss some fantastic places even if they are only two hundred meters away from these locations.

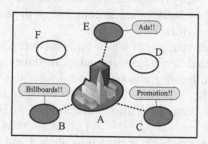

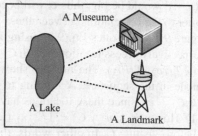

A) Put promotion information or ads. at correlated locations

B) Recommend places to tourists in terms of location correlation

Fig. 9.11 Some application scenarios of this location correlation

However, mining the correlation from people's location histories faces the following two challenges. First, the correlation between two locations does not only depend on the number of users visiting the two locations but also lies in these users' travel experiences. The locations sequentially accessed by the people with more travel knowledge would be more correlated than the locations visited by those having little idea about the region. For instance, some overseas tourists might randomly visit some places in Beijing because they are not familiar with the city. However, the local people of Beijing are more capable of determining the best itineraries for a visit there.

Second, the correlation between two locations, A and B, also depends on the sequences in which both locations have been visited. 1) This correlation between A and B, $Cor(A,B)$, is asymmetric; i.e., $Cor(A,B) \neq Cor(B,A)$. The semantic meaning of a travel sequence $A \to B$ might be quite different from $B \to A$. For example, on a one-way road, people would only go to B from A while never traveling to A from B. 2) The two locations continuously accessed by a user would be more correlated than those being visited discontinuously. Some users would reach B directly from A $(A \to B)$ while others would access another location C before arriving at B $(A \to C \to B)$. Intuitively, the $Cor(A,B)$ indicated by the two sequences might be different. Likewise, in a sequence $A \to C \to B$, $Cor(A,C)$ would be greater than $Cor(A,B)$, as the user consecutively accessed $A \to C$, but traveled to B after visiting C.

In short, the correlation between two locations can be calculated by integrating the travel experiences of the users visiting them on a trip in a weighted manner. Formally, the correlation between location A and B can be calculated as Eq. 9.21.

$$Cor(A,B) = \sum_{u_k \in U'} \alpha \cdot e_k, \qquad (9.21)$$

where U' is the collection of users who have visited A and B on a trip; e_k is u_k's travel experience, $u_k \in U'$, (Section 9.2.1.2 details the method for calculating a user's travel experience) . $0 < \alpha \leq 1$ is a dumping factor, decreasing as the interval between these two locations' index on a trip increases. For example, in the experiment of [43, 44], $\alpha = 2^{-(|j-i|-1)}$, where i and j are indices of A and B in the trip they pertain to. That is, the more discontinuously two locations being accessed by a user ($|i-j|$ would be big, thus α will become small), the less contribution the user can offer to the correlation between these two locations.

Figure 9.12 illustrates Eq. 9.21 using an example, in which three users (u_1, u_2, u_3) respectively access locations (A, B, C) in different manners and create three trips ($Trip_1$, $Trip_2$, $Trip_3$). The number shown below each node denotes the index of this node in the sequence. According to Eq. 9.21, for $Trip_1$, $Cor(A,B) = e_1$ and $Cor(B,C) = e_1$, since these locations have been consecutively accessed by u_1(i.e., $\alpha = 1$). However, $Cor(A,C) = 1/2 \cdot e_1$ (i.e., $\alpha = 2^{-(|2-0|-1)} = 1/2$) as u_1 traveled to B before visiting C. In other words, the correlation (between location A and C) that can be sensed from $Trip_1$ might not be that strong if they are not consecutively visited by u_1. Likewise, $Cor(A,C) = e_2$, $Cor(C,B) = e_2$, $Cor(A,B) = 1/2 \cdot e_2$ in terms of $Trip_2$, and $Cor(B,A) = e_3$, $Cor(A,C) = e_3$, $Cor(B,C) = 1/2 \cdot e_3$ by $Trip_3$. Later, the correlation inferred from each user's trips is integrated as follows.

$$Cor(A,B) = e_1 + \frac{1}{2} \cdot e_2; \quad Cor(A,C) = \frac{1}{2} \cdot e_1 + e_2 + e_3;$$

$$Cor(B,C) = e_1 + \frac{1}{2} \cdot e_3; \quad Cor(C,B) = e_2; \quad Cor(B,A) = e_3.$$

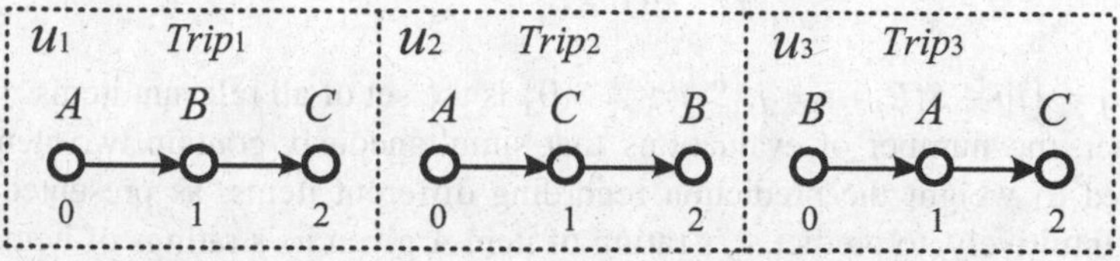

Fig. 9.12 An example calculating the correlation between locations

9.3.3.2 Rating Inference

The typical way to estimate the similarity between two items is calculating the Cosine similarity between two rating vectors that correspond to the two items. For instance, the similarity between location l_3 and l_5 can be represented by the Cosine similarity between the two rating vectors, $< 3,2,1,0 >^T$ and $< 4,2,0,1 >^T$. This rating-based similarity measure is fast, however it neglects the information of user mobility patterns among locations. As a result, the rating-based method is not a very effective similarity measure for an item-based location recommender, in which people's mobility patterns is a key factor determining the quality of recommendations.

Instead of using the rating-based similarity measure, the recommender presented in [44] integrates the location correlation (introduced in Section 9.3.3.1) into an item-based collaborative filtering (specifically, the Slope One algorithm) [19], thereby inferring the ratings of a particular user to some unvisited locations.

1)The Slope One algorithms

Notations: The ratings from user u_p, called an evaluation, are represented as array $R_p = < r_{p0}, r_{p1}, \ldots, r_{pn} >$, where r_{pj} is u_p's implicit ratings in location j. $S(R_p)$ is the subset of R_p, $\forall r_{pj} \in S(R_p)$, $r_{pj} \neq 0$. The collection of all evaluations in the training set is χ. $S_j(\chi)$ means the set of evaluations containing item j, $\forall R_p \in S_j(\chi)$, $j \in S(R_p)$. Likewise, $S_{i,j}(\chi)$ is the set of evaluations simultaneously containing item i and j.

The Slope One algorithms [19] are famous and representative item-based CF algorithms, which are easy to implement, efficient to query, and reasonably accurate. Given any two items i and j with ratings r_{pj} and r_{pi} respectively in some user evaluation $R_p \in S_{j,i}(\chi)$, the average deviation of item i with regard to item j is calculated as Eq. 9.22.

$$dev_{j,i} = \sum_{R_p \in S_{j,i}(\chi)} \frac{r_{pj} - r_{pi}}{S_{j,i}(\chi)}, \tag{9.22}$$

Given that $dev_{j,i} + r_{pi}$ is a prediction for r_{pj} based on r_{pi}, a reasonable predictor might be the average of all the predictions, as shown in Eq. 9.23.

$$P(r_{pj}) = \frac{1}{|w_j|} \sum_{i \in w_j} (dev_{j,i} + r_{pi}), \tag{9.23}$$

where $w_j = \{i | i \in S(R_p), i \neq j, |S_{j,i}(\chi)| > 0\}$ is the set of all relevant items.

Further, the number of evaluations that simultaneously contain two items has been used to weight the prediction regarding different items, as presented in Eq. 9.25. Intuitively, to predict u_p's rating of item A given u_p's ratings of item B and C, if 2000 users rated the pair of A and B whereas only 20 users rated pair of A and C, then u_p's ratings of item B is likely to be a far better predictor for item A than u_p's ratings of item C is.

$$P(r_{pj}) = \frac{\sum_{w_j} (dev_{j,i} + r_{pi}) \cdot |S_{j,i}(X)|}{\sum_{w_j} |S_{j,i}(X)|} \tag{9.24}$$

2) *The Slope One algorithm using the location correlation:* Intuitively, to predict u_p's rating of location A given u_p's ratings of location B and C, if location B is more related to A beyond C, then u_p's ratings of location B is likely to be a far better predictor for location A than u_p's ratings of location C is. Therefore, as shown in Eq. 9.25, the $|S_{j,i}(\chi)|$ in Eq. 9.24 is replaced by the correlation cor_{ji} (inferred in Section 9.3.3.1):

$$P(r_{pj}) = \frac{\sum_{i \in S(R_p) \wedge i \neq j} (dev_{j,i} + r_{pi}) \cdot cor_{ji}}{\sum_{i \in S(R_p) \wedge i \neq j} cor_{ji}}, \tag{9.25}$$

where cor_{ji} denotes the correlation between location i and j, and $dev_{j,i}$ is still calculated as Eq. 9.22.

In contrast to the number of observed ratings (i.e., $|S_{j,i}(\chi)|$) used by the weighted Slope One algorithm, the mined location correlation considers more human travel behavior, such as the travel sequence, user experience, and transition probability between locations. Using Eq. 9.25, an individual's ratings of locations they have not accessed can be inferred. Later, the top n locations with relatively high ratings can be recommended to the user.

9.3.4 Open Challenges

Although they are able to provide useful recommendations to an individual, personalized location recommender systems are faced with some open challenges during real implementation. These challenges include cold start problems, data sparseness, and scalability, which will be discussed individually.

9.3.4.1 Cold Start

A cold start is a prevalent problem in recommender systems, as a system cannot draw inferences for users or items when it has not yet gathered sufficient information. Generally, the problem is caused by new users or new items entering a recommender system. This sub-section discusses the possible solutions dealing with new locations and new users with respect to personalized location recommender systems.

New Location problem: When a new location is added to a location recommendation system, it usually has few ratings with which to determine the correlation (or similarity) between this new location and other places. As a consequence, the new location is hardly recommended to users even if it is a good place to visit. One possible solution is assigning the new location with a small number of users' ratings of existing places that are similar to the new location. Specifically, this method estimates the similarity between a new location and some existing locations (with an adequate number of ratings) according to their category information (e.g., the feature vector shown Fig. 9.10) [46]. For example, if both the new location and an existing place belong to the category of <restaurant, café, and bar>, the ratings that a user gave to a previous existing place might be similar to the new location. The locations in the system having a closer distance to the new location will be given a higher weight. As a result, the k most similar places can be selected for new locations. Then, the ratings of a few users to these similar places can be utilized to estimate their ratings to the new location, for instance, using the average mean of existing ratings. With these virtually generated ratings, the new location can be recommended to real users, thereby getting real ratings in return. The virtual ratings can be removed from the system once the new location has obtained enough ratings. Note that we can only select a few similar locations and users to generate the virtual ratings for a new location. Otherwise, these virtual ratings will dominate future inferences.

New User problem: When signing up in a recommender system, a new user has no location data accumulated in the system. Therefore, the recommender cannot offer her personalized recommendations effectively. One possible solution is providing an individual with some of the most popular recommendations at first regardless of her interests. If we have an individual's profile (e.g., likes movies), the popular locations with a category matching the individual's preferences can be recommended. Regarding the users who have only visited a very limited number of locations, some similarity-based mapping methods can be used to propagate a user's rating to a visited location to a few similar places that have not been accessed by the individual. Specifically, a similarity between two locations can be determined using either the location correlation (if both locations have sufficient ratings), or categories of these two locations as mentioned in the new location problem.

9.3.4.2 Data Sparseness

Intuitively, a user-location matrix is very sparse as a user can only visit a certain number of locations. The prediction of an individual's interest in a location is unable to be very accurate based on such a sparse matrix. Some approaches using additional information can improve the inferences. For example, the method introduced in Section 9.2.3 uses the category information of a location. Also, propagation and similarity-based mapping methods (mentioned in the above section) can be employed to reduce the empty entries in a matrix. Alternative methods can transfer this user-location matrix into a user-category matrix, which has a smaller number of columns and fewer empty entries than the former. As shown in Fig. 9.13, the similarity between users can be determined in terms of this user-category matrix with sufficient ratings, and then used in turn in the user-location matrix for location recommendation. However, this is still a very challenging problem that remains open to research and real implementation.

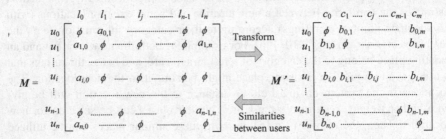

Fig. 9.13 Transforming a user-location matrix into a user-category matrix

9.3.4.3 Scalability

While the number of locations is limited in the real world and usually much smaller than that of users, the problem concerning the scalability of a location recommender arises in large part from the increasing number of users. Some approximation approaches can be used to reduce this problem to some extent.

For example, in the user-based CF model (introduced in Section 9.3.2), the mechanisms demonstrated in Fig. 8.10 can be employed to diminish the computations when a new user joins the system. First, existing users in a recommendation system can be clustered into groups according to the similarity between one another.

The users who are similar to a new user can then be found solely in the groups that the new user pertains to. Second, a location history can be quickly built for a new user by directly inserting her stay points into existing framework F. Later, we can compute the similarity between the new user and the representative user of a cluster based on this location history, thereby determining which group the new user belongs to. Third, the shared framework and user clusters can also be updated at a relatively low frequency, e.g., 1 update per month, as the arrival of a few users will not significantly change them. Fourth, once the similarity between the new users and other users in the cluster is computed, the similarity can be used for a time even if the new user has new data uploaded to the system. Adding a few trajectories will not change the similarity between two users significantly.

An alternative way to enhance the scalability of a location recommender system uses the item-based CF model presented in Section 9.3.3. The correlation (or similarity) between two locations can be updated at a very low frequency, e.g., once a month, as the arrival of a few new users does not change them significantly.

9.4 Summary

This chapter explores research topics in a location-based social network from the perspective of understanding locations with user-generated GPS trajectories. Using travel as a main application scenario, both generic and personalized travel recommendations are studied.

The generic travel recommender starts by finding interesting locations and travel sequences from a large amount of raw trajectories, and then offers itinerary and location-activity recommendations. By tapping into the collective social knowledge, these recommendations help people to travel to an unfamiliar place and plan their journey with minimal effort.

The personalized location recommender provides a particular user with locations matching her preferences, based on the location history of this user and that of others. Regarding a user's visits to a location as an implicit rating to that location, two kinds of collaborative filtering-based models are proposed to predict the user's interests in unvisited places. One is a user-based CF model, which incorporates the similarity between two different users (derived from their location histories) as a distance function between them. The other is a location-based CF model using the correlation between two different locations (inferred from many users' GPS trajectories) as a distance measure between them. The user-based CF model is able to accurately model an individual's behavior while it suffers from the increasing scale of users. The location-based CF model is efficient and reasonably accurate.

Some challenges are also discussed in the end of this chapter, aiming to encourage more research effort into this field.

References

1. Adomavicius, G., Tuzhilin, A.: Toward the next generation of recommender systems: A survey of the state-of-the-art and possible extensions. IEEE Trans. on Knowl. and Data Eng. **17**, 734–749 (2005)

2. Arase, Y., Xie, X., Hara, T., Nishio, S.: Mining people's trips from large scale geo-tagged photos. In: Proceedings of the international conference on Multimedia, MM '10, pp. 133–142. ACM, New York, NY, USA (2010)

3. Ardissono, L., Goy, A., Petrone, G., Segnan, M.: A multi-agent infrastructure for developing personalized web-based systems. ACM Trans. Internet Technol. **5**, 47–69 (2005)

4. Breese, J.S., Heckerman, D., Kadie, C.M.: Empirical analysis of predictive algorithms for collaborative filtering. In: Proceedings of the International 14th Conference on Uncertainty in Artificial Intelligence, pp. 43–52 (1998)

5. Cao, X., Cong, G., Jensen, C.S.: Mining significant semantic locations from gps data. Proc. VLDB Endow. **3**, 1009–1020 (2010)

6. Counts, S., Smith, M.: Where were we: communities for sharing space-time trails. In: Proceedings of the 15th annual ACM international symposium on Advances in geographic information systems, GIS '07, pp. 10:1–10:8. ACM, New York, NY, USA (2007)

7. Cranshaw, J., Toch, E., Hong, J., Kittur, A., Sadeh, N.: Bridging the gap between physical location and online social networks. In: Proceedings of the 12th ACM international conference on Ubiquitous computing, Ubicomp '10, pp. 119–128. ACM, New York, NY, USA (2010)

8. De Choudhury, M., Feldman, M., Amer-Yahia, S., Golbandi, N., Lempel, R., Yu, C.: Automatic construction of travel itineraries using social breadcrumbs. In: Proceedings of the 21st ACM conference on Hypertext and hypermedia, HT '10, pp. 35–44. ACM, New York, NY, USA (2010)

9. Dunstall, S., Horn, M.E.T., Kilby, P., Krishnamoorthy, M., Owens, B., Sier, D., Thiébaux, S.: An automated itinerary planning system for holiday travel. J. of IT & Tourism **6**(3), 195–210 (2003)

10. Getoor, L., Sahami, M.: Using probabilistic relational models for collaborative filtering. In: Working Notes of the KDD Workshop on Web Usage Analysis and User Profiling (1999)

11. Goldberg, D., Nichols, D., Oki, B.M., Terry, D.: Using collaborative filtering to weave an information tapestry. Commun. ACM **35**, 61–70 (1992)

12. Hofmann, T.: Collaborative filtering via gaussian probabilistic latent semantic analysis. In: Proceedings of the 26th annual international ACM SIGIR conference on Research and development in informaion retrieval, SIGIR '03, pp. 259–266. ACM, New York, NY, USA (2003)

13. Horozov, T., Narasimhan, N., Vasudevan, V.: Using location for personalized poi recommendations in mobile environments. In: Proceedings of the International Symposium on Applications on Internet, pp. 124–129. IEEE Computer Society, Washington, DC, USA (2006)

14. Huang, Y., Shekhar, S., Xiong, H.: Discovering colocation patterns from spatial data sets: A general approach. IEEE Trans. on Knowl. and Data Eng. **16**, 1472–1485 (2004)

15. Hung, C.C., Chang, C.W., Peng, W.C.: Mining trajectory profiles for discovering user communities. In: Proceedings of the 2009 International Workshop on Location Based Social Networks, LBSN '09, pp. 1–8. ACM, New York, NY, USA (2009)

16. Kim, J., Kim, H., Ryu, J.h.: Triptip: a trip planning service with tag-based recommendation. In: Proceedings of the 27th international conference extended abstracts on Human factors in computing systems, CHI EA '09, pp. 3467–3472. ACM, New York, NY, USA (2009)

17. Kleinberg, J.M.: Authoritative sources in a hyperlinked environment. In: Proceedings of the ninth annual ACM-SIAM symposium on Discrete algorithms, SODA '98, pp. 668–677. Society for Industrial and Applied Mathematics, Philadelphia, PA, USA (1998)

18. Kumar, P., Singh, V., Reddy, D.: Advanced traveler information system for hyderabad city. Intelligent Transportation Systems, IEEE Transactions on **6**(1), 26–37 (2005)

19. Lemire, D., Maclachlan, A.: Slope one predictors for online rating-based collaborative filtering. In: Proceedings of SIAM Data Mining. SIAM press (2005)

20. Li, Q., Zheng, Y., Xie, X., Chen, Y., Liu, W., Ma, W.Y.: Mining user similarity based on location history. In: Proceedings of the 16th ACM SIGSPATIAL international conference on Advances in geographic information systems, GIS '08, pp. 34:1–34:10. ACM, New York, NY, USA (2008)

21. Li, Q., Zheng, Y., Xie, X., Chen, Y., Liu, W., Ma, W.Y.: Mining user similarity based on location history. In: Proceedings of the 16th ACM SIGSPATIAL international conference on Advances in geographic information systems, GIS '08, pp. 34:1–34:10. ACM, New York, NY, USA (2008)

22. Linden, G., Smith, B., York, J.: Amazon.com recommendations: Item-to-item collaborative filtering. IEEE Internet Computing **7**, 76–80 (2003)

23. Lu, X., Wang, C., Yang, J.M., Pang, Y., Zhang, L.: Photo2trip: generating travel routes from geo-tagged photos for trip planning. In: Proceedings of the international conference on Multimedia, MM '10, pp. 143–152. ACM, New York, NY, USA (2010)

24. Nakamura, A., Abe, N.: Collaborative filtering using weighted majority prediction algorithms. In: Proceedings of the Fifteenth International Conference on Machine Learning, ICML '98, pp. 395–403. Morgan Kaufmann Publishers Inc., San Francisco, CA, USA (1998)

25. Resnick, P., Iacovou, N., Suchak, M., Bergstrom, P., Riedl, J.: Grouplens: an open architecture for collaborative filtering of netnews. In: Proceedings of the 1994 ACM conference on Computer supported cooperative work, CSCW '94, pp. 175–186. ACM, New York, NY, USA (1994)

26. Salton, G., Buckley, C.: Term-weighting approaches in automatic text retrieval. Inf. Process. Manage. **24**, 513–523 (1988)

27. Salton, G., Fox, E.A., Wu, H.: Extended boolean information retrieval. Commun. ACM **26**, 1022–1036 (1983)

28. Sarwar, B., Karypis, G., Konstan, J., Reidl, J.: Item-based collaborative filtering recommendation algorithms. In: Proceedings of the 10th international conference on World Wide Web, WWW '01, pp. 285–295. ACM, New York, NY, USA (2001)

29. Shardanand, U., Maes, P.: Social information filtering: algorithms for automating "Word of Mouth". In: Proceedings of the SIGCHI conference on Human factors in computing systems, CHI '95, pp. 210–217. ACM Press/Addison-Wesley Publishing Co., New York, NY, USA (1995)

30. Sheng, C., Zheng, Y., Hsu, W., Lee, M.L., Xie, X.: Answering top- similar region queries. In: Proceedings of Database Systems For Advanced Applications, vol. 5981, pp. 186–201. Springer (2010)

31. Singh, A.P., Gordon, G.J.: Relational learning via collective matrix factorization. In: Proceeding of the 14th ACM SIGKDD international conference on Knowledge discovery and data mining, KDD '08, pp. 650–658. ACM, New York, NY, USA (2008)

32. Spertus, E., Sahami, M., Buyukkokten, O.: Evaluating similarity measures: a large-scale study in the orkut social network. In: Proceedings of the eleventh ACM SIGKDD international conference on Knowledge discovery in data mining, KDD '05, pp. 678–684. ACM, New York, NY, USA (2005)

33. Takeuchi, Y., Sugimoto, M.: Cityvoyager: An outdoor recommendation system based on user location history. In: Proceedings of the 3rd International Conference Ubiquitous Intelligence and Computing, pp. 625–636. Springer press (2006)

34. Xiao, X., Zheng, Y., Luo, Q., Xie, X.: Finding similar users using category-based location history. In: Proceedings of the 18th SIGSPATIAL International Conference on Advances in Geographic Information Systems, GIS '10, pp. 442–445. ACM, New York, NY, USA (2010)

35. Xiao, X., Zheng, Y., Luo, Q., Xie, X.: Finding similar users using category-based location history. In: Proceedings of the 18th SIGSPATIAL International Conference on Advances in Geographic Information Systems, GIS '10, pp. 442–445. ACM, New York, NY, USA (2010)

36. Ying, J.J.C., Lu, E.H.C., Lee, W.C., Weng, T.C., Tseng, V.S.: Mining user similarity from semantic trajectories. In: Proceedings of the 2nd ACM SIGSPATIAL International Workshop on Location Based Social Networks, LBSN '10, pp. 19–26. ACM, New York, NY, USA (2010)

37. Yoon, H., Zheng, Y., Xie, X., Woo, W.: Smart itinerary recommendation based on user-generated gps trajectories. In: Proceedings of the 7th international conference on Ubiquitous intelligence and computing, UIC'10, pp. 19–34. Springer-Verlag, Berlin, Heidelberg (2010)

38. Yoon, H., Zheng, Y., Xie, X., Woo, W.: Social itinerary recommendation from user-generated digital trails. Personal and Ubiquitous Computing (2011)

39. Zheng, V.W., Cao, B., Zheng, Y., Xie, X., Yang, Q.: Collaborative filtering meets mobile recommendation: A user-centered approach. In: Proceedings of AAAI conference on Artificial Intelligence (AAAI 2010), pp. 236–241. ACM, New York, NY, USA (2010)

40. Zheng, V.W., Zheng, Y., Xie, X., Yang, Q.: Collaborative location and activity recommendations with gps history data. In: Proceedings of the 19th international conference on World wide web, WWW '10, pp. 1029–1038. ACM, New York, NY, USA (2010)

41. Zheng, Y., Chen, Y., Xie, X., Ma, W.Y.: Geolife2.0: A location-based social networking service. In: Proceedings of the 2009 Tenth International Conference on Mobile Data Management: Systems, Services and Middleware, MDM '09, pp. 357–358. IEEE Computer Society (2009)

42. Zheng, Y., Wang, L., Zhang, R., Xie, X., Ma, W.Y.: Geolife: Managing and understanding your past life over maps. In: Proceedings of the The Ninth International Conference on Mobile Data Management, pp. 211–212. IEEE Computer Society, Washington, DC, USA (2008)

43. Zheng, Y., Xie, X.: Learning location correlation from gps trajectories. In: Proceedings of the 2010 Eleventh International Conference on Mobile Data Management, MDM '10, pp. 27–32. IEEE Computer Society, Washington, DC, USA (2010)

44. Zheng, Y., Xie, X.: Learning travel recommendations from user-generated gps traces. ACM Trans. Intell. Syst. Technol. 2, 2:1–2:29 (2011)

45. Zheng, Y., Xie, X., Ma, W.Y.: Geolife: A collaborative social networking service among user, location and trajectory. IEEE Data Eng. Bull. 33(2), 32–39 (2010)

46. Zheng, Y., Zhang, L., Ma, Z., Xie, X., Ma, W.Y.: Recommending friends and locations based on individual location history. ACM Trans. Web 5, 5:1–5:44 (2011)

47. Zheng, Y., Zhang, L., Xie, X., Ma, W.Y.: Mining correlation between locations using human location history. In: Proceedings of the 17th ACM SIGSPATIAL International Conference on Advances in Geographic Information Systems, GIS '09, pp. 472–475. ACM, New York, NY, USA (2009)

48. Zheng, Y., Zhang, L., Xie, X., Ma, W.Y.: Mining interesting locations and travel sequences from gps trajectories. In: Proceedings of the 18th international conference on World wide web, WWW '09, pp. 791–800. ACM, New York, NY, USA (2009)